Frontiers in Polar Social Science

The Earth is Faster Now

Indigenous Observations of Arctic Environmental Change

Kunuk watches the ice. Photo © Bill Hess, Running Dog Publications.

The Earth is Faster Now

Indigenous Observations of Arctic Environmental Change

Edited by
Igor Krupnik and Dyanna Jolly

Published by
Arctic Research Consortium of the United States
in cooperation with the Arctic Studies Center, Smithsonian Institution
2002

Cover photo © 1980 James H. Barker: Nelson Island seal hunters from Toksook Bay study ice conditions during an April hunt.

This book may be cited as:
Krupnik, Igor, and Jolly, Dyanna (eds.). 2002. *The Earth is Faster Now: Indigenous Observations of Arctic Environmental Change.* Fairbanks, Alaska: Arctic Research Consortium of the United States. 384 pp. ISBN 0-9720449-0-6.

This book is published by ARCUS with funding provided by the National Science Foundation (NSF) Arctic Social Sciences Program under cooperative agreement OPP-0101279. Any opinions, findings, and conclusions, or recommendations expressed in this material are those of the authors and do not necessarily reflect the views of the NSF.

Additional funds were provided by the Arctic Studies Center, Smithsonian Institution, to increase distribution and availability.

ARCUS, 3535 College Road, Suite 101, Fairbanks, Alaska, USA 99709
907/474-1600 • fax 907/474-1604 • arcus@arcus.org • www.arcus.org

Printed in Canada

Foreword:
Where a Storm is a Symphony and Land and Ice are One

Jose A. Kusugak
President, Inuit Tapiriit Kanatami

I have precious memories, childhood memories of our home life in dwellings and shelters going back to when I was a kid in Naujaat (Repulse Bay), Nunavut. During one of those times, it was during a severe March arctic blizzard. My brothers Arvaarluk, Qunngaataq, and I exhausted all indoor things to do like listening to legends and stories, finding and marking capital "H"s on all the magazines that we recycled as wallpaper for our *qarmak,* sod house. We had exhausted our string games and were now resorting to *Aaqsiiq,* the Silence Game.

This was not a clever ploy by our parents to keep us quiet. *Aaqsiiq* has a clear and important purpose and that is to test and increase one's own sense of hearing.

We could hear the winds and blowing snow through our very basic air duct. It was crafted by fitting empty tin cans together end to end. (We were into recycling sand reusing long before blue boxes invention.)

Here was our own symphony of wind instruments. On the south side of our *qarmak* (sod house) you could hear rolling snow driving and building and shaping snowdrifts. On our roof, a different sound and a faster tempo with the swift movement of *natiruviaq* (flooring snow) and then the percussion, the wind itself beating its tempo through the air duct. If it were put on a music sheet, it would read, "The seals will be happy. They have snow in their own *natsiaqqitit;* those baby seal bearing holes have a fresh snow covering." For us, the drifting snow will smooth out the sea ice for easier and faster travel. New snow will bury old prints and reveal only the new and the fresh animal tracks.

The gradual softening of the north wind, guaranteeing many days of clear weather, sunshine, and endless blue skies lie ahead. And best of all, we will be free from the smothering *qarmak* and back outside to be part of the symphony of elements. I think it is fair to say that until colonization dragged Inuit into houses, we generally lived our lives as nomads, outside in the open and in sync with both our eco and solar systems.

Inuit have "moons" in a year roughly coinciding with the same months as *qablunaat,* but each may be its own season. The moon is both a warning and a gauge of when it is good timing to go to the floe edge. About three days before and after the full moon is considered *piturniq* (extra high tide) and often is windy. In the winter when Inuit go to the floe edge to hunt they are taught to be aware of this since it might cause the main ice to *aukkaq* (break off and float away). It is quite useless to go *sinaasiuq* (seal hunting at the edge of ice) or literally to "look for the edge of ice" at high tide since there is no edge. It is necessary to know when to go sealing as not to waste your time or even worse, lose your life. For Inuit, perhaps in the past more than now, there are times of the year when there is no difference between land and sea ice. If there is a difference, it's the fact that the sea ice is more important to our survival.

For Inuit, during most of the year, sea ice is really a large extension of land. In winter, it was rare to find igloos and camps built on land. The land was colder than building igloos on sea or lake ice. The radiant heat of the water made that much difference.

Climate change has real and serious implications for Inuit life because much of the traditional knowledge is based on the times of seasons and not traditionally on temperatures. In other words, one does "this" at "this time" of the year rather than when the temperature gets "like this." For example, caching caribou is done in the fall after flies stop flying; not only to prevent maggots but because the meat shouldn't be too fermented or for that matter, too fresh. It is called *pirujat* (cached) or fermented meat and fat is called *igunaq* (singular), *igunat* (plural). There are many grades of *igunat* from mild to green. Now with climate change and warmer temperatures, much meat is going to waste because of over fermentation and botulism is becoming a real hazard. *Iqalugjuat* and *kujjaitat* (fermented fish and "hung upside down" fish) are especially susceptible since fish rots easily.

I had better change the subject before I get too hungry.

I think it is important to note that at least for now global warming is not all bad. Many northerners, who love to boat, actually are enjoying longer boating seasons. Many Inuit fish with fish nets under ice and are happy that the ice is not as thick as it once was. Fish are fat and plentiful at the moment. Berry picking in the fall has never been better. In biblical terms, it would be the seven years of plenty. But life as we knew it and know it now, is changing fast. *Sirmik* (permanent ice) in many areas is melting causing lakes to drain. *Aniuvat* (permanent snow patches) are disappearing. *Aniuvat* produces our favorite tea water, and caribou frequent *aniuvat* patches to get away from mosquitoes and flies. What I fear is that lives will be lost, because of thinning of ice, and because after lake ice melts and snow on the land is gone in the late spring people are still travelling on sea ice to the beginning of July. Will the ice still be safe for them?

Finding examples of effects of climate change is easy and endless. I know Inuit, old and young, want to be informed of outside influences of global warming. Like acupuncture, they know that the pain is much in their homelands but the needles have to be inserted in the south, since that is where the disease really is. I also know that given the opportunity to partake in data keeping, Inuit are more than willing. We have been careful caretakers of the Arctic from Siberia to Alaska to Canada to Greenland and back to Siberia for a long time. We have one language with varying dialects. Our customs and laws were designed to ensure our survival, and there are reasons behind everything we did. For example, we never stayed in one place or used one campsite for more than a few years, to avoid disease. We were told and taught *anijaaq,* which means "go outside to greet the day and elements first thing in the morning in order to live long."

What it all came down to was respect for the Earth and doing your part in keeping the world as in its original state. Inuit see themselves as part of the ecosystem and want to be included: not as victims, but as a people who can help.

I believe Inuit can provide the rest of society with useful and timely information because we are at the forefront where the impacts and effects of climate change are felt first and may be the most severe. This book is a good example of "all of us earthlings" getting together to protect this bubble of atmosphere because it is a matter of life. Listen and look intently to the great outdoors as though it was a great symphony, where Land and Ice are One.

Contributors[1]

Fikret Berkes is professor of natural resources at the University of Manitoba, Winnipeg, Canada. He is an applied ecologist by background and works at the interface of natural and social sciences. Berkes has devoted most of his professional life to investigating the interrelations between societies and their resources, and to examining the conditions under which the "tragedy of the commons" may be avoided. His main area of expertise is common-property resources and community-based resource management. He has conducted research throughout the Canadian North on indigenous resource use systems, small-scale fisheries, co-management, traditional ecological knowledge, subsistence economies, and environmental assessment. He has worked with the Cree people of Quebec and Ontario (James Bay area) and Manitoba, and also with the Nishnabe (Ojibwa), Dene, Nishga, and Inuit.
Fikret Berkes, Natural Resources Institute, University of Manitoba, Winnipeg, Manitoba, Canada R3T 2N2. berkes@cc.umanitoba.ca.

Claudette Bradley is an associate professor of education at the University of Alaska Fairbanks. She is a member of the Schaghitcoke Tribe, whose reservation is located in Kent, Connecticut. Bradley has taught mathematics and education courses via distance delivery to rural students for the University of Alaska Fairbanks since 1989. Her research and publications address the development of culturally appropriate mathematics, science, and technology curriculum for Alaskan Natives and American Indian students. Since 1988, Claudette has been involved in teaching and coordinating precollege summer camps and science fairs for Alaska Native students. Her other projects include the development of an education module for grades five and six on Yup'ik Star Navigation Across the Tundra, which is part of National Science Foundation-funded Yupiaq Mathematics Project at the University of Alaska Fairbanks.
C. Bradley, University of Alaska Fairbanks, Interior-Aleutians Campus, P.O. Box 756720, Fairbanks, Alaska 99775 USA. ffceb@uaf.edu.

Shari Fox is currently a PhD student at the University of Colorado at Boulder and holds bachelor and master's degrees in environmental studies from the University of Waterloo. Fox is also the co-lead author, with Henry Huntington, of the Indigenous

1. This is a list of corresponding authors only. Affiliations of other contributors can be found in individual chapters.

Perspectives chapter of the Arctic Climate Impact Assessment, an international project organized under the auspices of the Arctic Council. Born and raised in the flat expanses of southern Ontario, these days Fox divides her time between studying and playing in the Rocky Mountains. Fox is committed to working on issues that concern indigenous people and hopes to continue visiting the North and working with Inuit.
S. Fox, Cooperative Institute for Research in Environmental Sciences (CIRES), University of Colorado at Boulder, Boulder, CO 80309-0449 USA. sfox@kryos.colorado.edu.

Chris Furgal is a researcher in the Public Health Research Unit, Centre Hospitalier Universitaire de Laval (CHUL) Research Centre, Université Laval, Québec. He conducts work in both the biological and social sciences on issues related to environmental contaminants, climate change, and other environmental health related issues and their management and impacts on people in the circumpolar North. Much of this work is conducted in cooperation with Aboriginal organizations in the Canadian North. He is a member of the Nunavik Nutrition and Health Committee and co-lead author of the Arctic Climate Impact Assessment (ACIA) chapter on health impacts in circumpolar arctic regions.
C. Furgal, Public Health Research Unit, CHUQ-Pavillon CHUL, 2400 rue d'Estimauville, Beauport, Québec G1S 1S7 Canada. christopher.furgal@crchul.ulaval.ca.

Henry P. Huntington is an independent researcher in Eagle River, Alaska, where he lives with his wife and two young sons. He received his PhD from the Scott Polar Research Institute at the University of Cambridge (UK), having done his research on the interactions of wildlife management institutions and Native hunting practices in northern Alaska. His subsequent research has examined traditional ecological knowledge, environmental contaminants, climate change, conservation in the Arctic, and other topics concerning the Arctic, its environment, and the peoples who live there. He is currently president of the Arctic Research Consortium of the United States.
H. Huntington, 23834 The Clearing Drive, Eagle River, AK 99577, USA. hph@alaska.net.

Dyanna Jolly is a Canadian researcher currently affiliated with the Centre for Maori and Indigenous Planning and Development at Lincoln University in New Zealand, and working on co-management issues. Before making the move to the southern hemisphere, Jolly did her graduate research at the University of Manitoba, exploring the contributions of traditional knowledge to understanding climate change in the Canadian Arctic. Her research was part of the project Inuit Observations on Climate Change, a joint effort between the International Institute for Sustainable Development (IISD) and the Western Arctic community of Sachs Harbour. She admits that she was born in (the dirty south of) Ontario, but brags about growing up in the Cree-Nakota-Saulteaux prairie community of Whitebear First Nations, in beautiful southern Saskatchewan.
D. Jolly, Centre for Maori and Indigenous Planning and Development, P.O. Box 84, Lincoln University, Canterbury, New Zealand. dyjolly@ihug.co.nz.

Gary Kofinas is a research assistant professor of public policy at the Institute of Social and Economic Research of the University of Alaska Anchorage and a senior fellow at the Institute of Arctic Studies of Dartmouth College. His research focuses on the evolution of co-management systems, the interface of local/traditional knowledge and science, and the sustainability of arctic communities. Gary received a PhD from the University of British Columbia in interdisciplinary studies/resource management

science. He serves as project leader for the Human Role in Reindeer/Caribou Systems initiative of the International Arctic Science Committee and advises on the design and implementation of the Arctic Borderlands Ecological Knowledge Co-op community monitoring program. Old Crow, Aklavik, Fort McPherson, and Arctic Village are partner communities in the Arctic Borderlands Knowledge Co-op program.
G. Kofinas, Institute of Arctic Studies, Dartmouth College, 6214 Fairchild, Hanover, New Hampshire 03755 USA. gary.kofinas@dartmouth.edu.

Jose Kusugak was born on May 2, 1950, in an iglu in Naujaat (Repulse Bay), located on the Arctic Circle. He is the second eldest of seven brothers and four sisters and attended school in Chesterfield Inlet and Churchill and high school in Saskatoon. Jose Kusugak first got involved with the Inuit Tapiriit Kanatami (then Inuit Tapirisat of Canada) in the early 1970s to work on the standardization of the Inuit writing system. Because project funding was delayed, Kusugak worked as an assistant to then-president Tagak Curley, introducing the concept of land claims to Inuit in the Arctic. In 1974, Kusugak went to Alaska to study the Alaska land claims and traveled the Inuvialuit region as part of the land use and occupancy study tour. From late 1974 to 1977, Kusugak chaired the standardization program of the Inuktitut language; from 1980 to 1990, he was the area manager of Canadian Broadcasting Corporation in the Kivalliq (Keewatin) region; and from 1994 to 2000 he was the president of Nunavut Tunngavik Incorporated, an affiliate of Inuit Tapirisat of Canada. He was elected president of Inuit Tapiriit Kanatami in June 2000. His wife Nellie works as a northern studies teacher at Arctic College in Rankin Inlet.
J. Kusugak, Inuit Tapiriit Kanatami, Suite 510, 170 Laurier Ave. W., Ottawa, Ontario K1P 5V5 Canada. jkusugak@tapirisat.ca.

Igor Krupnik is an ethnologist at the Arctic Studies Center of the National Museum of Natural History, Smithsonian Institution in Washington, D.C. Born and educated in Russia, he has done extensive fieldwork among the Siberian Native people in the Bering Strait area, in the Russian Far East, and recently on St. Lawrence Island in Alaska. He is currently coordinator of various international projects studying the impacts of global climate change and the preservation of the cultural heritage and ecological knowledge of Native peoples. He has published and co-authored several books and catalogs, and he writes extensively on arctic Native peoples, Native heritage resources, modernization and minority issues, and the history of anthropological research in the Arctic/North Pacific region.
I. Krupnik, Arctic Studies Centre, National Museum of Natural History, Smithsonian Institution, Washington, D.C. 20560-0112 USA. krupnik.igor@nmnh.si.edu.

Tero Mustonen is manager of the project Snowchange: Indigenous Observations of Climate Change at the Tampere Polytechnic in Finland, where he works as a lecturer and researcher. He is currently finishing his masters thesis at the University of Tampere on indigenous self-governance in the Russian, European and Canadian Arctic. He has visited and lived in many indigenous communities in Canada, Alaska, Finland, and Russia, including those of the Inuvialuit, Haida Gwaii, Tahltan Nation, Cree Nation on James Bay, and Sto:lo Nation, as well as the Sami communities of Ochejohka in Finland and Luujavre in the Murmansk region of Russia. Mustonen recently coordinated a three-year climate change-related online educational project Northern Environment Student Forum between institutions in British Columbia,

Murmansk and Tampere. When he has free time, he plans a kayaking trip to New Zealand and enjoys playing bad rock music real loud.
T. Mustonen, Tampere Polytechnic Teiskontie 33 FIN 33521, Tampere, Finland. tero.mustonen@tpu.fi.

Scot Nickels has many years of experience working with both Inuit and First Nations on environmental issues, from documenting Inuit knowledge of bowhead whales and investigating the environmental, cultural, and socioeconomic impacts of tourism in the North to the development of an integrated resource management plan for the Algonquin community of Barriere Lake. Nickels obtained his PhD at McGill University in February of 2000. His doctoral thesis is entitled *Ecological Knowledge and Experience of the Forest Environment: the Algonquins of Barriere Lake, Quebec,* and provides an analysis of how human knowledge and action are shaped through on-going interactions of people with each other and their environment. In 1998, Scot became the director of the Environment Department for Inuit Tapiriit Kanatami (ITK) (formerly ITC), the national organization dedicated to supporting Inuit in Canada. He is currently involved in policy and research issues such as species at risk, marine conservation, environmental assessment, contaminants, and climate change as they relate to Inuit.
S. Nickels, Environment Department, Inuit Tapiriit Kanatami, Suite 510, 170 Laurier Ave. W., Ottawa, Ontario K1P 5V5 Canada. snickels@tapirisat.ca.

Dave Norton has conducted research from Alaska's Prince William Sound to the Arctic, in fields as diverse as ornithology, human ecology, vertebrate (dinosaur) palaeontology, history, and fishery ecology. He has been called an "arctophile" and a generalist bucking trends of specialization. He credits his nine-year residency and teaching in Barrow with enthusing him for transdisciplinary and transcultural environmental science. After a sabbatical assignment to the University of Calgary in 1996-97, he assembled a trinational team of investigators to work on arctic marine contaminants for National Oceanic and Atmospheric Administration. Dave now leads another interdisciplinary research group for National Science Foundation's Office of Polar Programs, which encourages linkages between traditional knowledge and remote-sensing applications to analyze the responses of coastal sea ice to environmental variables. *Fifty More Years Below Zero,* which Dave edited for the Arctic Institute of North America, appeared in 2001 and is distributed by the University of Alaska Press. The book chronicles the fifty-year history of the Naval Arctic Research Laboratory at Barrow.
D. Norton, Arctic Rim Research, 1749 Red Fox Drive, Fairbanks Alaska 99709 USA arcrim@ptialaska.net or ffdwn@uaf.edu.

Natasha Thorpe has been fortunate to share tea and stories with Inuit throughout the Kitikmeot region of the Canadian Arctic where she has worked, travelled and camped for the last seven years. Natasha holds a masters degree in resource management from Simon Fraser University in Vancouver. Her most recent project involvement was as principal researcher for the Tuktu and Nogak Project (TNP), a community-driven effort to document and communicate Inuit knowledge of caribou and calving grounds for the Bathurst caribou herd. *Thunder on the Tundra: Inuit Qaujimajatuqangit of the Bathurst Caribou,* a compilation of the results of the five-year study, was published in late 2001 and co-authored by Thorpe, Naikak Hakongak, Sandra Eyegetok, and Kitikmeot elders. She now lives in Victoria where she continues to be committed to Inuit Qaujimajatuqangit projects.
N. Thorpe, Tuktu and Nogak Project, 231 Irving Road, Victoria, British Columbia V8S 4A1 Canada. nthorpe@telus.net.

Aklavik elders Anne B. and Danny A. Gordon watch the changing arctic sky. Photo by Deborah Robinson.

Table of Contents

List of Figures

List of Tables

Iñupiat whalers in spring near Barrow, Alaska, waiting on the sea ice by their umiaq as they scan the open lead for bowhead whales. Photo by Henry P. Huntington.

Preface:
Human Understanding and Understanding Humans in the Arctic System

Henry P. Huntington
President, Arctic Research Consortium of the United States
Director, Science Management Office,
Human Dimensions of the Arctic System Science Program, National Science Foundation

Social sciences in the Arctic enjoy many exciting and important opportunities. The series *Frontiers in Polar Social Science,* of which this is the first volume, explores and articulates those opportunities, the new frontiers of arctic research. It offers new insights for both social scientists and others to learn about developments in research concerning humans and their societies in northern regions. Although such a series cannot hope to encompass every branch of the social sciences, it can and should provide an introduction to several aspects of social science research in the North, stimulating further interest among specialists and non-specialists alike.

This volume, *The Earth is Faster Now,* addresses indigenous observations of arctic environmental change and the implications of such change for arctic peoples. Despite all the attention currently being given to climate change globally and in the Arctic, indigenous perspectives are all too frequently overlooked. As this volume shows, arctic residents have a great deal to say. Understanding and addressing climate change simply cannot be done without incorporating their specific and detailed views. The processes by which this can be done, however, take considerable time and effort of the part of both researchers and arctic residents. Applying this approach not just in a few isolated projects but as an integral part of arctic environmental research is an ambitious and important goal.

In making these points, the chapters in this volume also examine the ways in which social science methods and results contribute to our collective understanding of the arctic system—the combination and interactions within and among physical, biological, and social conditions—and its relationship with the rest of the world. In academic terms, an understanding of the way the system works must rest squarely on a solid foundation of research in and across the various disciplines that address the many aspects of a complex system such as the Arctic. But that overarching understanding requires thinking in broad terms, finding ways to encompass disparate disciplines. This obstacle does not appear in indigenous perspectives, which draw naturally on changes observed in the physical, biological, and social realms to develop a coherent view of the world.

Although this volume addresses primarily the process of adapting scientific perspectives to incorporate indigenous ones, it is important to keep in mind that information flows the other way as well. Indigenous communities wrestle with the ways in which they can use scientific research and information for their purposes. Such information can be understood and also misunderstood, sometimes with significant consequences. Effectively communicating the results of scientific research is no less important than helping scientists appreciate what they can learn from arctic residents.

Understanding the Connections Between Humans and Their Environment

Crossing cultures as well as disciplines is a vital challenge for arctic social science, and environmental change is an excellent way to explore that challenge. These changes have been observed by arctic residents and by scientists. They include thinning sea ice and changes in its characteristics, poor body condition in many animals, earlier growing seasons, a greater frequency of extreme weather events, rising temperatures in permafrost, and many others from the general and to the highly specific. Humans are connected to this changing environment in ways that are both similar and distinct from the ecological relationships that tie plants and animals to their ecosystems. Understanding the nature of those connections is essential to understanding the implications of environmental changes.

In the premodern era, arctic peoples were dependent largely on local resources but used a variety of means to influence their relationship with the living things on which they depended. Today, the use of living resources by arctic residents retains many ecological characteristics but is shaped also by regional, national, and international politics governing the allocation of harvests, the management of environmental impacts, and the influence of ideas such as animal rights. As human societies change, willingly or otherwise, their relationship to the environment is inevitably affected. If the environment is changing as well, we face an additional layer of complexity in trying to understand how arctic communities will respond to and cope with this challenge and how they will function in the future.

The connections of people with their environment must also be understood broadly. Typically, researchers focus on physical connections such as hunting, fishing, travel, and construction. There is also, however, the perceptual realm in which people understand their position relative to their surroundings in ethical and spiritual terms. Indigenous people often talk about this aspect of their relationship with their surroundings, but their listeners rarely take it in. When speaking of indigenous perspectives, however, it is essential to include the full range of the indigenous worldview and not just the subset that fits most neatly into the current scientific paradigm. Although the spiritual dimensions of the environment may be inaccessible to science, they nonetheless play a vital role in shaping perceptions and actions and thus must be taken into account when discussing appropriate responses to environmental change.

The nine papers collected in this volume focus primarily on documenting and understanding the nature of changes that are being seen by northern indigenous residents in their environment. The special emphasis of the papers is not simply on change but also on the ways arctic peoples perceive, influence, and are influenced by their surroundings. As we move towards broad, multidisciplinary attempts to characterize the arctic system, it becomes increasingly important to understand the nature of the human components of that system and how they are connected to the physical and the biological realms. The nature of those connections is especially important in the development of responses to environmental change.

Understanding the Arctic Environment and the Knowledge of Arctic Peoples

Scientific understanding of arctic environmental and climate change is based on records that are often short-term, fragmentary, or both. Weather records and sea ice data are available for some places, but rarely extending back before the twentieth century. Satellite monitoring of snow, ice, greenness, and other parameters covers most or all of the Arctic, but obviously only for recent decades, if not years. Process studies have been carried out in several locations around the Arctic, giving insight into the dynamics of ecological processes but leaving open the question of how to extrapolate those findings across the vast areas where no such studies have taken place. Examination of paleoenvironmental and archeological records reveals a great deal about certain indicators of climatic, environmental, and social conditions but requires careful and cautious interpretation. Into this mix is thrown the fact that the arctic environment is highly variable. Change, large and small, frequent or infrequent, is simply a basic characteristic of the region, compounding the difficulty of identifying trends and causal relationships.

Arctic residents have long known about, and had to cope with, this variability. Alternative strategies for finding food, a cultural acceptance of recurring hardship and privation, and migration were among the means—not always successful—of surviving and thriving. The intimate knowledge that arctic peoples have acquired about their surroundings is well known. The applicability of this knowledge to scientific studies of topics such as environmental change or the resilience of arctic social and ecological systems, however, is another matter. One key theme running through the papers in this volume is how to develop methods for documenting the knowledge of arctic peoples and how to present it in a form that is accessible both to the environmental sciences and to arctic indigenous communities themselves.

The documentation of knowledge is, on the surface, relatively straightforward. Through interviews, for example, the researcher can record what an experienced elder has to say about a given topic. There are, to be sure, many potential pitfalls to this process, but the basic methods are already established in the social sciences, and much has been written about the way such research can and should be conducted. Presenting the collected information is a separate step, but again, one that has been examined and improved over time. The difficulty comes in interpretation.

One hazard in the attempt to document indigenous knowledge about arctic environmental change is that what is collected often turns out to be little more than a body of accumulated facts, revealing what has been observed but not how it has been understood. This is particularly true when indigenous knowledge is collected separately from an appreciation of the system or framework in which it was originally generated, gathered, and held. Facts and observations, carefully collected and reviewed with those who provided them, can be immensely useful. Nonetheless, the difference between facts and a system of knowledge is akin to the difference between anatomy and physiology—knowing the parts does not tell you how they function and interact with one another in a living organism.

As scholars and researchers become more familiar with the environmental knowledge of arctic residents, the need to understand their system of knowing becomes ever more apparent. There is considerable overlap between the processes of observation employed by scientists and by arctic indigenous peoples. There is considerable interest in both groups about what the other has found. But there are also differences. Attempting a translation from one system to the other requires a certain level of understanding of how each system is constructed and how it works. Simply put, it requires understanding what people look for, how and why they look for it, and to what use they put the resulting information. The growing appreciation of the cultural basis of science is one manifestation of this line of inquiry. A similar appreciation of the cultural context of the knowledge of arctic indigenous peoples is necessary to move beyond acknowledging and using that knowledge solely as a source of raw data.

Researchers need, in other words, to understand why people see and interpret things as they do. Information is always collected and organized to suit a particular purpose or set of purposes, and certain biases are thus inherent in the body of knowledge that is created by human societies. Knowledge changes over time, and it is important to understand how it changes, how new information is incorporated, what are regarded as significant and reliable sources of information, and so on. In short, researchers need to understand the implicit method that lies beneath the knowledge of arctic peoples, just as they work diligently to follow the scientific method in other research. Only in this way can researchers appreciate not only the information that is generated by the environmental knowledge of the arctic residents, but also their perspective on the environment, their relationship with it, and what if any actions are needed to protect that relationship.

Addressing the Implications of Climate Change

This last point raises a related question. We have looked briefly at the ways in which researchers can better understand arctic peoples' knowledge and thus improve their common understanding of topics such as environmental change. If indeed the environment is changing, human actions may have to change, too, either as a passive consequence of environmental shift or in an active attempt to prevent, manage, or adapt to it. The former is simply a response to changes in environmental stimuli, something that can be watched but not controlled. The latter consists of the responses humans make to the prospect of environmental change, a conscious effort to manipulate society or the environment or both.

As is clear from the ongoing international debates concerning climate change and what, if anything, can and should be done about it, reaching consensus is probably an impossible task. Nonetheless, a broadly shared understanding of the nature, magnitude, and scope of expected environmental changes is an important starting point to determining what actions can be taken. When, as is the case in the arctic, those who must be included in that shared understanding are from various cultures and backgrounds, it is vitally important not only that they agree on the basic facts but that they understand how each group has acquired those facts and how they interpret them. This is a highly practical outcome of a thorough appreciation for the knowledge that arctic indigenous peoples have about their environment.

From a shared understanding of what is occurring in the arctic system, we can then design appropriate response strategies. Those strategies must also be based firmly on an understanding of how they themselves will affect the human and natural systems in which they will be applied. Response strategies that do not reflect the values, priorities, and needs of society will fail, either because they are not acted on or because they destroy the very thing they were supposed to help protect. The same is true for the ecological relevance of responses—they must take into account the environmental feedbacks they establish in order to avoid unintended problems that may be greater than the threat the response sought to address.

Incorporating the knowledge and perspectives of arctic peoples is the starting point for working together to address an issue from start to finish. That knowledge has been used at times to generate various hypotheses and models that are then subjected to the rigors of scientific research. Such an approach is one way to make use of the knowledge of arctic peoples, which by this means has contributed a great deal to modern understanding of the Arctic. When we are concerned, however, not just with academic understanding, but with real-world responses that have consequences of their own, it is essential that the use of the observations and knowledge accumulated by arctic residents is not limited to the start of the effort.

Similarly, it is not enough to take into account only their ecological expertise. The dynamics of their societies and cultures are every bit as important to the study of their knowledge and the collaborative design of response strategies. This leads us back to the critical role of the social sciences in understanding environmental change and its implications across the Arctic. Arctic residents can certainly speak for themselves, but social scientists have a great deal to offer in this discussion, particularly from a comparative vantage point that can nicely complement the detailed but local understanding that a community has about itself. Building on the comparative studies of environmental observations at many sites and in various areas around the North, we can and should establish true partnerships between social scientists, natural scientists, and arctic communities. In addition to researchers, such partnerships should include those within and outside arctic communities who are responsible for taking action to address the impacts of environmental change.

What Next?

The study of the interactions between humans and the arctic system is not, perhaps, in its infancy, but certainly in its youth. Researchers have already described, both for the past and for the present, some of the direct links, such as people as predators, or people as consumers of environmental conditions. But the indirect links and feedbacks and the dynamics of these together, especially in connection with other social, biological, and physical variables, require more work. We can measure many inputs and outputs, but the inner workings of the system have not yet been discerned, even in schematic form. The relationships between all these variables are complex and changing. We need to devise new approaches for identifying and elaborating particularly the human components of the arctic system. The papers in this volume point us in the right direction, and the next steps are to follow their lead and take these ideas further.

As we do so, we must keep in mind that there are other sources of social change than a changing environment, and other sources of environmental change than climate. For many arctic residents, particularly in the Russian Arctic, the immediate struggle for survival far outweighs an abstract concern about future effects of a changing climate. In other areas, industrial development, competition for fish and other resources, and environmental contaminants are among the drivers of environmental change of most concern to local residents. Throughout the Arctic, the process of modernization, though it takes many forms, has caused rapid and often painful social and cultural transitions and it continues to do so. This process is largely independent of environmental conditions but must nonetheless be taken into account because it has powerful implications for the human-environment relationship.

As we develop a better understanding of the dynamics of the arctic system, we inevitably begin to speak in terms of prediction. As a means of testing and pushing the limits of scientific understanding, this is a useful exercise. But for planning our responses to environmental change, "predictions" in the midst of complexity and incomplete understanding remain vague and unreliable. Instead, as we move from pure research to the application of what we have learned, perhaps we should think in terms of anticipating what may happen, rather than predicting what will happen. This approach lowers the standard for looking to the future, but by doing so it gives us a more realistic target. By working towards anticipation, we can include a number of possible outcomes, each of which can be evaluated for magnitude and likelihood of the threat, as is done in risk assessments. From a list of what we can anticipate, we can in turn develop response strategies based on the best current understanding of the threat and its implications, including the numerous implications of our responses.

Again, this must be undertaken as a collaborative enterprise. Linking social and natural systems together with arctic residents and those who are politically responsible for developing response strategies to environmental changes will not happen by accident. Nor will it occur simply because it seems to be the right thing to do. Instead, the challenge to polar social scientists is to demonstrate not only the relevance of their research and results but also their ability to work with others and to use social understanding to improve the collective understanding of the dynamics of the arctic system. Solid disciplinary research is needed to meet this challenge, but it must be matched by a strong commitment by the social science community as a whole to reach outwards as well. This, truly, is a frontier of arctic social science today.

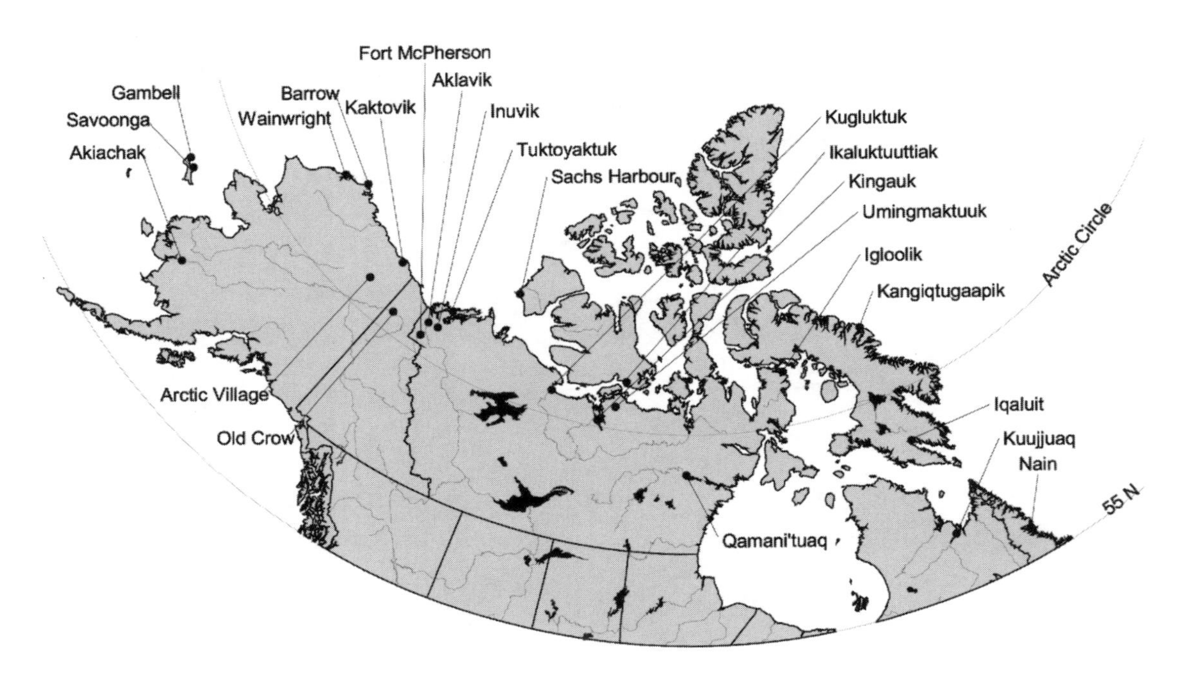

Northern Native communities that participated in the arctic environmental change documentation projects discussed in this volume (illustration by Philippa McNeil).

Introduction

Igor Krupnik and Dyanna Jolly

This book has two main messages to its readers and to the polar science community at large. First, arctic residents are witnessing far-reaching changes in their environment, and they are ready to create partnerships with scientists, to document their observations and to make their voices heard. Second, interaction between local experts and academic scholars will require other patterns of collaboration than between, let's say, physical and social scientists. Familiar ways of doing research—scanning earlier data, disputing other people's concepts, and borrowing references across disciplines—will be inadequate for this new unfolding collaboration. We rather have to learn to act through sharing, listening, and accommodation to others' ways of observing and "knowing." This is why we put this issue as the starting point for a new publication initiative, Frontiers in Polar Social Science, advanced by the Arctic Research Consortium of the United States (ARCUS).

As with many other pioneer ventures, this volume is a product of an uncommon though highly successful collaboration. It represents a coming together of researchers, communities, and organizations, but in many different ways. Each of the volume's several chapters describes a long-term research and collaborative effort, with a potential to be a book project in itself, as some already are. Despite our separate projects and focuses, we share a common vision and, basically, we put the same message on paper. To us, this volume is neither a "silver bullet" nor a handbook for our colleagues of what indigenous people know about arctic change. First and foremost, it is a collection of studies and reflections on how indigenous peoples see changes and what they say about changes around them.

What gave this book its specific edge? We believe we can disclose some driving forces behind our joint venture in this volume introduction. Our first "secret" is that this initiative found an immediate and enthusiastic response from a small, though growing community of Native experts and researchers who already work together in documenting Native observations of environmental changes occurring across the Arctic. Our second resource was what we call "the magic of the science frontier." It looks increasingly obvious to polar scientists, arctic residents, and the general public alike that arctic indigenous people have a special stake in modern studies of global environmental change. They also have a lot to contribute—when and if they are given the chance and the appropriate means to participate fully in the ongoing global change discourse.

Records of local observations are created in dozens of indigenous communities across the circumpolar zone, by human inquisitiveness and people's interaction with each other and the environment. Such records are constantly reinforced and immediately tested in discussions with neighbors, fellow hunters, and experienced elders. This observation process is nonstop, daily, and intergenerational, without any granting agencies and science planning involved. This is indeed an exciting impetus for partnership and THE new research frontier to guide the course of today's environmental change research and to draw public attention and debate.

Last but not least, the "frontier" paradigm is a specific pattern of human vision as well as the powerful and universal drive to expand the boundaries of the known into the unknown. This drive pushes researchers' explorations in remote arctic villages and at university labs alike. In fact, the lure of the frontier existed ages before modern professional scholarship was established. As humans, we are always anxious to know what is beyond the horizon but also what is—or will be—our next challenge to face.

Current environmental change in the Arctic offers a testing challenge to our ability to observe, to analyze, and to respond, by using the tools and resources of today's society. If this change is coming (and some of us believe that the challenge of rapid environmental shift is already here), then this volume may well become one of the first coordinated efforts in response. By focusing on local observations and interpretations of change, it opens paths to the next "frontiers" in science analysis and public actions. This was the third major appeal that worked on behalf of our venture at every stage. It was also the critical bond that helped generate our project design, recruit volume contributors, and bridge several individual papers into a common message.

Why "The Frontiers"?

Several project papers collected in this volume document various aspects of Native observations of current environmental change across the Arctic—from weather to sea ice to caribou to marine mammals to permafrost to plant communities. Our purpose, of course, is to bring to light the richness of local expertise of northern residents as they witness and interpret shifts, transitions and/or abnormal events in their familiar habitats. What people actually know is closely connected to both historical and current land use and occupancy. It reflects their various daily encounters with the arctic environment—where they hunt, fish, and travel; when they do it; and what factors are significant in framing the scope of their seasonal activity. As will be illustrated below, such a broad observational base contributes to a unique reliability of local environmental monitoring, where change is often traced by and related through personal life histories and experiences. Individual living memory is then extended through storytelling and information sharing. Such an exchange of oral traditions takes place daily in the family and community setting, and it expands the time-depth of personal expertise and observations.

By bringing together a series of projects in documenting what people really see and talk about these days in their communities, we argue that indigenous arctic residents are clearly noticing changes across their regions. Many common themes and stories arise from voices to be heard in this volume. These are themes of increasing variability and unpredictability of the weather and seasonal patterns and thus of the need to be more careful when hunting, travelling, and forecasting the weather. These are stories of the extensive loss of multiyear sea ice, the appearance of species never seen in living memory, and of a growing concern for the health of game populations upon which the communities depend.

We see our mission here in terms of finding new ways to document things that are already emerging as the issues of public concern at both elders' meetings in distant northern villages and at scientific symposia. These should be better ways, creative ways, more appropriate ways. The chapters in this volume explore new methods and research tools for learning about and sharing data generated through observations in "another kind of science," which is indigenous environmental knowledge. They address new models of community partnerships and research cooperation that go well beyond the now-dominant pattern of "us" scientists informing "them" (that is, Native residents and public at large) on what we learn through "our" scientific research in "their" areas. This is our vision of the frontier of polar social science, in terms of the science's philosophy, research goals, and ethics.

Still, there is much more in local observations that justifies this volume's headline as "Frontiers in Polar Social Science." Tied to the themes of complexity, unpredictability, and increased variability are many indications of how current environmental change is part of a larger group of challenges and changes that people face as northern residents. To many people on the ground, their daily concerns about weather and sea ice shifts are hardly separated from other critical issues, such as oil and mineral exploration, contaminants, animal rights campaigns, and land-claim negotiations. Scientists often forget to acknowledge that their cause-driven research models are, nevertheless, rather simplistic approximations of a real life. Whereas scientists love to talk about "independent links," "critical factors," and "stratified impacts" in their abstract scenarios of change, arctic residents have their own vision of recent change, which is always a multifaceted process. There is no need for any special "interdisciplinary" dimension, because complexity is a phenomenon of daily existence. To any scholarly approach, this local perspective, if properly understood and accurately documented, is an invaluable reality check.

There is one more critical aspect of environmental observations by Native elders and subsistence experts. Their understanding of what is happening with the weather, the lands, and the oceans is often articulated as community-based assessments of change. Such assessments translate the global-scale process of change into local-scale evidence. Hence, the observations described in this volume are the best arguments for the value (and the urgency) of doing local-scale, place-based research and modelling of the global process. This is what current science-based understandings of climate change are often missing—an indication of how changes are affecting places. The discourse of global climate change so far is largely on modeling studies, global predictions, and international policy. Little is known about local places—communities, ice floes used for spring sea hunts, caribou calving grounds, historical sites and protected areas, airports, travel routes and camping areas, the spring nesting sites of snow geese, and the river mouth where the fish spawn.

A common (though usually unspoken) question within the community of polar physical scientists is, what is the reason to be engaged in listening to some hunters' stories and elders' recollections in the era of global modelling, supercomputers, and satellite imagery? Why should we bother to document indigenous knowledge and perspectives of environmental change in the first place? To social scholars (at least, to some of us working in the Arctic), the answer is simple—because indigenous people really want this done. A striking motive that can be seen in almost every paper in this volume is the amount of community initiative and support for these types of projects across the Arctic. Rarely, if ever, does polar science enjoy such an unprecedented level of public backing and readiness to share data and expertise from local residents.

These and other similar projects also highlight the human context of environmental change—another "frontier" topic in global change science and in related public debates. The stories collected in this volume are not just about changes to the sea ice, weather or caribou. They are about how people see these changes in the context of their lives. As the global change issue is going to shift from being the subject of scientific research to the matter of political discussions to the focus of public actions, there is a growing need for transitional mechanisms from science to practical policies. Such a need to translate local, place-based pictures of change into public policy finds an almost natural venue through community-based observation projects like those presented in this volume. It also provides additional impetus of the value and urgency of documenting Native observations of environmental change.

Raising awareness and reaching wider audiences will be better facilitated through documentation within the context of collaborative, community-supported initiatives—rather than through the more conventional scholarly studies generated by the governmental science agencies and introduced "from without." The papers collected here illustrate the power and ingenuity behind such innovative projects, since their authors pioneered many creative ways to do this—through community workshops, bilingual CD-ROM's, videos, school classes, and community knowledge sourcebooks.

Finally, there is a growing emphasis on recording Native observations of change as part of a larger effort to alter the status of knowledge shared by elders and local experts vis-à-vis the samples of data collected by scientists. The "frontier" zone here is, actually, in interpretation rather than documentation, as the need to document what locals see in their environment becomes gradually acknowledged by scholars and policy-makers alike. There is already an established respect among many northern specialists of the expertise of indigenous arctic residents, particularly in areas such as wildlife management. In some cases, such as the Northwest Territories in northern Canada, indigenous knowledge now holds a government-mandated place in any developmental assessment and decision-making. Still, documenting Native observations of an issue as potentially controversial as climate change is only just emerging and has yet to reach its full potential. We all have a long way to go before the paradigms and interpretations of change advanced by Native experts and community scholars are given the credits and attention they deserve in research planning as well as in science funding.

As with science-based understandings of arctic environmental change, community-based assessments are evolving. The difference, however, is that while much of the science research is focused upon predicting what might happen, local experts are talking about what is happening. They also want to do something about it—and to take action quickly. Arctic residents, through projects such as described in this volume, are clear in saying that monitoring of changes needs to start immediately and that monitoring should be facilitated by both science and indigenous knowledge, in genuine sharing and full cooperation.

How This Volume Originated

This volume came out of an almost fortuitous meeting of the two present co-editors (as well as a few other key backers—see Acknowledgements) at the thirteenth annual ARCUS meeting in May 2001. At the time, both of us were actively engaged in independent collaborative efforts in documentation of indigenous knowledge on environmental change in Alaska and the Canadian Arctic, respectively. Similarly, we had both been publicly advocating the value and the importance of indigenous scholarship to the interdisciplinary study of arctic climate and environmental change at several meetings and in our previous publications. Between the two of us, we had a good network of colleagues—Americans, Canadians, and Europeans—who were fast exploring the same field and who shared the same passion and ideology (though none of them was present at that same meeting). When the key volume's headline, "Frontiers in Polar Social Science" was introduced to our discussions, it came through as a lightning rod.

Once advanced, this book progressed along some unconventional lines in science cooperation. It took exactly a full year from the moment the possibility of such a "science frontiers" collection was discussed for the first time at the ARCUS meeting in Washington, D.C., to the day this volume is to be released at the fourteenth annual ARCUS meeting in May 2002. Such a year-term is a remarkably short lifespan for a full-size international collection of papers to materialize, by any standards of modern scholarly publication. Maybe it was the very title of our volume, *The Earth is Faster Now,* that worked on its behalf.

We believe it was this magic of the science "frontier" that pushed us to shelve many other personal commitments and to vow to the ARCUS leadership that we could organize a collection of project papers about the documentation of indigenous knowledge of arctic environmental change in one year's time. This was a risky pledge to make, since most of the potential contributors we were thinking of had never met each other. There was not enough time to organize one of the usual symposiums that both brings authors together and produces a symposium volume. In fact, in a full reversal of the usual science practice, we instead considered the issue of a symposium, but as a post-volume venture(!). Nevertheless, we soon succeeded in commissioning pledges for volume papers from a dozen enthusiastic colleagues—in less than three months. The magic of the "science frontier" worked again; modern technology of e-mail communication made it technically possible.

If, in accordance with the "frontier" theory, the frontier acts as an advancing zone of activity driven by the high level of energy and the concentration of human resources, then we struck at the true "frontier" community in science research. Beyond shared enthusiasm, most of the projects presented in this volume are the outcomes of many years of work and of elaborate planning and development processes. They are collaborative scientific and public enterprises that engage the energy and resources of many people and agencies. They are also products of long-term and trusted partnership between researchers and arctic communities.

The opportunity and importance of translating research projects (documentation of indigenous knowledge) into the next stage (public action, policy, etc.) was immediately recognized by all the people we contacted as potential volume contributors. That, we believe, was one of the key reasons why they came on board with such enthusiasm. With each of us fully aware of the pioneer nature of the projects we were doing in distant arctic communities, we saw an opportunity for strength in numbers. Science frontiers may be launched by a few isolated pioneers, but they take shape and advance with the steady number of people involved and a certain level of (previously) accumulated activity only. Therefore, we consider the swiftness of this volume's preparation as one more demonstration of a highly developed status, if not maturity, of our common field of research.

The "frontier" paradigm also advocates a high level of communication within the advancing zone of interaction and change. Our "frontier" was driven and supported by today's most universal means of communication, the Internet, and it became possible because of the technology that was unimaginable even a decade ago. Whereas the volume's basic outlines were discussed in person during a few days of the ARCUS meetings, almost everything else was accomplished via e-mail. Stories of distant lands and images of isolated arctic villages that once took months to reach, traveled freely through cyberspace. They also came out of desk printers by mere computer clicks. All of this, of course, is a familiar practice in today's science. However, we think we probably pushed the envelope slightly. When one volume editor relocated from North America to New Zealand, a new virtual arctic research node was set in motion eighteen time zones, a full hemisphere, and a whole new day away. In daily communications between Washington, D.C., New Zealand, and the ARCUS headquarters in Fairbanks (where our volume's publication office was located), it often created a twenty-hour-long business day. It also made us to acknowledge almost physically that this is truly a small world—which is another critical function of any frontier experience.

The Focus and the Structure of This Volume

Communicating environmental change from the perspective of indigenous arctic people is a central theme to this volume. What emerges in the chapters to follow is a group of local, place-based stories that—taken together—reveal a much more extensive record of events to be put into circumpolar or, at least, in the North American continental perspective.

The projects presented here report ground observations from some twenty-three indigenous communities (see map, page xxviii). They cover thousands of miles—almost the entire stretch—of the North American polar zone, from the Bering Strait area to north Alaska to the central Canadian Arctic to Baffin Island and the Labrador eastern coast. Despite their differences in local geography, patterns of community involvement, and researchers' approach, the authors have structured their papers around the observations, stories and personal reflections of the people they work with and learn from. In this way, these papers offer Native residents various venues to speak for themselves, using their own words and explanations. They also all recognize the need to encourage more unconventional and better working relationships between researchers and local experts. Each chapter thus offers its own perspective on finding ways to make this kind of collaboration possible.

The title of the book, *The Earth is Faster Now,* also carries our common message. It comes from a comment of a local elder, first shared by Caleb Pungowiyi, then-the president of the Alaska Eskimo Walrus Commission. While talking to elders in his native village of Savoonga on St. Lawrence Island, Alaska, he recorded the following statement from Mabel Toolie (Legraaghaq, born in 1912). This is how Pungowiyi explained this statement himself:[1]

> My aunt, Mabel Toolie, said [to me]: "The Earth is faster now." She was not meaning that the time is moving fast these days or that the events are going faster. But she was talking about how all this weather is changing. Back in the old days they could predict the weather by observing the stars, the sky, and other events. The old people think that back then they could predict the weather pattern for a few days in advance. Not anymore! And my aunt was saying that because the weather patterns are [changing] so fast now, those predictions cannot be made anymore. The weather patterns are changing so quickly she could think the Earth is moving faster now.

The contributions in this volume thus respond to a common urgent issue—the need to come to grips with what is happening in the North, as explained by the people who live there and who are experiencing changes firsthand. The first three chapters in the volume tell the stories of building a whole network (actually, three different types of networks) of engaging communities in climate/environmental change documentation. The change is obvious to the people on the ground—the question is, what is the best way to reveal it and to send the message to those agencies who make management and policy decisions. In the opening chapter by Shari Fox, this task has been achieved by building a climate change observational record via a combination of techniques: from meetings, presentations, and informal discussions to semidirected interviews, focus groups, mapping exercises, and videography in four communities in Nunavut. The second paper by Gary Kofinas et al. presents a different model—when a whole system of community-based environmental monitoring was created across the U.S.-Canadian Arctic Borderlands via a cooperative of five northern communities, government agencies, and university researchers. The main purpose here was to build a long-term database and process of communication, through which evidences of change could be traced along many parameters and at many different levels. The third paper, by Jolly et al., describes how residents from the community of Sachs Harbour on Banks Island in the western Canadian Arctic decided to format their observations of and ideas about the recent changes into a powerful media message—by making the video *Sila Alangotok* to raise awareness in more southern regions and to make links with other communities.

The two next papers reveal documentation projects that are both heavy in researchers' and observers' reflections about their partnership, though not framed directly by the "arctic climate change" paradigm. David Norton expresses his own reservations about the "linear" nature of climate change scenario, a perspective which is shared by his native collaborators in northern Alaska. Instead, it was rather a concern for safety of hunters on the sea ice that worked as a key factor encouraging local participation in a comparative documentation of a series of abnormal ice events at the recent Barrow Symposium on Sea Ice and beyond. Krupnik's paper documents how the desire to preserve ancestors' knowledge on ice and weather monitoring practices became the key factor for local participation in sea ice observation project on St. Lawrence Island, Alaska. Thus, the St. Lawrence Island project was rather an experiment in creating a record of Native ice and weather observations by local residents themselves and by the way "we see it."

The next paper by Thorpe et al. presents a specific cross-section from a long-term collaborative effort in documenting Native knowledge related to the biology, behavior, and changes in arctic caribou population in arctic Canada. Here, again, the main message is that of urgency, because the Quitirmuit in the Nunavut Territory are concerned that changes in the weather, environment, and wildlife populations seem to be happening too quickly for people and the environment to adapt.

In her paper on traditional navigation skills, Claudette Bradley illustrates how the sharing of knowledge always comes to the issue of change, even when the topic is the sky, the stars, and land orientation techniques. What originally looked (at least to an outside researcher) like a pure knowledge documentation effort to build a classroom curriculum in a Yup'ik community in western Alaska eventually emerged as a discussion of environmental change, as elders observed their old techniques being altered by the new environmental realities.

The next two papers in the volume also focus on the people side of the human-environment equation. Both contributions describe the impacts of environmental change on indigenous people; Furgal et al. in terms of human health, and Nickels et al. with the need to support communities in finding ways to cope with the changes they are experiencing. For Furgal et al., the link between human health and environmental change is critical; links seen, for example, in the effects of environmental changes on the abundance or availability of country food, or the ability of Inuit to access these resources. In the next paper, Nickels et al. describe a definite role for national organizations such as Inuit Tapiriit Kanatami (ITK) in "putting the human face" on climate change. The authors show how ITK is partnering with local, regional, and national organizations to bring indigenous voices into climate change science and policy through a series of community workshops.

We were also very lucky to commission three of our colleagues with extensive experience and involvement in indigenous knowledge and climate change issues to contribute their more general perspectives on the subject of this volume. In his opening Foreword, Jose A. Kusugak, the president of the Inuit Tapiriit Kanatami, addresses the critical importance of the ongoing environmental change in the Arctic, both from his personal life experience and from the position of the organized Inuit community in general. In the following Preface, Henry Huntington, the current president of ARCUS and key author for the Indigenous Perspectives chapter in the forthcoming Arctic Climate Impact Assessment (ACIA) document, shares his remarks on understanding the Arctic environment system through the richness of knowledge of arctic people. Beyond the obvious value of such an approach, he also writes of the many challenges that lie on the paths that seem so smooth and straightforward from the outside. Finally, in the volume's Epilogue, Fikret Berkes looks at arctic environmental change through the framework of a sustainability science that includes local knowledge and observations. He argues that traditional knowledge, and civil science in general, are essential ingredients of sustainability science because more conventional scientific approaches are limited in their ability to deal with complex systems problems such as climate change. His thoughts offer a challenging test to current stereotypes on how indigenous knowledge can be matched with, checked by, and recorded along the practices of modern science. We hope that those readers looking for "silver bullets" in comparative data from other sciences and fields will read this Epilogue with the intensity it deserves.

The last (and the most recently joined) component of this volume is a report by a group of our European colleagues established in Tampere, Finland. Their project, Snowchange, is a multiyear education-oriented effort to document indigenous observations of climate change in northern regions, particularly across the Nordic countries, arctic Russia, and Siberia. Tero Mustonen, who chairs the Snowchange project group, contributed his short report on their activities during 2001 and early 2002, which complements the primarily North American focus of this volume.

As our readers will see, by putting together this volume we bring under a common cover a network of researchers doing similar kinds of work and the communities who are supporting them. The ability of individual voices and projects to raise awareness about what is happening is thus multiplied by sharing the stories all across the North: from Alaska to the central Arctic to Nunavut to Labrador, and, finally, to the Eurasian Arctic. As these individual stories report, as (and if) the climate and landscape continue to change, many arctic residents wonder "what then?" Neither they nor us scientists have a definite answer, but both parties represented in this volume see it as extremely urgent to start the discussion.

Acknowledgements

First of all, we are very grateful to all our volume contributors, whose work made this international collection of papers possible. There was no other magic in the integrity and swiftness of the editorial process other than your high discipline, dedication, and perseverance. It was a real pleasure working with you and we thank you all.

Despite all these daring efforts in scientific writing and analysis, there would never be a volume on documentation of indigenous knowledge if not for cooperation and goodwill of so many northern communities and dozens of individual local collaborators. Although each volume paper acknowledges this contribution and cites the names of particular Native participants, this book in general is a tribute to the scholarly explorations and research expertise of the arctic residents. Such knowledge is their special treasure, the best of their scholarship, and a pinnacle of generations of inquiries and achievements in mastering their beloved though often unforgiving environment. Many projects described here were, in fact, generated by the Native communities rather than by the intellectual curiosity and research agendas of individual scholars. In return, this contributed to the unique level of knowledge sharing among local experts and academic researchers, which, we believe, is the true mission and spirit of our common enterprise.

As volume editors, we are particularly grateful to Fikret Berkes and Henry Huntington for their many insights, friendly advice, and the willingness to share their perspectives on the role of indigenous knowledge in the study of arctic environmental change (eventually summarized in this volume's Epilogue and Preface, respectively). This was often done on a very short notice, and we appreciate your help. Also helpful on short notice was the very skillful Philippa McNeil, from the Canadian Wildlife Service in Whitehorse, Yukon Territory, who produced our general North American Arctic map showing the communities covered in the individual volume papers. We are also thankful to our home institutions—the Smithsonian Arctic Studies Center (ASC) and the Centre for Maori and Indigenous Planning and Development at Lincoln University in New Zealand—for providing logistical support and other resources to work on this project. ASC's timely financial contribution to this project also allowed us to increase distribution of this volume to share with Native communities and colleagues in northern studies.

Three people made critical contributions to this venture and they deserve special thanks. Wendy Warnick, ARCUS executive director, embraced the idea of our "frontiers" volume on indigenous knowledge from its very beginning; she also backed it with the full support and resources of the ARCUS "powerhouse" office in Fairbanks. Fae Korsmo, former program manager of the Arctic Social Sciences Program at the National Science Foundation's Office of Polar Programs, was always a source of inspiration and encouragement. Finally, this volume is very much the tribute to Sue Mitchell's (our ARCUS editor's) skills in style editing and design as well as to her patience in bringing several papers in different styles and colors, with dozens of illustrations, delivered in every format possible, under a common cover. We hope the readers will appreciate this final product of our "frontier" teamwork. We also hope that more similar "frontiers" volumes in other fields of polar social research will eventually come out of this initiative through the ongoing support of the ARCUS and the NSF Arctic Social Sciences Program.

1. Caleb Pungowiyi (from Kotzebue) speaking at the Girdwood Workshop on Sea Ice and Environmental Change, February 15, 2000 (quoted in Krupnik 2000: 26). We are grateful to Caleb for his kind permission to use his story for the title of this book.

Figure 1-1. Flying near Clyde River, August 2000. Dramatic fjords are characteristic of the regions around Clyde.

1

These are Things That are Really Happening:

Inuit Perspectives on the Evidence and Impacts of Climate Change in Nunavut

Shari Fox
Cooperative Institute for Research in Environmental Sciences (CIRES) and the Department of Geography, University of Colorado at Boulder, USA

Resume ᐸᐃᐸᑦ ᐃᓱᒪᒋᔭᐅᔪᑦ

ᐅᐊᑦᑎᐊᕈᓂᐸᐊᓗᒃ ᑕᐃᒪᓐᖓᓂᑦ, ᐃᓄᐃᑦ ᓯᕗᓕᖏᓪᓗ ᐊᓐᓇᐅᒪᓇᓱᐊᕐᓯᒪᕗᑦ ᓄᓇᖕᖓᓂᑦ ᑕᕆᐅᖕᖓᓂᓪᓗ
ᐅᑭᐅᖅᑕᖅᐳᑐᑉ ᐃᓗᐊᓂᑦ. ᐊᑐᖅᓯᒪᔪᒥᒍᑦ ᐅᔾᔨᕆᓯᒪᔪᒥᒍᓪᓗ ᑲᑎᖅᓱᐃᓯᒪᕗᑦ ᐊᖏᔪᒻᒪᕆᖕᒥᒃ ᐊᑐᕈᓐᓇᖅᑕᒥᓄᑦ
ᐊᕙᑎᒥᓄᑦ ᐊᒻᒪᓗ ᓯᓚᐅᑉ ᓄᓇᐅᓪᓗ ᖃᓄᐃᓕᖅᐸᓪᓕᐊᓂᖓᓄᑦ ᐃᓕᖅᑯᓯᕆᔭᖏᓄᓪᓗ. ᑕᒪᓐᓇ ᖃᐅᔨᒪᔭ
ᑐᖃᐅᔪᖅ ᐃᓚᖃᕆᕗᖅ ᐅᑭᐅᖅᑕᖅᑐᒥᑦ ᓯᓚᐅᑉ ᖃᓄᐃᑦᐳᑐᓂᖓᓂᒃ ᖃᓄᕐᓗ ᐊᓯᔾᔨᐅᑎᓯᒪᓕᕐᒪᐸᖓᑦ ᐅᓪᓗᒥᒍᑦ.
ᑕᒪᓐᓇ ᑐᑭᓯᐅᒪᓇᓱᐊᕐᓂᖅ ᓯᓚᐅᑉ ᖃᓄᐃᓕᕐᓂᖓᓂᒃ ᐊᐳᑎᖃᑦᓚᓇᖕᒪᔅᓯᓕ ᐃᓄᐃᑦ ᐃᓅᓯᖏᓐᓄᑦ
ᑕᒪᒃᑯᓇᓐᖓᑦ ᐅᒃᐱᓇᔭᐅᔪᓂᒃ ᑕᒪᒃᑯᓄᖕᖓ ᐊᓐᓇᐅᒪᓇᓱᐊᕐᓂᕐᒧᑦ ᐊᔪᓐᖏᔭᔪᑎᑦ. ᓱᓕ ᐅᓪᓗᒥᒧᑦ ᐃᓄᐃᑦ ᐊᒥᓱᑦ
ᐊᑐᖅᐸᖕᒪᑕ ᖃᐅᔨᒪᔭᑐᖃᕐᓂᒃ ᓯᓚᐅᑉ ᖃᓄᐃᓕᕐᓂᐊᕐᓂᖓᓄᑦ ᓇᓚᐅᑦᑕᕈᑎᓂᒃ, ᐊᖅᑯᑉ ᐃᓗᐊᒍᑦ ᒪᓕᒃᓗᒍ
ᓯᓚᐅᑉ ᐃᓚᓴᓯᐅᒪᓂᖏᑦ ᖃᓄᐃᑦᑐᐊᕐᓂᒃᓴᖓᓂᓪᓗ. ᑕᐃᒪᓕ ᑕᒪᑐᐸᒍᓇ ᖃᐅᔨᒪᓂᕐᒥᒍᑦ ᓯᓚᐅᑉ
ᐃᓕᖅᑯᓯᖏᓐᓂᑦ ᐊᑐᖅᑕᖏᓐᓄᓪᓗ, ᐃᓄᐃᑦ ᖃᐅᔨᓴᕋᐃᓐᓂᖅᓴᐅᖕᒪᑕ ᐊᓯᔾᔨᖅᐸᓪᓕᐊᓂᖕᒪᓂᑦ
ᐃᓕᖅᑯᓯᐅᐸᕙᓪᓗᐊᑦ ᐊᓯᐊᒍᑦ, ᓱᕐᓗ ᐅᐱᕐᖓᒃᐳᓴᖅ ᐊᓯᐅᓪᓗᓂ ᐅᐱᕐᖓᑲᐅᑎᒥᖃᑦᑕᓕᕐᓂᖓ, ᓯᓚᐅᑉ
ᐃᓕᖅᑯᓯᖅᓱᖏᓐᓂᖓ, ᐊᐳᑎᐅᑉ ᓂᓚᐅᓪᓗ ᐊᓯᔾᔨᕐᓂᖏᑦ, ᓂᓪᓚᓱᖃᑦᑕᕐᓂᖓ ᐊᒻᒪᓗ ᐱᕈᖅᑐᑦ
ᓱᒃᑲᓂᖅᓴᐅᓕᖅᑐᑎᒃ. ᑕᐃᒪᓕ ᓄᓇᒥᓂᒃ ᐊᑐᐃᓐᓇᕋᒥᒃ ᐊᒃᑐᐊᖏᓐᓇᖅᑐᑎᓪᓗ ᐃᓄᐃᑦ ᖃᐅᔨᓴᕋᐃᒻᒪᑕ
ᐊᓯᔾᔨᐅᑎᔪᓂᒃ ᓯᓚᒥᓄᑦ ᐊᕙᑎᒥᓄᓪᓗ ᐊᒻᒪᓗ ᖃᐅᔨᒪᔭᑎᖃᑦ ᑕᑭᔭᐅᑦᑕᑎᒃ ᑕᒪᒃᑯᐊ ᐊᒃᑐᐃᕙᓪᓕᐊᔪᑦ
ᐅᑭᐅᖅᑕᖅᑐᒥᑦ ᐊᕙᑎᒍᑦ ᓂᕐᔪᑎᓄᑦ ᖃᐅᔨᕙᖕᒪᒥᑦ. ᑕᒪᓐᓇ ᖃᐅᔨᓴᖅᑎᓪᓚᕆᖕᓂᑦ ᐃᓚᓴᕆᔭᐅᓯᒪᖏᒻᒪᑦ
ᐃᓄᐃᑦ ᖃᐅᔨᒪᔭᑐᖃᓄᓯᒃᓯᒪᖏᑦᑐᖅ, ᑭᓯᐊᓂᓕ ᐅᖃᐅᓯᐅᓂᖓ ᐊᒻᒪᓗ ᓇᓗᓇᐃᖅᑕᐅᓂᖓ
ᑕᑯᒍᑦᑎᒍᑦ ᑖᕙᓂ ᑎᑎᕋᖅᓯᒪᔪᓂᒃ, ᐃᓄᐃᑦ ᐃᓚᐅᔭᕆᐊᖃᑦᑕᕆᒃᐳᑦ ᓂᐱᖃᖅᑎᑕᐅᓗᑎᓪᓗ ᐅᑭᐅᖅᑕᖅᐳᑐᑉ
ᓯᓚᖓᑦ ᐅᖃᐅᓯᐅᓂᐊᕈᓂ ᖃᐅᔨᓴᖅᑕᐅᓂᐊᕈᓂᓗ.

ᑕᒪᓐᓇ ᑎᑎᕋᖅᑕᐅᓯᒪᔪᖅ ᑕᒪᐃᓐᓂᒃ ᑕᐅᑦᑐᒐᖃᖅᑐᓂ ᕿᒥᕐᕈᐃᔪᖅ ᖃᐅᔨᓴᕐᓂᐅᔪᓂᒃ ᐃᓄᖕᓂᑦ
ᐃᓚᖃᖅᑐᑎᒃ ᐃᖃᓗᖕᓂᑦ, ᐃᒡᓗᓕᖕᒥᑦ, ᖃᒪᓂᑦᑐᐊᒥᑦ ᐊᒻᒪᓗ ᖃᖏᖅᑐᒑᐱᑉᒥᑦ, ᐊᑯᓐᓂᖏᓂᑦ 1995-2001
ᒧᑦ (ᓱᓕ ᖃᐅᔨᓴᕐᓂᖅ ᑲᔪᓯᕗᖅ), ᖃᐅᔨᒪᓂᖏᑦ ᐅᔾᔨᕆᓯᒪᔭᖏᓪᓗ ᓯᓚᐅᑉ ᐊᕙᑕᑕᓗ
ᐊᓯᔾᔨᖅᐸᓪᓕᐊᓯᒪᓂᖏᓐᓂᒃ.

Introduction

There is no question about it, you can tell the big difference in weather patterns from the past. In the years past, there used to be long periods of fine weather. There would be a period of bad weather, which was usually pretty bad, but once it cleared up it would be a long time before another storm would get to us. Change in the weather patterns is very much noticeable. The earth, and the natural environment, are so much different now from what it was in the past. (Z. Uqalik, Igloolik, 1991)

Inuit in the Canadian Arctic are experiencing the effects of climate change in their communities and hunting grounds, and these environmental changes are affecting Inuit livelihoods in a number of ways. For example: changes in sea ice characteristics and distribution have made travel more difficult and dangerous; increasingly unpredictable weather has caused a number of accidents on the land as people get lost or caught unprepared in storms; dried up rivers and lakes block access to traditional summer hunting grounds; hunters who rely on igloos for emergency and temporary shelters in winter can no longer build them due to poor snow conditions caused by changing wind patterns; poor vegetation growth is effecting the health of caribou, and Inuit are finding that the meat tastes different and the skins are of poor quality.

Despite a wealth of empirical knowledge held by Inuit regarding climate change, scientific research on this topic, which has been going on with much vigour in the Arctic for decades, has rarely included observations of local people. Mainly, this is because science has been unable (and often unwilling) to recognize and accept the knowledge of indigenous people (frequently termed "traditional ecological knowledge" or TEK) as valid and functional. During the last decade in Canada, however, as a number of research projects (e.g., Ferguson and Messier, 1997) and government agencies (e.g., Nunavut Wildlife Management Board) began to demonstrate the value and utility of TEK, it began to gain acceptance. Some northern scientists and researchers began to recommend that TEK be incorporated in climate change studies (e.g., Cohen, 1997; Peterson and Johnson, 1995) but it was unclear how to go about doing so. However, in the last few years, efforts to engage indigenous people in arctic climate change research have been undertaken by a small, but growing number of projects (e.g., Fox, 1998, 2000; Huntington et al., 2001; McDonald et al., 1997; Riedlinger and Berkes, 2001; Thorpe, 2000).

Since 1995, I have been working with Inuit in Nunavut to document and examine their knowledge of climate processes and change. Over the course of this study, the methods and research design have adapted to meet the changing needs and interests of the communities and scientists involved, and each phase of the study has had a different focus, with each phase growing out of the last. The overarching goal of this research is the documentation and analysis of Inuit observations, knowledge, and perspectives of climate variability and change, with an aim to help develop frameworks and tools for reciprocal communication with scientists and decision-makers. The research is meant to benefit both Inuit and science by (a) helping Inuit preserve traditional knowledge through documentation, (b) helping communities communicate and preserve traditional knowledge *in situ* through activities that pass knowledge between communities, community members, elders and youth, (c) providing local knowledge and information to decision-makers at various levels for climate change strategies and policies, and (d) developing conceptual and practical products, tools and frameworks for communication and collaboration between Inuit and scientists.

In order to carry out this study, I have been living for different periods of time in four communities in Nunavut; Iqaluit, Igloolik, Baker Lake *(Qamani'tuaq),* and Clyde

River *(Kangiqtugaapik)* to work primarily with Inuit elders and experienced active hunters. A multimethod approach drives data collection and includes interviews, participatory mapping, focus groups, participatory observation, and videography. These complementary methods strengthen the project by providing flexibility, adaptability and means to crosscheck information, while multicommunity involvement provides opportunities for comparisons and elements of representativeness (e.g., differences in local climates and cultures).

This study illustrates the detailed knowledge and sensitivity of Inuit to patterns, variability and changes in the environment. Inuit observations of climate and environmental change cover a wide range of phenomena and are often housed within complex understandings of ecological and environmental processes. As a result, observations of change often come with detailed interpretations that draw on knowledge of webs, links, and feedbacks within ecosystems and landscapes. The impacts of environmental change on lives and livelihoods in the North are also an important component of this work as is community vulnerability, coping strategies and adaptability related to these impacts. Inuit knowledge and responses to climate variability and change are contextualized within family, community and Nunavut life. Links to scientific observations, analyses and methods are made and tools for communication between indigenous and scientific knowledge are identified. This aspect is aided by in-depth interviews with climatologists and other scientists to help find links in ways of learning, knowing and communicating between these researchers and Inuit.

This chapter draws on experiences and results of a long-term, ongoing study with Inuit on their knowledge of climate and environmental change. To try and include the details of the entire project here would be impossible, therefore, this paper focuses on a few selected aspects of the study. In keeping with the theme of the volume, an overview of the project design and methodology (and at times their evolution) are provided. The primary goal of this paper, however, is to present examples of Inuit observations and perspectives on the impacts of climate change. This is done through the transmission of local knowledge—through quotes, photos, an example of a participatory map, discussions of research experiences and findings, and an in-depth look into one community, Baker Lake.

Four Nunavut Communities

This study takes place in four Nunavut communities; Iqaluit, Igloolik, Clyde River *(Kangiqtugaapik)* and Baker Lake *(Qamani'tuaq)* (Figure 1-2). The participation of these four different communities adds several unique aspects to the project:

- Each community tells their own story about climate and climate change. The research is grounded in four different communities (case studies) that have their own particular history and experiences.
- Comparisons between communities of local cultures, language, histories, resources, livelihood strategies and climates, shed light on how Inuit knowledge and perspectives of climate change vary, and are shared, between communities.

Inuit in different regions have unique ways of observing climate and weather, validating knowledge and assessing changes and impacts; they also have many techniques and systems in common. Through these comparisons, better recommendations for information collection and sharing can be made to local and territorial decision-makers.

- This project helps to set up communication between Inuit in different communities. Throughout the project, Inuit have expressed interest in the knowledge and perspectives of climate change from people in other communities, as well as from scientists. Through radio shows, community presentations and smaller group meetings, observations from different communities in Nunavut are shared and discussed.
- Each community has unique climatic and environmental characteristics and is experiencing climate change differently, as noted by both scientific and Inuit observations. Undertaking individual and comparative analyses of communities can yield a variety of useful recommendations for further climate change research and adaptation strategies. For example, for scientists, arctic climate change is usually addressed on a scale that does not allow for regional or local analysis. Often, the kind of data needed for this is not available. Inuit knowledge and observations from different communities can help inform scientific analysis on a smaller (local and regional) scale, as well as provide knowledge on the variation of climate change and its impacts around Nunavut. For policymakers, the study helps to elucidate what actions may be appropriate in order to address the needs of specific communities, as well as Nunavut as a whole.

Iqaluit

Iqaluit (Figure 1-3) is the capital of Nunavut, located approximately 2,000 air kilometres North of Montreal on the southern end of Baffin Island (63° 43' N, 68° 30' W). With a population of 4,220 as of the last census (Census of Canada, 1996), Iqaluit is Nunavut's largest community. Approximately 60% of the residents here are Inuit, compared to other Nunavut communities where many are over 90% Inuit (Ipellie and Rigby, 1999). Since 1999, when Nunavut based its legislature in Iqaluit, government jobs have attracted many people from all over the territory and Canada. Combined with an increase in supporting infrastructure and businesses, Iqaluit has been growing rapidly in recent years.

Traditionally, nomadic Inuit hunters and their ancestors frequented Iqaluit and its surrounding areas which are characterized by rolling, rocky hills, and many lakes. It was not until 1942, during the Second World War, that Iqaluit became a settlement, when the U.S. Air Force built an airstrip there in order to more conveniently support troops and allies in Europe. Inuit in the area, who were employed to construct the airstrip and hangers, settled a small village nearby and named the growing settlement Iqaluit, "a school of fish [sic]" (Ipellie and Rigby, 1999) although the community would be named Frobisher Bay on maps until it was changed back to Iqaluit in 1987.

Like other Northern communities, Iqaluit was also influenced by the presence of explorers, whalers, missionaries, the Hudson Bay Company (1950), and a DEW (Distant Early Warning) Line site (1955).

Today, Iqaluit is a busy town complete with hotels, museum, library, coffee shops, restaurants, bars, as well as a hospital, racquet club and rapidly expanding tourism industry. However, many Iqaluit residents still live at outpost camps for parts of the year and land-based activities, such as hunting and fishing, remain an important part of life here. A strong Elders' Society and Hunters and Trappers Organization help to teach

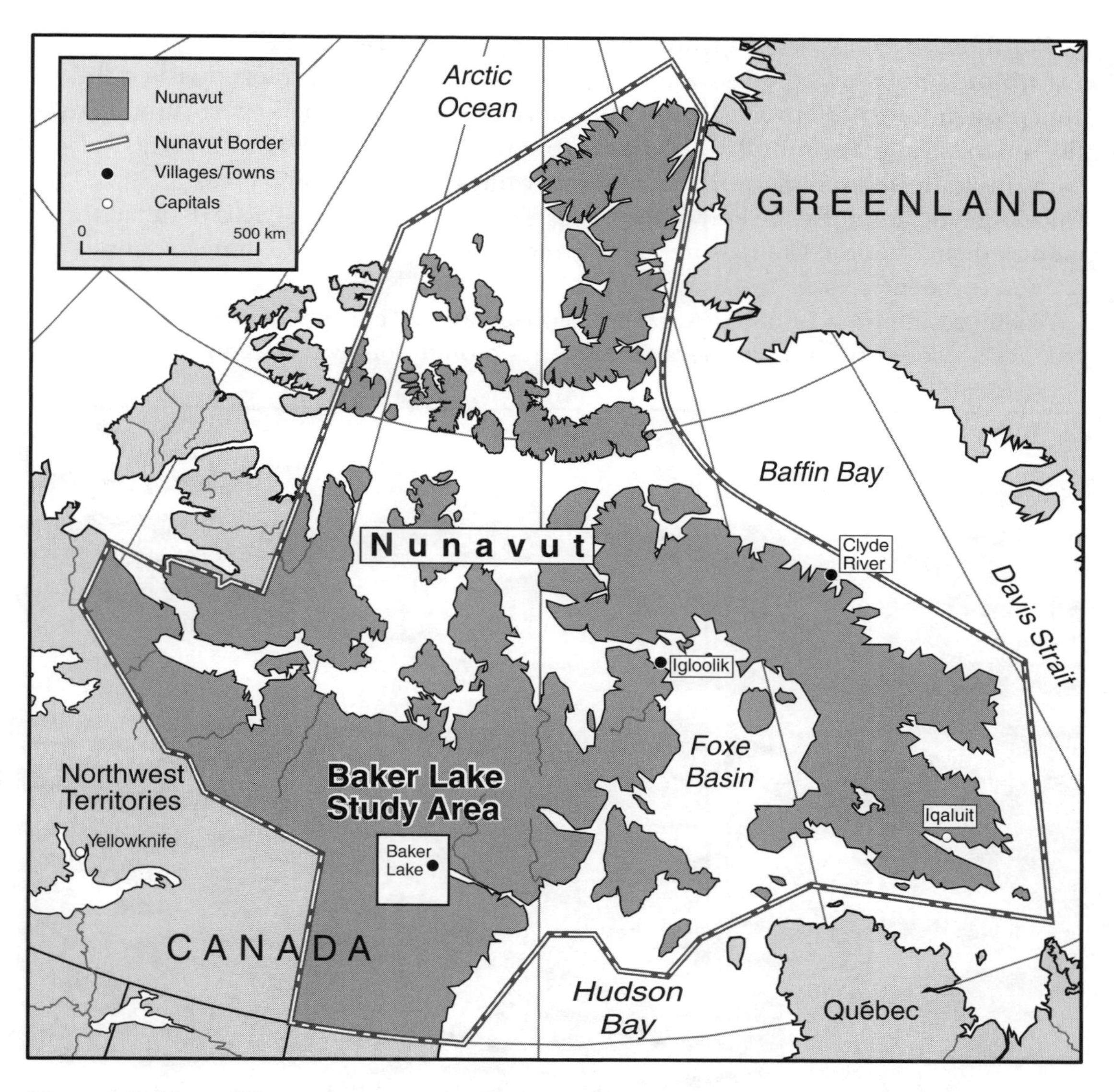

Figure 1-2. Map of Nunavut showing Iqaluit, Igloolik, Baker Lake, and Clyde River. The "Baker Lake study area" provides the location for Figure 1-14.

traditional skills and often host various cultural events such as feasts and traditional games.

Igloolik

Located on a small, relatively flat island in the northern Foxe Basin, next to the eastern coast of Melville Peninsula, Igloolik (Figure 1-4; 69° 22' N, 81° 40' W) is a community of 1,174 people, 93% of whom are Inuit (Census of Canada, 1996). Plentiful wildlife on the land and in the ocean are central to Inuit culture and identity in Igloolik, and these resources continue to support community economic, social and spiritual well-being (MacDonald, 1999).

For thousands of years, Inuit and their ancestors have lived in the resource-rich areas around Igloolik. In the 1800s and early 1900s, a number of explorers visited the area though Roman Catholic missionaries were the first outsiders to permanently reside on the island beginning in the 1930s, followed shortly by the Hudson Bay Company. Inuit, however, remained in traditional camps in surrounding areas (MacDonald, 1999). By the end of the 1960s, Igloolik had a school, RCMP detachment, nursing station, Co-op store and a growing number of Inuit families settling around these services.

Today, residents of Igloolik (widely recognized as the "cultural centre of Nunavut"), balance a strong cultural heritage with changes brought on by modernization

Figure 1-3. View overlooking Iqaluit and Koojessee Inlet, July 2001.[1]

(MacDonald, 1999). The Inullariit Society, Igloolik's elders group, takes an active role in teaching Igloolik youth traditional land skills and sewing, while some also take GPS (Global Positioning System) classes at the Igloolik Research Centre. The Inullariit Society, along with the Research Centre, also manages an ongoing oral history project which documents traditional knowledge of local elders. A local film production company, Igloolik Isuma Productions Inc., has made several films about the local culture and recently produced the internationally acclaimed film, *Atanarjuat: The Fast Runner,* based on an Inuit legend.

Kangiqtugaapik ***(Clyde River)***

With a population of just over 700 (94% Inuit) (Census of Canada, 1996), Clyde River (Figures 1-1 and 1-5) is the smallest community involved in this project. Located on the northeast coast of Baffin Island (70° 27' N, 68° 34' W), the community of Clyde River sits inside Clyde Inlet, neighbour to some of the most dramatic fjords and cliffs in the world. Spectacular glaciers and a seemingly endless supply of icebergs add to the remarkable scenery of this region.

It is thought that Vikings visited Clyde River 1,000 years ago, then later, in 1616, it was explored and mapped by Robert Bylot and William Baffin (Illauq, 1999). Traditionally, Inuit had always moved through Clyde to search for game or visit other

Figure 1-4. Igloolik in February 1996. The round building near the centre of the photo is the Igloolik Research Centre (Nunavut Research Institute).

groups. A permanent settlement began to grow, like many other arctic communities, around various services and trading opportunities that moved to the area (e.g., Hudson's Bay Company, 1924, U.S. Coast Guard weather station, 1940s, federal school, 1960) (Illauq, 1999).

Livelihoods in Clyde still centre on land-based activities. Camping and hunting remain important activities and many people harvest animals, particularly seals and whales, throughout the year for food. Subsistence harvesting is part of a mixed economy in Clyde, where residents also collect an income from arts and crafts, government jobs and tourism ventures.

Qamani'tuaq *(Baker Lake)*

In several ways, Baker Lake (Figure 1-6) is a sharp contrast to Clyde River. Located much farther south (64° 19' N, 96° 1' W) in the Kivalliq region of Nunavut, Baker Lake is the only inland Inuit community in Nunavut. The landscape here is flat tundra, dotted with countless lakes and ponds, as far as the eye can see. Having worked primarily in the Qikiqtani (North Baffin Region of Nunavut) for several years before coming to Baker Lake, I will never forget my first visit to this area where plants can grow knee-high! The settlement is on the north shore of Baker Lake, which is connected to Hudson Bay, approximately 300 km away, via Chesterfield Inlet. Only a day or two travel from the treeline (by dog team), Inuit in Baker Lake traditionally made

Figure 1-5. An inuksuk *overlooks the community of Clyde River, August 2001.*

teepee-style *tupiit* (tents) out of wood and caribous skins (Figure 1-7). Unlike Inuit from elsewhere in Nunavut, the people from this region are not marine based. Caribou and fish are the most abundant and important animals here.

Baker Lake has a unique cultural history. According to Baker Lake elders, nine major cultural groups who were in the area prior to settlement are represented in the community[2] (Webster, 1999). These groups lived in their respective areas until the mid-1950s. Around this time, the nursing station was established, as well as the school. As children were brought to school, this, in combination with a recent period of starvation, led Inuit into the settlement to stay (Mannik, 1998; Keith, 1999). Since then, Baker Lake has becoming a steadily growing community of 1,385 (90% Inuit) (Census of Canada, 1996) known internationally for its fine art. Several galleries and artists' organizations in Baker Lake support a large number of artisans, some world-renowned, who produce carvings, wallhangings, prints, and other crafts.

Research Design

Reflections on the Research Approach

In this paper, I speak to the main component of this study, the documentation and analysis of Inuit observations, knowledge, and perspectives of climate change. In order to collect this information, a research approach was needed that allowed the project to be flexible and responsive to the time, place, and situation at hand. A

Figure 1-6. Baker Lake, July 2001.

multiphase, multimethod framework was used to implement a number of complementary methods, including interviews, participatory mapping, participant observation, focus groups, and videography. These methods will be discussed in detail in the next section. Here, I believe it is important to address aspects of the research, and the researcher, that are significant to understanding the project in terms of its foundations, strengths, and limitations.

This is a community-supported research project. In some ways, this is a distinction from other projects described in this volume that are "community-based"—where it was the community that initiated and carried out the research. In the case of this study, the idea came from me, as a university student in 1995. However, since the very beginning, the communities I approached in regards to starting up this kind of research were extremely supportive and have worked with me to develop the research design over the years. In this respect, this project may be considered "community-based" in that it has always incorporated many of the philosophies and techniques of "participatory action research" and "participatory rural appraisal", both forms of research commonly used in community-based projects (see Chambers, 1994a, 1994b, 1994c; Grenier, 1998; St. Denis, 1992). The flexible, adaptive, and collaborative nature of these approaches is reflected in the methods that have been designed and applied throughout this study.

Figure 1-7. Elders make caribou skin tupiit *(tents) as part of a "traditional day" celebration in Baker Lake, August 2000.*

A key element to this study is time. The project is based on long, repeat visits. Over the course of this study, I will have made seven visits to Nunavut, between one and five months each time and during different seasons of the year. The long-term commitment has benefited the research in a number of ways. First, repeated visits and over a year spent in the field have helped to develop many close relationships in the communities, providing the trust needed to undertake this type of research. Second, this time in the field has helped me to learn the language, through one-on-one tutoring, informal coaching and simply being there. Though my abilities remain only slightly better than *kutaq*—("baby talk"), and my vocabulary is limited, it is clear that my interest and efforts in learning the language have a role to play in how some people react to me and the project, and indeed it has helped develop some of my relationships. Also, methodologically speaking, even a basic understanding of Inuktitut has helped me tremendously to follow participants during interviews and ask better questions.

I would be remiss if I did not acknowledge that this study is, for me, a very personal experience. More and more, I have come to feel a part of people's lives and a part of these places I visit. Now, when I visit Nunavut I do not feel like I am "in the field" anymore, in many ways it is like coming home. I often find myself at different times babysitter, cook, confidant, travel companion, or in one of a number of other family or community roles. Without a long-term approach and these personal experiences and relationships, I would have a very different perspective on this study. For example, I would not be able to contextualize the "data" that I have collected. Also, I would not have a sense of how the context can change, between locations and over time, or an understanding of how climate change fits into a larger picture—individual lives, family life, community, and Nunavut, and how the concern over climate change fits with other (often more important and certainly more urgent) problems and issues facing Inuit communities.

The personal aspects of this project also inject a number of shortcomings and biases into the research. My experiences and relationships in the North have undoubtedly shaped the questions I have asked and, even with local help, the way I have interpreted the answers. Also, my personal circumstances of being a young, female, non-Inuit, university student must also have an influence. And, as a final caveat, quoting a friend, "the researcher always goes and investigates what he/she finds most fascinating, and this is, of course, a subjective process" (Kolff, 2000: 8).

Methods

To date, this study incorporates seven periods of fieldwork. The last three periods, perhaps the most pertinent to this paper, are summarized in Table 1-1. Following, I provide a more detailed discussion of the methods used.

Interviews

Interviewing is the cornerstone method of this study. Two types of interviewing styles are used, semidirected interviews (the majority of those conducted in this study) and unstructured interviews. In semidirected interviews, the participants are guided in the discussions by the interviewer, but the direction and scope of the interview are allowed to follow the participants' train of thought (Huntington, 2000: 1270). During the interviews for this study, I had a list of questions ready, but it was much more effective to let the person speak freely, with me playing off their comments and prompting them with a new question only when necessary. Usually, I would begin these interviews with an open question such as, "I am doing a project about Inuit knowledge of climate change. I wonder if that brings anything to mind for you?" Nine times out of ten, it would, and the participants would be free to begin with what they felt was important to discuss related to this topic. Leading questions and statements were carefully avoided, such as "do you think it is getting warmer?", or "people say that the ice is melting earlier." Both of these statements, for Inuit, would imply that someone else held this knowledge, therefore it must be true, and one would not want to disrespect the knowledge of someone else. Ferguson and Messier (1997) sternly warn that leading questions are especially inappropriate for Inuit, and can lead to incorrect conclusions, since Inuit will not disagree openly if they think that there might be a sound basis for the question, in fact, they will usually respond positively even without personal knowledge. In addition, leading questions are impolite and if an informant suspects that the interviewer is using this tactic intentionally, the interviewer may lose credibility with all informants (Ferguson and Messier, 1997). By using the semi-directed method of interviewing, and keeping the conversation as "interviewee-centred" as possible, leading questions, and thus misinterpreted answers, can be avoided.

Unstructured interviews are slightly different. This method involves minimum control over the participant's responses (Bernard, 1995) and is rather a listening technique whereby the informant is encouraged to freely discuss the topic the way they choose, with very little guidance from the interviewer. I have found this method useful with elders who are most comfortable telling stories. Rather than directing the elder with prepared questions, they were simply asked to share any knowledge they had on the topic, and I tried to make it clear that I simply wanted to know what they think. What often resulted was a wealth of information, but that which needed to be realized through the context of various stories of hunting, travelling, hardships and happy times. Certainly, this method is most useful if one has a good deal of time in the field since it may take several interviews like this to thoroughly cover a topic. This method has also been effective for gathering information about context. By allowing a person to respond with little probing, there were many times when that discussion provided context or meaning to knowledge and information gathered elsewhere. Sometimes, it involved just letting people talk, or rant, but often, people say what they do for a reason and letting them do so can often provide insights that might otherwise be missed.

Over the course of this study, I have conducted 135 interviews (Figure 1-8). Many people were interviewed once; however, some were interviewed a second time to clarify information. Most of the key informants have been interviewed three or four

Table 1-1. Fieldwork phases and methods

Phase	Purpose	Methods Used
Information Sharing and Research Design (4 months, 2000)	To disseminate results of previous work through community presentations, media interviews, and local radio shows	Meetings, radio shows, presentations
	To establish common goals, research strategies and research assistants for the new project phase	Meetings, informal discussion
	To conduct a small set of preliminary interviews in each of the communities to help decide which would be appropriate for in-depth study	Semidirected interviews
In-depth Study (5 months, 2001)	To conduct in-depth studies in Clyde River and Baker Lake	Semidirected interviews, focus groups, participant observation, participatory mapping, videography
	To clarify and verify information with participants	Semidirected interviews, informal discussion, focus groups
Project Follow-up (3–4 months, 2002)	To clarify and verify any unclear information before the final report. To give back to all communities products from the study such as videos, written summaries (in Inuktitut), and maps. Schools and heritage centres will receive summary CD-ROM. Radio shows and presentations will also help to disseminate results	Meetings, informal discussion, presentations, radio shows

times each since they have so much knowledge it is impossible to address it all in one interview,[3] and for some, even after four, we were still only scratching the surface.

All of the interviews were conducted in Inuktitut with the help of interpreters. Fortunately, I have been able to work with the same interpreters over a number of years. In addition to being supportive friends and skillful translators, they have helped enormously with research design and helping me to understand various aspects of local culture. They are valuable research assistants and having their help has added consistency to the interview process and the project as a whole, since I frequently contact them for advice while I am at home. All of the interviews were audiotaped, translated and transcribed. Ten interviews each in Baker Lake and Clyde River were videotaped. This will be discussed in upcoming sections.

Almost all of the interviews have been conducted in people's homes, though sometimes at outpost camps or out on the land and, of course, always over tea. In Baker Lake, the elders' room in the Inuit Heritage Centre also provided a comfortable, quiet place for discussions. All participants were taken through an informed-consent process and encouraged to ask questions before the interview. They were asked to sign a consent form unless verbal consent was more appropriate (only in a few cases). Also, following the Igloolik Oral History Project, participants were paid.

Mapping

During the in-depth study period in 2001, mapping was an important component of many interviews and focus groups (Figure 1-9). Using topographic maps on a variety of scales, mylar overlays and coloured pencils, participants could draw where they have observed environmental changes. Usually it was the older, active hunters who chose to use maps to help explain their observations since these hunters sometimes use maps and GPS units for travel. Even if they were not familiar with maps, once they got oriented, many participants found the maps useful for communicating and places on the maps would jog their memories about some story or piece of information to share. This was not always the case, and some participants chose to sketch or use storytelling to relay their knowledge.

Maps were particularly useful in focus groups. As with the interviews, participants made decisions about when and how to use maps, including when it was useful to make their own. As noted by Flavelle (1995), what is interesting about mapping with groups is not only the map created, but also the process itself. With a number of local experts gathered around a map on a table, people were able to point out particular features they wanted to talk about (e.g., lakes or glaciers) or areas they considered sensitive and in need of monitoring. People were interested in the observations that others had about their respective camps, hunting grounds, travel routes and places where they originally came from. From these starting points, the group could discuss and map what observations they had individually, in common, and various ideas on how to deal with the impacts of various environmental changes. The map was at once a conversation piece and a means to bring together a variety of observations. Sketches, air photos, and photographs also facilitated these discussions.

Participant Observation

Participant observation is one of the principal qualitative methods for conducting research on a particular group or locality. It is an observational method in which the researcher becomes part of the events being observed—often having to make decisions about how much they are observing versus participating (Dane, 1990; Ellen, 1984). For this study, this method translated into being an active member of community life and participating in one of the most important aspects of Inuit life—going out on the land. Being *nunami* (on the land) with friends and family is often when Inuit are happiest (Otak, personal communication 1997) and as expressed by Quaqtaq Inuit, is how one has the best chance to become, and remain, an *inutuinnaq,* a genuine Inuk (Dorais 1997:89).

Several times I had the opportunity to travel with local families and friends on camping trips or to outpost camps (Figures 1-10 through 1-12). These experiences were key to my understanding of Inuit climate knowledge, since I could ask questions in context. Hunters, elders, and others could discuss their observations of environ-

Figure 1-8 (left). An interview with Peter Kunuliusie, an elder from Clyde River, in 2001.

Figure 1-9 (below). Margaret Kaluraq and Shari Fox listen as Silas Aittauq, an elder from Baker Lake, discusses and maps water level changes around Baker Lake (2001).

mental change and indicate examples on the landscape.

Participant observation has been important for developing trust and relationships. My participation in community events or in the chores of packing, fishing and camping are key to being a meaningful member of the group. My willingness to try new things and trust other people has returned that same trust and respect back to me.

Focus Groups

Focus groups can be useful tools as the combined elements of both individual interviews and participant observation (Morgan, 1988). In focus groups, the researcher acts as a facilitator, directing the flow of conversation rather than asking specific questions. The facilitator is interested in the discussion, but also in the interaction between members of the group.

To date, I have conducted three focus groups in Clyde River and three in Baker Lake. In both cases, I worked with my interpreters to select key informants from people we had already interviewed. By facilitating focus groups after we completed all the interviews, we avoided influencing individual answers through a group discussion. We chose people based on their interest in the project and the areas with which they were familiar. Some people wanted to work in groups that had people who were familiar with areas different from their own (e.g., hunting grounds or areas of origin), while others wanted to work with people familiar with the same areas. There were between four and six people per focus group. The sessions lasted approximately three hours.

In Baker Lake, high school students also participated in one of the focus groups. Elders took turns sharing their knowledge of climate and climate change and students were able to ask questions and help facilitate the session. The focus group was incorporated into class material, and students wrote reports on the discussion. This prompted a school-based project where students did their own independent studies, interviewing community elders about their traditional knowledge of weather and climate and incorporated that with what they were learning in science classes about climate change.

Videography

In 2001, twenty interviews and several trips on the land were digitally videotaped in Baker Lake and Clyde River. The idea behind the videotaping was to provide an alternative means to document and communicate Inuit knowledge of environmental change (besides through writing). The video is being used to support a supplemental pilot project to my dissertation, an interactive, multimedia CD-ROM on Inuit observations of environmental change (currently in production).

The CD-ROM incorporates video (clips, as well as an introductory video), maps (both topographic and those made with participants), photos, audio, music, and text (Inuktitut and English) to help communicate Inuit observations and perspectives on environmental change. The CD-ROM is meant to be for anyone interested in Inuit knowledge of climate and environmental change. It has aspects appealing to both community members and academics. Copies of the CD-ROM will be available for

Figure 1-10. Interviewing Apak Qaqqsiq of Clyde River at his outpost camp, 2001.

Figure 1-11. Camping near Igloolik, 1997.

Figure 1-12. Ice fishing at Mogg Bay, near Igloolik, 1997.

schools and cultural centres in the communities involved. More accessible VHS copies of the video portion, plus hard copies of the maps produced, will be available to all participants (both video and text will be in both English and Inuktitut). A copy of the CD-ROM will be included as part of my dissertation and will be accessible to climate researchers at my institution. I will also present the CD-ROM at a variety of conferences and public meetings. At this time, as a pilot project, it will not be available for wide distribution.

Inuit Observations and Perspectives of Environmental Change

> There is a big change. The snow is almost like the ice now. The wind blows a lot harder and when the wind blows that hard it packs the snow really hard. It packs it so hard that it's almost like ice and it's not good now, you can't make igloo shelters with it. These are things that are really happening. These are not things that I think are happening, but these are really happening (M. Aqigaaq, Baker Lake, 2001).

> Inuit have a traditional juggling game. The weather is sort of like that now. The weather is being juggled, it is changing so quickly and drastically (N. Attungala, Baker Lake, 2001).

Inuit are observing changes in their environment and having to make adjustments to their usual activities as a result. Participants in all four communities report changes in such things as weather patterns, ice conditions, vegetation growth, and wildlife, all of which fall outside of what they expect based on oral histories and a lifetime of experience on the land. In many cases, elders, experienced hunters and others relate these changes to a changing climate. Specific observations and concerns can vary between communities and between community members, however, there are also key themes that span communities and come up consistently in interviews.

In this section, I outline examples of observations from Iqaluit, Igloolik and Clyde River providing some general discussion from these communities. Baker Lake is then used as a case study to look more closely at observations of environmental change, their linkages and impacts in one community. Lastly, I discuss the different ways Inuit are reacting to and coping with the impacts of climate change, how climate change fits into the bigger picture of Inuit life and how Inuit knowledge of climate and environmental change can make an important contribution to both climate change research and local and territorial policies that attempt to deal with the effects of climate change.

A Note on Interpreting Inuit knowledge

Each community, and indeed each Inuk, has their own important story to tell about their relationship with the environment and how they relate to changes. To summarize observations and knowledge of environmental change only tells a portion of those stories. We might consider that among Inuit, "general statements are viewed as vague and confusing, whereas specific statements are seen as providing much more interesting information" (Kublu et al., in Csonka, 2000: 4). At the same time, summarizations and generalizations about Inuit observations are important in order to provide a picture of how climate change is affecting Nunavummiut.[4] Through careful collection and interpretation of climate knowledge with Inuit, representative statements can be made which are useful to help communicate knowledge and concerns about climate change to decision-makers, researchers, other Nunavummiut, and the general public. However, it is important to remember that there is more to the source of this knowledge—it comes from individual accounts. In the more detailed description of Baker Lake, presented in the next section, we see examples of the richness of Inuit knowledge through quotes and maps made by various people. At the individual scale, we can learn more about subtle linkages between the environment and local and personal histories, identities, and livelihoods, as well as how knowledge of the environment and its changes is variously constructed, validated and communicated. This gives us a more complex and nuanced look at the Arctic and environmental change and helps us to further understand Inuit knowledge and culture, which in turn may also contribute to the initiation of meaningful and sensitive research and policy-making. Both presentations of Inuit knowledge (in summaries and individual quotes) are complimentary (Csonka, 2000) and form the basis of this section.

Iqaluit, Igloolik and Clyde River

Community members from Iqaluit, Igloolik and Clyde River have unique observations regarding change, while sharing others at the same time. For example, all three communities observe that the weather has become more variable and unpredictable, mainly within the last ten years. Elders in all three communities have expressed disappointment and frustration in the fact that their traditional methods of predicting the weather are not working.

> Even today with the clouds around [since they are used in weather prediction], it is hard to predict what kind of weather we are going to have because it has been getting windier and [changing] more rapidly. Because of that, it has been hard to predict. In the past, it was easier [to predict the weather] because it would be uncommon for winds to come around suddenly, it wouldn't be as windy, and it would be easier to predict with clouds. But now it becomes more difficult to predict because the weather changes [so quickly]. (T. Qillaq, Clyde River, 2000).

> Right now the weather is unpredictable. In the older days, the elders used to predict the weather and they were always right, but right now, when they try to predict the weather, it's always something different. It's very unpredictable right now (Z. Aqqiaruq, Igloolik, 2000).

> The weather has changed. For instance, elders will predict that it might be windy, but then it doesn't become windy. And then it often seems like it's going to be very calm and then it suddenly becomes windy. So their predictions are never correct anymore, the predictions according to what they see haven't been true (P. Kunuliusie, Clyde River, 2000).

In all three communities, the sun feels like it has been getting stronger. Inuit residents are sustaining sunburns and strange skin rashes (attributed to sun exposure), both conditions they have not experienced in the past. Older hunters who have spent long periods of time out on the sea ice in twenty-four-hour sunshine for much of their lives have rarely had their skin burn. Sunburns and rashes seem to be an increasing phenomena of the last nine years or so (though health centres may have also contributed to increased awareness of risks to sun exposure and skin problems, thus more recognition of these conditions). Many residents describe the sun as "stinging" or "sharp feeling." They also complain that the sun in these recent years hurts their eyes.

In Iqaluit and Clyde River, *aniuvat* (permanent snow patches) are disappearing. In Clyde River, the disappearance of these snow patches has meant the loss of some good areas for caching meat. In Iqaluit, it has meant the loss of drinking water for some residents who still prefer to collect and melt this ice instead of using water delivered by the municipality. In Iqaluit, it is also interesting that this melting has been occurring even though some residents claim that there has been a cooling trend—a phenomenon not contradictory if one considers, from a scientific standpoint, explanations involving changes in precipitation or mean relative humidity (Seimon, personal communication, 1999). This example from Iqaluit illustrates how Inuit knowledge is localized and at this scale can identify problems that scientists might not be aware of. It is also an example of how environmental changes may be connected to climate change in ways not always obvious.

Related to temperature change, Inuit did not identify any clear trends in these communities. While participants still considered the environmental changes in Table 1-2 to be related to climate, most people noted that annual and seasonal temperatures were part of a cycle—they are always different from year to year. Again, we can draw some linkages with science. Both Inuit and scientists note that environmental changes are not always connected to temperature, but to other changes in the climate system, such as we saw with melting *aniuvat* in Iqaluit (Figure 1-13). Also, Inuit and scientific observations of climate both show that Iqaluit, Igloolik and Clyde River all fall within areas of the Arctic that have been experiencing a cooling or little to no changes in temperature trends (with annual and seasonal variability), unlike the Western Arctic which is experiencing a pronounced warming (IASC, 1999; Serreze et al., 2000; Riedlinger and Berkes, 2001; Thorpe, 2000). Long term observational records, both Inuit and scientific, are also in agreement that weather variability has increased over the last five year period (Fox and Daniels, in prep.).

Table 1-2. Examples of environmental changes observed by Inuit in Iqaluit, Igloolik, and Clyde River

Iqaluit	• winds change suddenly, weather changes used to be more subtle • weather unpredictable since 1990s • sun's rays feel stronger • sky is hazy, not as blue • birds arrive earlier and new species are arriving, e.g. robins since late 1990s	• *aniuvat* (permanent snow patches) are melting in the hills around the community • ice conditions becoming more unpredictable with several accidents occurring in the last few years • though some residents cited evidence of a cooling trend in the last few years, more residents noted they could not identify any temperature trends with their knowledge • more unusually hot days in summer
Igloolik	• weather increasingly unpredictable in recent years • sun's rays feel stronger • sky is hazy, not as blue	• less periods of extended clear weather • some residents claimed a warming or a cooling trend (opinion split) but more emphasis placed on variability from year to year and that weather and climate follow cycles
Clyde River	• *aniuvat* are melting all around the community • *auyuittuq* (glaciers) are changing—many are melting, though some advancing • increase in weather variability • weather increasingly difficult to predict in recent years • changes in snow distribution, depth and colour	• winds have changed in direction and strength and change suddenly • sea ice has changed—usual leads do not form and new ones open in unusual areas; ice thinner and dangerous for travel in some areas • more iceburgs • warmer springs • sun's rays feel stronger

As a tangent but important note, the impacts of climate change in Nunavut are not always negative. An example can be made from Clyde River. Peter Kunuliusie mentioned during one of our focus groups (2001) that maybe climate change could be a good thing. As many of the glaciers in the region are receding, they are exposing ground that has been covered for a long time. Old animal bones and antlers are to be found on this ground, and Peter, being a carver, is happy to find new resource materials for his artwork.

Baker Lake

In this section, I present observations and experiences from Baker Lake in order to give a more detailed picture of how Inuit are discussing and experiencing the impacts of climate change. Table 1-3 helps to organize this information so that one can see examples of the topics people chose to talk about, the specific observed changes they noted, and some of the evidence, linkages, and impacts associated with these observations. Though compartmentalizing and summarizing this knowledge in chart form is useful to give an overview of how people are experiencing change in Baker Lake, it cannot adequately illustrate the richness of Inuit knowledge. The sections following Table 1-3 attempt to address this by selecting two topics for more in-depth discussion, using the words of Inuit themselves and an example of how hunters used mapping as a tool for articulating their observations of change.

As seen in Table 1-3, Inuit in Baker Lake are seeing climate change impact their environment in a number of ways. Two phenomena cause particular concern and were repeatedly discussed in interviews, focus groups and even casual conversation—lowering water levels and changing winds. These changes carry special significance for residents due to their impacts on wildlife and land-based activities.

Figure 1-13. Scars left on the landscape near Clyde River where an area of aniuvat used to be, 2001.

Table 1-3: Evidence of climate change as discussed by Inuit in Baker Lake

Topic	Observed Change	Evidence, Linkages and Impacts
Climate and seasons	Winter colder in past	• caribou clothing no longer needed to keep one warm, "southern" clothing good enough • nose and cheeks used to turn white (frostbite) much sooner when outside in winter (pre-1960) • dogs used to get steamy in Jan/Feb before 1960s
	Shorter winters	• used to have enough snow November 1 for dog team travel, now comes later • ice expected in September, now comes later
	Early and quick spring melt	• lakes break up early, snow disappears quickly off land • snow gets "shiny" (melt layer) earlier (March instead of April), this is noticed particularly by dog team drivers • some people wait to go out for spring hunting—last two years melted so fast people didn't get to go
	Warmer springs	• people used to go out in the evenings for a quick hunt when ice and snow hardens again—now little to no refreezing at night
	Warmer summers	• poor vegetation growth • thin caribou due to poor vegetation and more activity (running around) possibly due to being agitated by the heat • less mosquitoes due to less standing water
	Longer summers	• people cache meat at the usual time in late summer/fall, but recently it's been too warm and meat is spoiling
Weather	Increased weather variability	• frequent inability of experienced hunters and elders to predict the weather • storms used to build up gradually, now storms occur without warning • direction of wind changes suddenly • many people remember longer periods of nice, clear weather in the past. These days there are more clouds. • unpredictability of weather has caused a number of people to be stranded on the land. • difficult to dry meat, weather too variable to do it

Topic	Observed Change	Evidence, Linkages and Impacts
Rain	Not as much rain in summer Rains harder but longer periods between rains	• poor vegetation growth, thin caribou • used to rain just enough to replenish the land, now rains harder and not as frequently
Snow	Not as much snow	• traditional fox trapping areas no longer have enough snow to bury traps • tops of hummocks show through snow in winter, hard to travel
	Snow layering changed	• usual layers not found when digging to make igloos • snow feels different when poked with a snowknife
Ice (Freshwater)	Freeze up is later	• lakes used to begin freeze up in August, now later in year
	Spring melt process is different	• used to be spring freeze/thaw cycle; now ice melts and does not refreeze
	Areas of thicker ice	• thicker as seen through drilled fishing holes—less snow on the ice (due to stronger winds) reduces insulation and causes thicker ice
	Areas of thinner ice	• some areas used to be too thick for ice fishing, now one can fish here because ice has thinned up
	Rougher ice	• more winds change ice formation process
Lakes and Rivers	Water levels down in lakes and rivers	• cannot navigate through usual lake and river boating routes due to shallow water • blocked routes have made some caribou summer hunting grounds inaccessible • islands have grown bigger and shorelines have extended • smaller lakes and streams have dried up • Thelon River used to run strong enough to push ice from the lake, now too shallow, not as strong • one woman visited the rocks she used to jump across to get to the other side of the Kazan river—now there is less water there
Winds	Winds blow stronger	• snow packed unusually hard, cannot make igloos, makes for unsafe travel when emergency shelter (igloo) cannot be made—a number of accidents on the land have been blamed on inability to find good snow to make a shelter • hard-packed snow from wind makes for bumpy, uncomfortable travel

Topic	Observed Change	Evidence, Linkages and Impacts
	Change in dominant wind	• winds used to blow mainly from the west, but now often from the northwest and north[5] • clouds used to appear from the west now seem to appear from nowhere • tents are still set up in the same place year after year. Originally set up so wind would be to back of the tent, now often hits the side • one must be very careful when using ridges in snow for navigation during winter. Changes in the wind have made these ridges less reliable
Caribou	Caribou are thin	• not as much fat on the body when butchered
	Caribou are not as tasty	• meat is tough because caribou are running too much (perhaps due to heat or being chased more by machines)
	Other changes in body condition	• more liquid in joints • neck areas of skins rip easily • white pustules on meat, more than usual
Insects	Less mosquitoes	• too hot, less standing water to breed these insects
	More of some insects	• more black flies—collect more in summer cabins. Some note that even though more of some insects, fish are still doing poorly
	New species of insects	• new insects arriving in Baker Lake area—some never seen, some familiar to a few people who come from treeline area
Fish	Less Fish	• not as many fish in waters where they are expected
	Poor health	• fish now often too small, skinny, strange colour (trout are darker) and different smell (smell "like earth") • when boiled, fish meat splits—these days one does not see the white fat layer between meat
Birds	Poor health	• seem smaller and "not as happy" • more common redpolls and white throated sparrows, hardly see lapland longspurs anymore (names identified using bird field book)

Lower Water Levels

> I have always lived around Baker Lake and traveled to the area's rivers and lakes, so I have noticed that there seems to be less water in the Kazan and Thelon [rivers], and here in Baker Lake itself. It is not just the rivers and streams that seem to be drying up or having less water than previous years . . . there also was a big pond between the town and the airport, but it isn't there anymore, because there is less water everywhere. (B. Peryouar, Baker Lake, 2000)

> The Thelon river used to be a very strong flowing river and it used to be so strong that it would push the ice from the lake, but in the past few years, when I've traveled along that river, it seems more shallow and a lot of the little lakes and rivers all around Baker Lake have gotten shallow and some places, like the little ponds and little lakes, have even dried up. (B. Peryouar, Baker Lake, 2001)

> The lakes and rivers do not have as much water as they have had in previous years. There is usually a lot of fish in a lot of the lakes but probably because there is not as much water in the lakes or the rivers that probably is why there is not a lot of fish in the waters. The fish move to places where there is deeper water, where there is more water in the lakes or rivers. (I. Kaluraq, Baker Lake, 2001)

> There is a lot less water, around all these islands [in Baker Lake]. The shore is getting closer to Sadluq Island, they are almost joined together. There used to be a lot of water. We could go through with our outboard motors and boats, but now there is getting to be less and less water all over. . . At the mouth of Prince River there used to be a lot of fish and you used to be able to get char. There's been a lot less fish because there's not as much water anymore. And we used to be able to get a lot of fish all the time at Qikiqtaujaq and all the other places where you can get fish. The fish were more plentiful and they used to be bigger. Now you hardly get char anymore at Prince River or any of these fishing places because the water level has gone down. (L. Arngaa'naaq, Baker Lake, 2001)

The heritage of Inuit in Baker Lake is tied very closely to the lakes and rivers. Many of the groups who settled in Baker Lake (see "Baker Lake" in previous section) depended upon the region's waters for their survival. For Ukkuhiksalingmiut, they survived mainly off of fishing in the Back River area (Webster, 1999). For Harvaqtuurmiuit, survival was primarily dependent on caribou, primarily hunting during the fall caribou crossings. Groups of hunters in kayaks would wait for caribou at the places they crossed rivers and intercept them with their spears (Webster, 1999). After settling in the community in the 1950s and '60s, these and Inuit from other groups continued to hunt and fish in their traditional waters and also in the waters of Baker Lake and the rivers and lakes close to it.

Knowledge of lakes and rivers is closely linked with the knowledge of other elements and processes in the environment. When elders and hunters discuss their assessments of lake and river conditions they do so with consideration to such things as ice, precipitation, vegetation, fish health and populations, temperature, and the history of the water body. In explaining the dramatic drop in water levels of lakes and rivers since the late 1990s, people make links to other observations of change.

> This summer, a lot of the lakes and little ponds have dried up. There is usually a small stream that goes around Baker Lake, the community, down to the lake, but it's all dried up right now. This summer I've noticed there is less water this year in the lakes and ponds. I noticed that last year [too]; we did have a very hot summer. Usually I go out walking or hunting, I have a cabin just a short distance from the community and I usually walk there during the week. When I was travelling between my cabin and the community I noticed that there was not as much mosquitoes around because it was so hot last summer. This year there seems to be a bit more mosquitoes around because it's not as hot this summer. . . . It's [the lower water levels] probably because this last spring there has been hardly any rain at all, it's just finally starting to rain a lot later in the summer and there might be more water now that it's just starting to rain, but there was hardly any rain at all during the spring and that's probably why. . . . It is unusual [that] this past spring there has been hardly any rain. I have a dog team and the place where I keep them there is usually some puddles and little streams that are right near where I keep my dogs but all the little streams and ponds are all dried up because, for some reason, there hasn't been a lot of rain. That's not usually the way it is. (S. Aittauq, Baker Lake, 2001)

Shallow water in lakes and rivers is having an impact on community life primarily by restricting travel and access to hunting areas. In the 1940s and 1950s, hunters could take their boats (with outboard motors) without any difficulty up the rivers that connect to Baker Lake. Since the 1960s, many people have noticed that the water has been getting slightly shallower (with variability) but never reaching the extremes that have been occurring since the 1990s. In the last four years, usual travel routes along rivers have been completely blocked due to shallow water. Portaging is not possible because the blocked areas are too big to cross and because it is too difficult to carry all the equipment people transport for camping and hunting on the land. Many hunting parties have had to turn back. Matthew Aqigaaq (Baker Lake, 2001) stressed the seriousness of this situation when he described how blocked boat routes have prevented hunters access to their usual caribou summer hunting grounds (see Figure 1-14). This area has good caribou hunting, but is only accessible by boat in the summer as the ground is too marshy and hummocky to travel by foot or four-wheeler.

Through interviews and focus groups, hunters like Aqigaaq told similar stories and mapped their observations and concerns about changing water levels. Figure 1-14 provides an example of how maps were used to communicate information and support discussion. The map shows how Aqigaaq delineated the boat route and hunting areas he is concerned about and where Silas Aittauq has observed a dried up river that he usually has to cross on his walking route to his cabin.[6]

Changing Winds. In Baker Lake, winds and wind shifts were once useful for making weather predictions and navigating the tundra. These days, they are a hazard. In the last decade, linked to increased weather variability, wind patterns in Baker Lake have changed. These changes have been associated mainly with changes in direction, frequency and strength.

Changes in wind direction and frequency have been observed through a number of problems they cause. For example, Inuit have traditionally used the dominant winds to navigate the land, especially in winter. Winds blow across the snow surface and create ridges. Before weather and wind patterns started to change, these ridges always ran in the direction of the dominant wind and were a reliable navigation tool—one knew which way the ridges ran in relation to where they were going and used the ridges to stay on course. Now, one has to be extremely careful when using these ridges for navigation since unpredictable shifting winds can throw one off course. Though some residents now use GPS units for travel on the land, these devices are not always dependable (e.g., they can break or freeze), and many Inuit stress the need to still know about and use the traditional methods of following snow ridges when travelling.

The increased strength of winds has also been observed through a variety of impacts to the land and land-based activities and linked to a number of changes in the landscape. For example, less snow accumulating on the ice due to stronger winds means some thicker areas of ice. However, most people are concerned about the increased strength of winds in terms of significant modifications to snow, primarily the impact on igloo building.

> There used to be different layers of snow back then. The wind would not blow as hard, not make the snow as hard as it is now. Nowadays, the snow gets very hard, but back then, there used to be the ground snow and then another layer of snow just above the ground snow and we used to be able to use those layers of snow to make our shelters. Above these two layers there would be the surface snow and you would be able to get at the surface snow. Just remove that snow to get to the snow that's good to make shelters with. But nowadays, the snow gets really hard and it's really hard to tell [the layers] and it's really hard to make shelters with that kind of snow because it's usually way too hard right to the ground. ... We would use these long sticks to see where there was a really good spot to make our shelter. We weren't trying to be fussy, but we had to find the right type of snow because the right type of snow had different layers and the different layers would be soft on the top and a little bit harder, but not too

hard for the next two layers. That was the kind of snow you wanted to make a good shelter. (T. Qaqimat, Baker Lake, 2001)

Some winter I would really like to make an igloo for you to see, but I can't find the right snow, so I can't make an igloo, the snow is not right for it. (S. Quinangnaq, Baker Lake, 2000).

ᖃᒪᓂᑦᑐᐊᖅ ᑕᓯᖕᒪᑕ
ᐃᒪᖃᕐᓂᖕᒪᑕ ᐊᓯᔾᔨᖅᑕᕐᓂᖕᒪᓄᑦ

Changing Water Levels in the Baker Lake Area

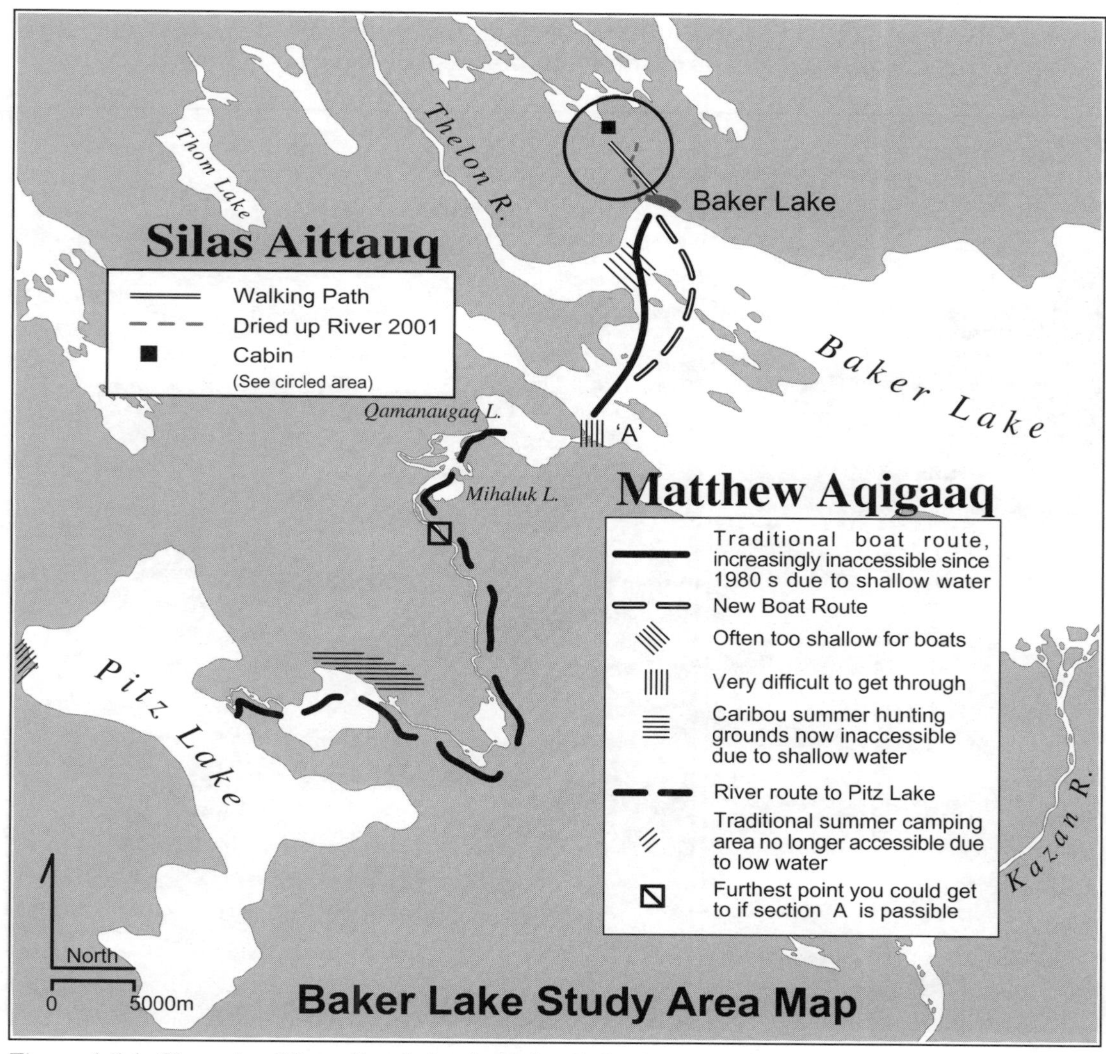

Figure 1-14: Changing Water Levels in the Baker Lake Area.

> I have a lot of experience with igloos. But I don't like to handle the snow anymore. The snow is too hard, the wind packs it. (P. Attutuvaa, Baker Lake, 2001)

Igloos are still used by Inuit today as temporary and emergency shelters while they are travelling on the land. The inability of Inuit in Baker Lake to build igloos due to stronger winds packing the snow has some residents very worried. Some residents think that a number of accidents and injuries on the land in the past few years may have been prevented had there been good snow for the individuals to build igloos. Some people are more nervous about travelling on the land these days, knowing they may not find good snow for a temporary shelter if they need one (Figure 1-15).

Reacting to and Coping with Impacts of Climate Change

Inuit in Nunavut are coping with the impacts of climate change in a number of different ways and on a number of different levels. One way Inuit are coping is by changing their patterns of subsistence. For example, in both Baker Lake and Clyde River, hunt-

Figure 1-15. Man building igloo, Kivalliq region, circa 1970 (photo by David Webster).

ing parties pack for several extra days when they travel since they now expect to get caught in unpredictable weather. Before, hunters could plan to be out for a certain number of days and pack accordingly. In Baker, travelling, for some, now includes packing extra tents, or staying near cabins since they can no longer depend on building igloos for shelter. In Clyde, hunters have had to find new locations for caching meat due to melting *aniuvat,* and in Baker, caching meat has to wait until later in the fall when temperatures cool. Also, hunting locations are being adjusted in both of these communities. Hunters in Baker Lake dealing with inaccessible traditional hunting grounds (Figure 1-14) now look for caribou in other areas. Hunters in Clyde River stay away from some usual travel routes due to dangerous conditions of thinning ice and unusual leads and have moved traditional campsites that have been washed away from surging rivers, thought to be caused by melting glaciers.

Since many of the environmental changes having an impact on the Arctic are relatively recent, Inuit communities are focussed primarily on day-to-day coping mechanisms, such as those described above. However, Inuit are seeking to extend their coping strategies into adaptive ones that may be appropriate for their increasingly unpredictable environment. For example, at the local level, meetings at community Hunters and Trappers Organizations are used as a venue to discuss observations and concerns about environmental change and how people are dealing with these changes. Collectively, under the auspices of various Inuit organizations (e.g., the Inuit Circumpolar Conference), Inuit are addressing issues of adaptive strategies at higher levels by seeking partnerships with the federal government to address various aspects of climate change including research, impacts, and adaptations (Fenge, 2001).

Emotional and Cultural Aspects of Environmental Change

Coping with the impacts of climate change does not go without some stress. How Inuit relate to a changing environment emotionally, culturally, and spiritually is an important aspect of coping and adapting that has not been adequately addressed in research. For example, consider how extremely skilled elders and hunters can no longer predict the weather as they have in the past. No longer able to be confident in their predictions, some elders and hunters are genuinely distressed, not only because they can no longer advise travel parties with assurance, but because their personal relationship with the weather itself has changed. This change in elders' relationships with the weather was expressed poignantly in a statement made by Zacharias Aqqiaruq (2000), an elder in Igloolik. After a discussion of his knowledge of climate and environmental changes, Aqqiaruq stated, "the weather has been called *uggianaqtuq*." Iyerak (personal communication, 2000) explained the meaning of *uggianaqtuq* in this way:

> For example, I am very close with my sister. Say I wasn't feeling myself one day and I went to go visit her. As soon as I walk in the room, or say something, she would know right away that something is wrong. She would ask me, "is there

> something wrong with you"? She would say I was *uggianaqtuq*. I was not myself, acting unexpectedly or in an unfamiliar way.

Aqqiaruq's use of *uggianaqtuq* characterizes the personal nature of Inuit knowledge and how deeply connections to the environment can be felt. In Nunavut, the climate is becoming unfamiliar, and its behaviour is unexpected and out of the norm to those who have known it intimately for so long.

Emotional responses to environmental change are often tied to feelings of identity. For example, in 1996, I was in Igloolik studying the role of walrus in community life and how this species, a major local resource, might be impacted by climate change. In an interview with Kotierk (1996), I learned how hunting walrus was central to Inuit identity in this community—hunting walrus was how one made oneself a "real Inuk." In our discussions of environmental change and potential negative impacts to walrus populations, Kotierk was passionate about how any loss in hunting would affect his life. He related this to dealing with government-imposed quotas on hunting. The restriction or loss of hunting can create deep resentment, anger, and depression for some hunters—they *are hunters* and need to hunt in order to fulfill something deep inside themselves (Kotierk, 1996). Also consider some examples from earlier in this paper. Though hunters and their families will change hunting strategies or camping areas if they are restricted by hazards such as shallow water or thin ice, they miss the areas they are used to. In some cases, these places have special significance to family heritage or hold meaningful memories for individuals. Many hunters and elders are extremely attached to the places they come from and travel. They are tied to the land through their intimate knowledge of its paths and processes, but also through emotions and a sense of identity.

As Inuit deal with the impacts of climate change, their way of life is affected, as seen above, through changes in subsistence strategies and practices. This study also found that culture can be affected, for example, through changes in the way people assemble and communicate knowledge about the environment. For example, in Igloolik, it was interesting to discover that a new (or very old) Inuktitut word was being used to describe the increasing incidence of rainfall observed in winter and early spring (Otak, personal communication, 1997). George Kappianaq (1997), an Igloolik elder, used *misullijuq,* an unknown word at the time, during an interview to describe a mix of rain and snow observed that spring. It seemed that either this word was created to describe a new phenomenon, or it was a very old word being used to describe an event that had not been observed in a very long time.

Inuit are extremely adaptive people and have dealt with both environmental and social change throughout their history. A flexible, ongoing process, Inuit knowledge will continue to blend traditions of *substance* (knowledge and skills handed down over generations which people incorporate into personal practice) with traditions of *process* (forming knowledge through practical engagement with the environment) (see

Ingold and Kurtilla, 2000) in order to construct knowledge that will help them best deal with their environment.

The emotional and cultural aspects of environmental change for Inuit are as complex as their observations. This was only a small venture into this discussion. Here, I wanted to make the point that although many northern scholars acknowledge how deeply Inuit are tied to the land in terms of subsistence, knowledge, culture, and spirituality, they often fail to address the latter two themes in discussions related to environmental change. In order to more fully understand Inuit perceptions and knowledge of environmental change, it is important to address these in the context of Inuit culture.

Climate Change and Community Life—The Need to Recognize a Bigger Picture

While climate change is clearly a concern for Inuit, it is important to recognize that there are other, often more pressing problems facing communities. Issues such as substance abuse, domestic violence, suicide, cultural preservation, housing and infrastructure improvements, contaminants, and poverty often top the priority list of things people want to address both inside the community and with help from outside.

These issues came up at various times during interviews conducted for this study. Participants would sometimes use interviews as an opportunity to express their distress and anxieties about problems in the community. We (my interpreters and I) learned early on the value of letting people speak their minds. It was interesting, and important, to let people speak about what mattered to them at the time. We heard concerns over disinterested youth, how the animals are affected by pollution and community garbage, how the latest government policies seem impractical for Nunavummiut, etc.

Time spent living in the communities also highlighted for me the importance of other issues. During casual conversations or evening visits, discussion did at times revolve around observations of environmental change, but more frequently, it was about other family and community concerns like those noted above. In addition, I was asked by community members to help them with tasks such as filling out housing applications, legal aid forms, grant applications, or organizing youth activities. I am sure that this is not unusual for researchers working in the North, however, it pointed out to me that there are many other issues besides climate change that concern Inuit and that my time sometimes better served communities through activities that were not related to this study. Projecting this idea into a bigger picture, it gives further meaning to the need for collaborative research approaches where communities determine what research they want to conduct and how they would like to do so.

Studies that address climate change and impacts to local communities in the North are an important contribution to arctic and climate change studies. However, recent media attention and research in northern communities on climate change has sometimes overshadowed other local issues that could also benefit from raised awareness, application of resources (i.e. funding) and research. As researchers, we must take care not to push our concerns onto communities and pay attention to what concerns people do have and how they prioritize these within the context of community life.

These concerns will be different from place to place. By addressing these concerns with community members through collaborative research processes, we will help to create more meaningful research.

Study Contributions

This study set out to benefit both Inuit and scientists. This is being accomplished in a variety of ways, through contributions to research, policy, and cultural preservation.

Research

The study contributes to research on indigenous knowledge, as well as arctic climate change, in a number of ways. For example, the study helps to develop frameworks for sharing information between scientific and indigenous knowledge. A number of previous works have developed conceptual frameworks that assist communication between scientific and indigenous knowledge using elements such as hypotheses formulation, community monitoring and indigenous knowledge as climate history (e.g., Ferguson and Messier, 1997; Riedlinger and Berkes, 2001; Thorpe, 2000); however, there has been little done on developing practical frameworks. This project directly addresses this gap by developing practical, functional methods and tools for communicating indigenous knowledge to science and vice versa. Participatory mapping exercises, for example, are a way through which scientists and Inuit can discuss their knowledge and ultimately, together, produce a product that is useful to both parties. The CD-ROM pilot project is another example, an experiment into presenting Inuit climate knowledge in an alternative, integrative way based on visual media that are accessible and useful to both Inuit and scientists, with hard copy products from the CD-ROM also available to wider Inuit and public audiences. Also, in-depth interviews with both Inuit and scientists in this project help to provide better insight into how each group constructs and communicates knowledge, thus leading to better informed frameworks and tools for cooperation.

It is key that the project addresses practical and conceptual frameworks in the spirit of reciprocity and collaboration. While the study is heavily weighted to how Inuit knowledge can be documented and communicated to scientists and decision-makers (to counter a status quo heavily weighted to science), it is also important to recognize that Inuit are interested in receiving scientific information. Therefore, this project also works with Inuit and scientists to development frameworks and tools that help transfer scientific information to Inuit in a meaningful manner.

The study makes contributions to climate change studies by providing large collections of observations from four different regions in Nunavut. These observations are being linked to scientific observations through an in-depth analysis of both Inuit and scientific observational data (Fox and Daniels, in prep). This is a key contribution to climate change studies, since although climate scientists and decision-makers are becoming more and more interested and accepting of indigenous knowledge, what may help break the reluctance of a dominant "western science" paradigm in research and policy is a direct analysis of how Inuit and scientific knowledge bases may be linked on issues such as data (observations) and issues of scale (temporal and spatial).

It is hoped this kind of analysis will also serve as a catalyst to encourage further collaboration between northern and scientific communities.

Lastly, but perhaps most importantly, this study highlights the need to go beyond the collection of indigenous observations. Local observations are valuable contributions to understanding the nature of environmental changes and their impacts to local ways of life. However, we (researchers) need to be sensitive to the contexts, social and cultural, in which observations are made and the changes are experienced. For example, while many researchers document the impacts of environmental change on subsistence practices, they do not take it any further to investigate how community members *feel* about these changes. As was discussed earlier in this paper, there are emotional and cultural aspects of environmental change that need attention.

Policy

This study contributes to policy in two ways, a) the documentation and communication of Inuit knowledge of environmental change, and b) documenting and communicating Inuit experiences of coping with, and adapting to, the impacts of environmental changes.

At present, the Government of Nunavut (GN) is developing a *Climate Change Strategy* that will implement policies related to climate change in Nunavut. The Strategy is guided by the following themes: encouraging action, promoting technology development and innovation, enhancing awareness and understanding, government leading by example, and investing in knowledge and building a foundation (GN, Dept. Sustainable Development, 2002). As part of this policy-building effort, the GN is incorporating Inuit knowledge and experiences. This study, through the documentation of Inuit knowledge and coping strategies related to climate change, will provide key information from four different communities in Nunavut. This knowledge will contribute to "investing in knowledge and building a foundation," thus directly helping to achieve the goal of the Strategy, to "concentrate on mitigating impacts, determining methods of adaptation to potential impacts of climate change and implementation of potential mitigation measures" (GN, Dept. Sustainable Development 2002). This study may also influence the strategy by presenting observations and experiences from the community level that the GN is not aware of.

Finally, through various channels such as participating in international conferences and academic writing, this study can raise awareness about the Arctic and about Inuit experiences of climate change to wider audiences.

Cultural Preservation

Lastly, this study helps to support the preservation of traditional knowledge and cultural heritage, an endeavour important to many Inuit communities, as well as many people outside of Nunavut. Copies of interview transcripts, videos and maps will be kept in the communities at heritage centres and research centres. The CD-ROM pilot project will be used in community schools and libraries. Participants receive copies of videotaped interviews, photographs, research summaries, and hard copies of maps, which document the knowledge of members of their families. Certainly, more is needed than documentation to help preserve Inuit culture and knowledge. This must be done in-situ, keeping in mind that writing down traditional knowledge cannot be a substitute for demonstrating or verbally communicating it (Berkes, 1999). This study tried to address this through participatory activities, local radio shows and elder-youth interactions that helped to transmit knowledge within the community and promote discussion of knowledge and skills.

Conclusions

This chapter provides an overview of a long-term study working with several Inuit communities in Nunavut to learn about and document their observations, knowledge, and perspectives of climate variability and change with an aim to help develop frameworks and tools for reciprocal communication with science. Embedded in a personal, relationship-based approach, the research utilizes an adaptive, multimethod research design to engage Inuit communities and document their knowledge. Long and repeated visits to communities over many years, project updates in accessible formats, and the continual incorporation of community and participant feedback strengthen the project design.

The results of the study include a large collection of Inuit observations and perspectives related to changes in the arctic environment. Maps and video technology provide some alternative means to collecting and expressing this knowledge and a CD-ROM pilot project attempts to provide an integrative, visual, interactive way for Inuit, scientists, decision-makers, students and others to learn about this knowledge. Hard copies of video, Inuktitut text and maps are also products from the research that are more accessible to a wider Inuit audience.

A unique aspect of this project was trying to understand the impacts of climate change on emotions, identity, and cultural change. The linkages between climate change and cultural change are an important, though often overlooked, aspect of indigenous knowledge and climate change research. Experiences from this project highlight the need to go beyond documenting observations of environmental change to understanding the community context within which these changes are occurring, and how climate change fits into the diversity of issues and concerns facing northern residents. A community-based approach can help to ensure that community needs and priorities for research are addressed.

Finally, the study makes direct contributions to science, policy and cultural preservation efforts through the documentation of Inuit climate knowledge, linking empirical observations of both scientists and Inuit, and developing conceptual and practical frameworks for communication and collaboration between these groups that will help facilitate activities to foster research and preservation of knowledge *in situ.*

It is clear that Inuit in Nunavut are observing changes in their environment and these changes are having an impact on their lives. Inuit have always been able to cope with and adapt to the impacts of extremes, variability and changes, whether environmental or social. However, as (and if) the climate and landscape continue to change out of the range of variability that locals expect, the question seems to be for many, "what then?" (Tatyakputumiraqtuq, 2001). Inuit, scientists, and decision-makers need to address this uncertainty together through meaningful collaboration that seeks to analyze observational data and perspectives on impacts and identify options for adaptation strategies that are appropriate to varying community situations.

Acknowledgements

I am extremely grateful to the elders, hunters, and other community members of Iqaluit, Igloolik, Kangiqtugaapik, and Qamani'tuaq who have been part of this study over the years. This work would not have been possible without your generosity and patience for so long. *Qujannamiik ammalu matna* for teaching me about your knowledge of climate and so much more. To my interpreters and teachers in the North, Shuvinai Mike, David Audlakiak, Leah Otak, John MacDonald, Ruby Irngaut, Jukeepa Hainnu, Geela Tigullaraq, Margaret Kaluraq, David Webster and Hattie Mannik, without your continual support and critiques, this research would have been impossible!

I am also grateful for the generous funding of the Royal Canadian Geographical Society, the Northern Scientific Training Program, the Cooperative Institute for Research in Environmental Sciences (CIRES), the University of Colorado at Boulder, the Social Sciences and Humanities Research Council of Canada (SSHRC) and the National Science Foundation (OPP-9906740).

Finally, thanks to Malachi Arreak at Tusaavut for help with the Inuktitut abstract, Jim Robb for his excellent maps, and to the editors of this volume and Natasha Thorpe for thoughtful comments and encouragement through several drafts of this paper.

Notes

1. All photos by Shari Fox unless otherwise noted.
2. The Iluiliqmiut, Kihlirnirmiut, Hanningayuqmiut, Ukkuhiksalingmiut, Qairnirmiut, Hauniqturmiut, Akilinirmiut, Harvaqtuurmiut and Paalirmiut (Webster, 1999).
3. Interviews lasted between one hour and two hours depending on how the person felt.
4. Nunavummiut refers to people who live in Nunavut.
5. Further analysis is needed to confirm the exact change in direction, however, it is clear from informants that the dominant wind direction has shifted.

6. Several other participants mapped observations in this area. Only two are included here, however, due to page size/legibility.
7. IOHP denotes Igloolik Oral History Project.

References[7]

Aittauq, S. 2001. Interview by S. Fox and M. Kaluraq, Baker Lake.

Aqigaaq, M. 2001. Interview by S. Fox and M. Kaluraq, Baker Lake.

Aqqiaruq, Z. 2000. Interview by S. Fox and R. Irngaut, Igloolik, IOHP File # unknown.

Arngna'naaq, L. 2001. Interview by S. Fox and M. Kaluraq, Baker Lake.

Attungala, N. 2001. Interview by S. Fox and M. Kaluraq, Baker Lake.

Attutuvaa, P. 2001. Interview by S. Fox and H. Mannik, Baker Lake.

Berkes, F. 1999. *Sacred ecology: Traditional ecological knowledge and resource management.* Philadelphia: Taylor and Francis.

Bernard, H.R. 1995. *Research methods in anthropology: Qualitative and quantitative approaches.* Second Edition. Walnut Creek: AltaMira Press.

Census of Canada. 1996. Northwest Territories Bureau of Statistics, T-Stat (Territorial Statistics On-Line). www.stats.gov.nt.ca

Chambers, R. 1994a. The origins and practice of participatory rural appraisal. *World Development 22*(7): 953–969.

Chambers, R. 1994b. Participatory rural appraisal (PRA): Analysis of experience. *World Development 22*(9): 1253–1268.

Chambers, R. 1994c. Participatory rural appraisal (PRA): Challenges, potentials and paradigm. *World Development 22*(10): 1437–1454.

Cohen, S.J. 1997. What if and so what in northwest canada: Could climate change make a difference to the future of the Mackenzie Basin: *Arctic 50*(4): 293–307.

Csonka, Y. 2000. Interpreting oral data in ethnohistory: A transposition of cultures? Paper presented at the twelfth Inuit Studies Conference, Aberdeen, Scotland, August, 2000. Referenced with permission from the author.

Dane, F.C. 1990. *Research methods.* Pacific Grove: Brooks/Cole Publishing.

Dorais, L. 1997. *Quaqtaq: Modernity and identity in an Inuit community.* Toronto: University of Toronto Press.

Ellen, R.F. 1984. *Ethnographic research: A guide to general conduct.* London: Academic Press.

Fenge, T. 2001. The Inuit and Climate Change. *Isuma: Canadian Journal of Policy Research 2*(4): 79–85.

Ferguson, M. A. D., and F. Messier. 1997. Collection and analysis of traditional ecological knowledge about a population of arctic tundra caribou. *Arctic 50*(1): 17–28.

Flavelle, A. 1995. Community-based mapping in Southeast Asia. *Cultural Survival* (winter): 72–73.

Fox, S., and Daniels, K. A. Climate variability in the eastern canadian arctic: Linking Inuit knowledge and meteorological data, in prep.

Fox, S. 2000. Arctic climate change: Observations of Inuit in the Eastern Canadian Arctic. In: F. Fetterer and V. Radionov, Eds. *Arctic Climatology Project, Arctic Meteorology and Climate Atlas, CD-ROM.* Boulder, CO: National Snow and Ice Data Center.

Fox, S. 1998. *Inuit knowledge of climate and climate change.* Unpublished thesis, master of environmental studies (geography), University of Waterloo, Canada.

Grenier, L. 1998. *Working with indigenous knowledge: A guide for researchers.* Ottawa: International Development Research Centre.

Government of Nunavut, Department of Sustainable Development. 2002. *Inuit qaujimajatuqangit on climate change in Nunavut* (in press).

Huntington, H. P. 2000. Using traditional ecological knowledge in science: Methods and applications. *Ecological Applications 10*(5):1270–1274.

Huntington, H. P., Brower Jr., H., and D. W. Norton. 2001. The Barrow Symposium on Sea Ice, 2000: Evaluation of one means of exchanging information between subsistence whalers and scientists. *Arctic 54*(2): 201–204.

International Arctic Science Committee (IASC). 1999. *Impacts of global climate change in the arctic regions: Report from a workshop on the impacts of global change, 25–26 April 1999, Tromsø, Norway.* Fairbanks: Center for Global Change and Arctic Systems Research.

Illauq, B. 1999. Clyde River. In: M. Soublière, ed. *The 1999 Nunavut handbook,* pp. 331–337. Iqaluit: Nortext Multimedia, Inc.

Ingold, T. and Kurtilla, T. 2000. Perceiving the Environment in Finnish Lapland. *Body and Society 6*(3-4): 183–196.

Ipellie, A. and Rigby, C. 1999. Iqaluit. In: M. Soublière, ed. *The 1999 Nunavut handbook,* pp. 357–372. Iqaluit: Nortext Multimedia, Inc.

Iyerak, T. 2000. personal communication, Igloolik Research Centre, Igloolik.

Kaluraq, I. 2001. Interview by S. Fox and M. Kaluraq, Baker Lake.

Kappianaq, G. 1997. Interview by S. Fox and L. Otak, Igloolik, NWT, Igloolik Oral History Project File #IE 397.

Keith, D. 1999. Baker Lake. In: M. Soublière, ed. *The 1999 Nunavut handbook,* pp. 232–239. Iqaluit: Nortext Multimedia, Inc.

Kolff, A. C. 2000. *The political ecology of mining and marginalization in the Peruvian Andes: A case study of the Cordillera Huayhuash.* Unpublished thesis, master of arts (geography), University of Colorado at Boulder.

Kotierk, A. 1996. Interview by S. Fox and L. Otak, Igloolik, NWT, IOHP File # unassigned.

Kublu, A., Laugrand, F., Oosten, J. 1999. Introduction. In: Saullu Nakasuk et al., *Interviewing Inuit elders,* Vol. 1. Iqaluit: Nunavut Arctic College.

Kunuliusie, P. 2001. Interview by S. Fox and G. Tigullaraq, Clyde River.

Kunuliusie, P. 2000. Interview by S. Fox and J. Hainnu, Clyde River.

MacDonald, J. 1999. Igloolik. In: M. Soublière, ed. *The 1999 Nunavut handbook,* pp. 248–253. Iqaluit: Nortext Multimedia, Inc.

Mannik, H. (editor) 1998. *Inuit Nunamiut: Inland Inuit.* Altona: Friesen.

McDonald, M., Arragutainaq, L. and Z. Novalinga. 1997. *Voices from the bay: Traditional ecological knowledge of Inuit and Cree in the Hudson Bay bioregion.* Ottawa: Canadian Arctic Resources Committee.

Morgan, D. L. 1988. *Focus groups as qualitative research.* Newbury Park: SAGE.

Otak, L. 1997. Personal communication, Igloolik Research Centre, Igloolik.

Peryouar, B. 2000. Interview by S. Fox and M. Kaluraq, Baker Lake.

Peryouar, B. 2001. Interview by S. Fox and M. Kaluraq, Baker Lake.

Peterson, D. L., and D. R. Johnson. 1995. An action plan for an uncertain future in the far north. In: D. L. Peterson and D. R. Johnson, Eds. *Human ecology and climate change: People and resources in the far north.* Washingon: Taylor and Francis.

Qaqimat, T. 2001. Interview by S. Fox and M. Kaluraq, Baker Lake.

Qillaq, T. 2000. Interview, S. Fox and J. Hainnu, Clyde River.

Quinangnaq, S. 2000. Interview, S. Fox and D. Jorah, Baker Lake.

Riedlinger, D. and Berkes, F. 2001. Contributions of Traditional Knowledge to Understanding Climate Change in the Canadian Arctic. *Polar Record, 37*(203): 315–328.

Seimon, A. 1999. personal communication, Environment and Societal Impacts Group, National Center for Atmospheric Research, Boulder, CO.

Serreze, M.C., Walsh, J.E., Chapin III, F.S., Osterkamp, T., Dyurgerov, M., Romanovsky, V., Oechel, W.C., Morison, J., Zhang, T., and R.G. Barry. 2000. Observational Evidence of Recent Change in the Northern High-Latitude Environment. *Climatic Change 46*(1-2): 159–207.

St. Denis, V. 1992. Community-based Participatory Research: Aspects of the Concept Relevant for Practice. *Native Studies Review 8*(2): 51–73.

Tatyakputumiraqtuq, W. 2001. Interview by S. Fox and H. Mannik, Baker Lake.

Thorpe, N. 2000. *Contributions of Inuit ecological knowledge to understanding the impact of climate change on the Bathurst Caribou Herd in the Kitikmeot region, Nunavut.* Unpublished thesis, masters of resource management, Simon Fraser University, British Columbia.

Thorpe, N., Hakongak, N., Eyegetok, S., and the Kitikmeot Elders. 2001. *Thunder on the tundra: Inuit Qaujimajatuqangit of the Bathurst caribou.* Vancouver: generation Printing.

Uqalik, Z. 1991. IOHP File #IE 209.

Webster, D. (with the Inuit Heritage Centre and Harvaqtuurmiut Elders) 1999. *Harvaqtuurmiut heritage: The heritage of the Inuit of the lower Kazan River.* Yellowknife: Artisan Press Ltd.

Figure 2-1. Moses Sam, an esteemed Neets'aii Gwich'in elder and avid fisher, checks his fish net on a pond near Arctic Village. Moses Sam passed away in 2001 and is missed by all.

2

Community Contributions to Ecological Monitoring: Knowledge Co-production in the U.S.-Canada Arctic Borderlands

Gary Kofinas with the communities of Aklavik, Arctic Village, Old Crow, and Fort McPherson

Indigenous hunters, fishers, and gatherers of the Arctic have traditionally been keen observers of their ecosystems. In former times, an acute awareness of one's environment was critical to survival. Today's northern peoples face challenges similar to those of their grandparents, although the overall complexity of conditions has increased and the forces for change are novel. To meet the challenges of community sustainability, indigenous people have stated their insistence on being involved in all functions of resource management, and in that process they expect their cultural perspectives to be respected. While significant challenges remain in realizing this goal, new models for community involvement are developing. Ecological monitoring and assessment is emerging as one of the most promising areas for the inclusion of local knowledge in regional resource management functions (Stevenson, 1996; Heiman, 1997; Brydges and Lumb, 1998; Russell et al., 2000).

During the past decade, the idea of involving local indigenous communities and scientists in a knowledge co-production process has received considerable attention (Freeman and Carbyn, 1988; Freeman, 1992; Inglis, 1993; Berkes, 1994; Ford and Martinez, 2000; Huntington, 2000; Kruse et al., 2000; Usher, 2000). Many argue that the combined perspectives of science and local/traditional knowledge have potential for generating a more holistic understanding of conditions than either can alone (Pinkerton, 1994; Berkes, 1999). In some jurisdictions of the North (e.g., Yukon Territory and Nunavut Territory of Canada), the inclusion of local and traditional knowledge in research and decision-making is now mandated by law, leaving agency managers wondering how to meet their legal obligations. As well, some approaches for engaging communities in a knowledge co-production process have been criticized for ignoring power dynamics (Nadasdy, 1999), fragmenting indigenous perspectives in a way that reduces knowledge to information (Cruikshank, 1998), and imposing western constructs (Anderson, 2000). Thus, the current state of affairs raises a methodological question—what approaches that offer the benefits of knowledge co-production address the substantive concerns on global change, while at the same time remaining sensitive to local cultures?

This paper addresses that question by describing and analyzing our experience in an ongoing ecological regional monitoring program. Our case study is the Arctic Borderlands Ecological Knowledge Co-op, a collaborative alliance of indigenous communities, First Nations, Inuvialuit organizations, co-management boards, government agencies, and university researchers.[1] The geographic focus of the Arctic Borderlands Co-op (ABC) is the U.S.-Canada Arctic Borderlands—a region defined by the range of the internationally migratory Porcupine Caribou Herd and its near-shore environment (see Figure 2-2). The ABC addresses concern for climate change, regional development, and contaminants with the central question, What is changing and why? First established in 1996 as the Northern Yukon Ecological Knowledge Co-op, in 2000 the Arctic Borderlands Ecological Knowledge Co-op was broadened to include Alaska. Today, the ABC has emerged as a model for collaboration in regional ecological monitor-

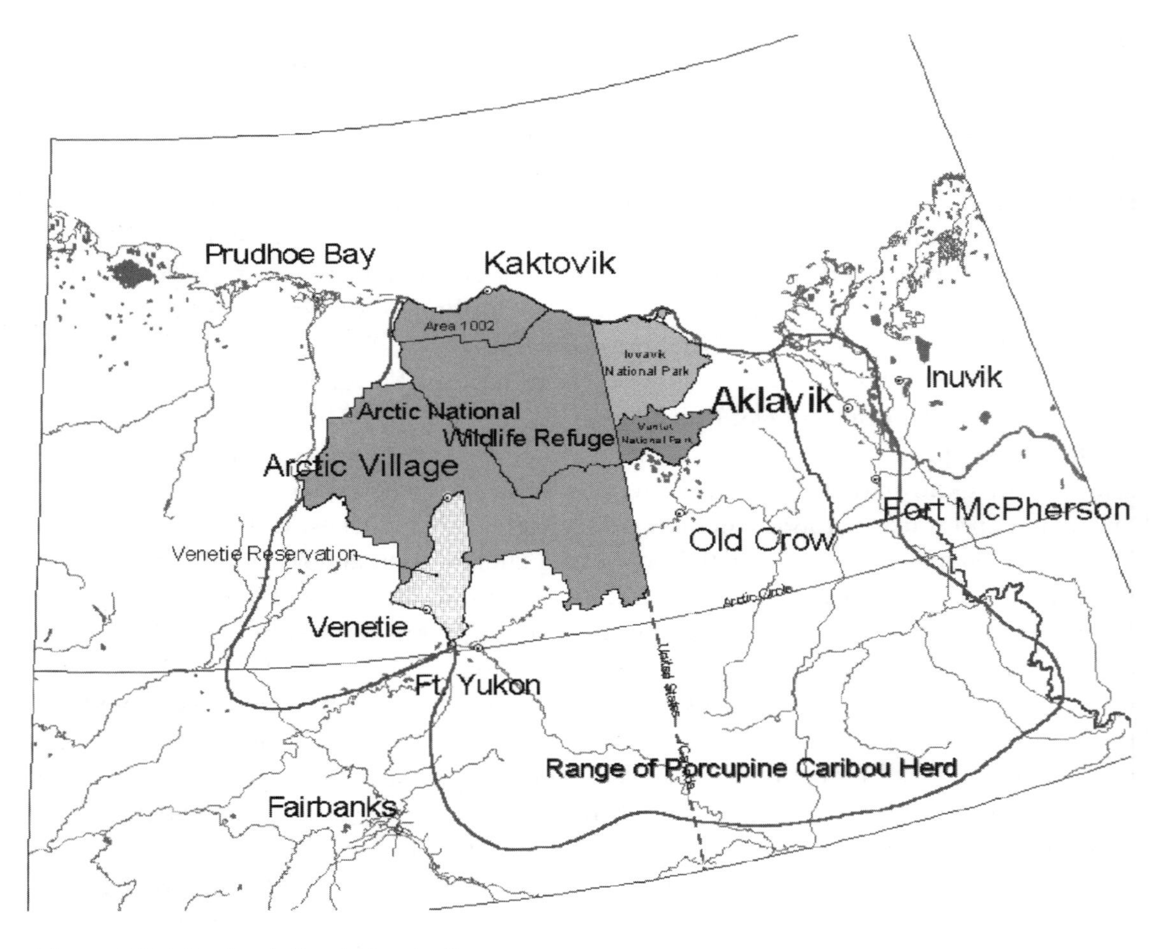

Figure 2-2. The Arctic Borderlands Co-op area.

ing that gives voice, respect, and legitimacy to the local knowledge of subsistence-based communities (Figure 2-3).

Community partners of the monitoring program include Old Crow, Yukon Territory (pop. 275); Aklavik, Northwest Territories (pop. 950); Fort McPherson, Northwest Territories (pop. 975); and more recently, Arctic Village (pop 140, joined in 2000); and Kaktovik, Alaska (pop. 250, joined in 2002). These five communities, along with Venetie, Alaska, are the primary users of Porcupine Caribou. All have vibrant mixed subsistence-cash economies and intimate relations with their homelands and living resources. As Native claimant groups, these communities are the Vuntut Gwitchin First Nation (Old Crow), the Gwich'in of Northwest Territories, (Aklavik and Fort McPherson), and the Inuvialuit Final Agreement beneficiaries (Aklavik). Because the population of Aklavik includes two Native claimant groups (Ehdiitat Gwich'in and Inuvialuit), each with its own traditional territory, its population is treated as two communities in the monitoring program. Arctic Village, situated on the Venetie Indian Reservation, is governed under the Venetie Traditional Council and the local Arctic Village Council. Kaktovik is part of the North Slope Borough, and has a newly created traditional village council. The vast majority of the residents in the five communities are indigenous people. Culturally, these communities are Iñupiat (Kaktovik),

Figure 2-3. Allan Benjamin of Old Crow scans the autumn landscape from above his cabin on the Crow River. Hunters like Allan are aware of changes in their landscape, both through the stories of their elders and from spending a lifetime travelling community homelands.

Inuvialuit (Aklavik), Vuntut Gwitchin (Old Crow), Tetl'it Gwich'in (Fort McPherson), Neets'aii Gwich'in (Arctic Village), and Ehdiitat Gwich'in (Aklavik). Together they have experience with the Arctic Borderlands extending back to time immemorial, interacting with a wide range of arctic environments from the coastal plain tundra and the Mackenzie Delta system, to the south slope taiga of the Brooks Range and northern tundra-taiga complex of Alaska, Yukon, and Northwest Territories. Since the goal of the monitoring project is to examine the status and trends of ecological change, local renewable resource committees and councils serve as the community contact organizations in the monitoring program.

As a region, the Arctic Borderlands is known internationally for its history of gas and oil development proposals. The Road to Resources Program of the Canadian federal government, launched in 1959, led to the construction of the Dempster Highway, the only major industrial development of the region. The Mackenzie Valley Pipeline and Arctic Gas Pipeline projects were proposed in the mid seventies and later dropped, but prompted the settlement of several Native land claims and the protection of sensitive caribou habitat in Canada (Berger, 1977; Page, 1986). Currently, a proposal for oil and gas exploration and development in the concentrated calving grounds of the Porcupine herd in Area 1002 of Alaska's Arctic National Wildlife Refuge is being debated as the most high-profile environmental issue of North America.

The U.S.-Canada Arctic Borderlands is also considered a hot spot of recent and projected future climate change scenarios (Overpeck et al., 1997; Zhang et al., 2000). Results from global climate model simulations for the region predict increases in the frequency of years with climate conditions that result in earlier spring green-up, deeper winter snows, 2°C to 4°C warmer summer temperatures, and more summer precipitation (Maxwell, 1992; Rowntree, 1997). Or as Annie B. Gordon, an ABC monitoring associate from Aklavik, has put it, elders of her youth told of a future time when Gwich'in people would eat food from trees, and that time is coming.

Several unique characteristics of the Arctic Borderlands monitoring program case study set it apart from other experiences presented in this volume. The Borderlands Co-op is not a one-shot research project but an effort to establish an ongoing monitoring program that will be maintained into the future. The focus of the monitoring is on several issue areas and not just on climate change or development. Also, the ABC is established in an unusually complex political landscape that involves two nation states, three state/territorial governments, and five Native claimant groups. Adding to the complexity is the uncertainty of keystone species like the Porcupine Caribou Herd—a resource known to scientists and locals alike for its mysterious ways and unpredictability (Klein, 1991; Russell et al., 1993; Kofinas, 1998: 120–171). The Borderlands Co-op's informal, interorganizational collaborative approach to monitoring is also noteworthy. This informal cooperative for monitoring has grown out of relations developed through formal co-management arrangements (e.g., Wildlife Management Advisory Council of the Yukon North Slope and the Canadian Porcupine Caribou Management Agreement) and includes a conscious effort by its participants to avoid bureaucratization and stay responsive to local communities.

Believing that local community contributions in the co-production of knowledge can and should infuse both the means and the results of an ecological monitoring program, we focus on the findings, methods, and experience of the Arctic Borderlands Ecological Knowledge Co-op's community monitoring program.

Philosophy and Rationale for Community Involvement

What is meant by local knowledge and why include it in an ecological monitoring program? The unique characteristics of indigenous peoples' perspective on land and animals have long been noted by some anthropologists and natural scientists (Figure 2-4) (Berkes, 1981; Feit, 1973, 1988; Irving, 1958; Levi-Strauss, 1966; Nelson, 1983; Nelson et al., 1982; Slobodin, 1962, 1981). In the development of the ABC program, we realized at the outset that there are clearly differences in local and science-based perspectives (Feit, 1986; Fienup-Riordan, 1990). In spite of the differences, there are also important similarities in the two systems of knowing (Scott, 1996). As we believe,

Figure 2-4. Sitting quietly for hours and observing is part of the Vuntut Gwitchin's autumn hunt for Tyso Denjik Vutzuii *("caribou of the Quill River," mistranslated by western scientists as the Porcupine Herd). August and September brings migrating caribou to traditional river crossings, long known by the Vuntut Gwitchin. These months are the most important time for harvesting caribou.*

the contributions of communities and the work of good science-based monitoring are not mutually exclusive. Going a step further, we initiated the project with the premise that while there are countless problems associated with power relations in the management of natural resources, these power dynamics do not suggest, as some have argued (Nadasdy, 1999), the need for communities to isolate themselves from management agencies. New approaches to knowledge co-production that are mutually respectful are consistent with the sustainability goals of the communities. Randall Tetlichi, 2001–2002 ABC Old Crow monitoring associate and former Vuntut Gwitchin First Nation Chief, talked about his knowledge and what he describes as the need to "double understand."

> The indigenous way is to use what we have—the natural resources and the environment we're given it to look after. It's our responsibility.
>
> When the white people first came here, the Native people said, "How" and the white people said, "Why?" I spent 90% of my time growing up with my grandparents. They taught me about a lot. I notice in the traditional world, it's all based on how am I going to do this. We never asked, why? People of my age are the last ones to say how. Traditional knowledge is passed on through generations—experiences from elders that they're passing down to me and to other people. I didn't ask why I use snowshoes to get a moose, it's by the knowledge that was passed down. Traditional knowledge is using the knowledge that we have. Traditional knowledge is all connected with the universe. Science wants to know why it works.
>
> A good thing today is that people have to come together. We have to know why and how. We have to double understand. I have to train my mind to remember, but I also have to train my mind to understand science. Now I have to double understand and pass the knowledge on. The young people have to double understand, use that knowledge—how and why.
>
> I have to engage myself both in what science and elders are saying. I have to engage in both. What about scientists? They have to also. That's one of the new rules.

Community monitoring, as defined in the ABC program, is not a process of hiring and training local community members to serve as technicians in the collection of data. The focus here is on local knowledge, defined as an interacting system of individually and collectively held observations, theories, and local preferences, underpinned with an ideological perspective. We recognize that local culture brings a distinct mode of learning and thought, intimate relations with land and resources, and time-tested strategies of human survival. We also acknowledge that local knowledge of indigenous communities, like all knowledge, is limited, dynamic, and to a great extent bound to the social institutions from which it is derived. Given the breadth of topic areas generally encompassed by local knowledge (vs. the narrow scope of disciplinary science), we expected that local knowledge would address a range of subjects relevant to community, regional, and global change. We also expected that local experts of the community would be in an ideal position to contribute to topics beyond those typically in the purview of ecology and to address social and economic questions as well.

Three elements found in all knowledge systems serve as our framework for the inclusion of local communities in monitoring. These elements are values and preferences, theories and explanations, observations and first-hand experience (see Table 2-1). The links between science-based and community monitoring are through the association between local observations and "data," local explanations and the identifica-

Table 2-1: Examples of "observations," "theories," and "values" documented from interviews with Old Crow, Aklavik, and Fort McPherson hunters

	Old Crow	Aklavik	Fort McPherson
Local observation:	In years of deep spring snow pack and spring icing, northbound caribou move to select lakes and feed on "muskrat push ups."	Hunters repeatedly observe migrating caribou pausing at the freshwater creeks draining into Ptarmigan Bay.	In one year caribou came into area, people did not bother them, and caribou over wintered near community. In another year caribou leaders were hunted upon first arrival, herd was redirected in new direction, and caribou wintered in Alaska.
Rule or theory:	Green vegetation of "push ups" provides important nutrition source for poor northbound caribou; muskrats in lakes add to health of herd; muskrat trapping can increase number of rats in lakes.	Caribou select local routes in coastal areas based, in part, on quality of fresh-water streams.	Caribou selection of migratory routes and winter habitat are based on movements (collective knowledge) of southbound vanguard. If approximately 500 caribou are allowed to pass through border area first, then caribou are more likely to use Caribou Mountain area.
Value/choice:	Traditional on-the-land pursuits can help to maintain a healthy ecosystem and should remain a part of the community's lifestyle. People need to keep trapping.	People should avoid disturbance and pollution of area.	Temporary hunting closures are needed seasonally to maintain local winter supply of fresh meat.

tion of monitoring indicators, and the common interest in preferences and policy recommendations. Thus, the ABC was launched with several assumptions about the possible benefits of indigenous community involvement in a regional monitoring program. These include:

- *Addressing complex problems:* The need to track change and assess the status of ecological conditions is especially important as people of a region face new situations like climate change, greater industrial development, and rapid cultural change. Viewed together, these forces for change bring new levels of uncertainty and perceived risk. Collective action is needed to address these unprecedented challenges.
- *Providing a unique perspective:* Local knowledge provides insights not available from many scientific studies—both in terms of the information reported as well as a way of learning, knowing, and communicating about the environment. The inclusion of local knowledge speaks to the problem of scale and patchiness of conditions found across a region. Indigenous community history and ongoing experience with traditional homelands suggest that local hunters, fishers, and gatherers can serve as the eyes and ears of a regional monitoring program to enhance the monitoring capacity for early warning of change.
- *Broadening the co-management process:* Although co-management arrangements have been established across the Arctic Borderlands Region, and co-management bodies regularly address issues of resource sustainability, the agendas of those bodies are often overwhelmed with discussions on policy, leaving little time for in-depth discussion about ecosystem process. As well, community members sitting on co-management boards are often not representative of those who spend considerable time on the land, harvesting resources. Local people most actively involved in subsistence activities are typically distant to the discussions at co-management board level.
- *Acknowledging legitimacy and value of traditional practice:* Giving standing to local knowledge in a monitoring program and providing honoraria to local experts who participate is one way of giving legitimacy to community perspectives on land and resources, acknowledging the value of traditional ways of life, and encouraging their ongoing practice.
- *Building organizational capacity:* Hiring and training of local research associates to work with the community monitoring program builds the human and organizational capacity of communities that strive to assume a strong role in functions of co-management. Greater community capacity is achieved, in part, through skill development and education in both the formal and traditional ways.
- *Capturing knowledge for future generations:* Documenting local knowledge preserves a record of community peoples' perceptions of conditions for future generations. If organized properly, the record provides community control over others' access to local knowledge information.
- *Sharing observations between communities and with scientists:* Local knowledge has long been exchanged through extensive communication networks among indigenous

communities of the North. Sharing of community monitoring reports on the status and trends of conditions by several communities reveals patterns at the regional scale. Sharing between local communities and scientists enhances this process and has the potential to cultivate mutual respect and trust among parties.

- *Promoting awareness and stewardship:* Encouraging local people to be active observers and keep a watchful eye on what is happening on the land promotes stewardship of living resources.

Regional and Community Monitoring as Process

Below we describe the roots, evolution, structure, operating principles, and methods of the Arctic Borderlands Ecological Knowledge Co-op and its community monitoring program. These elements of the program, all significantly shaped by the contributions of local communities, provide a model for the formation and implementation of a northern knowledge co-production process through regional ecological monitoring.

Problem Definition and Direction Setting

Successful collaboration among players generally passes through stages of development, beginning with a group's definition of a common problem, the identification of key participants, the establishment of operating principles, and the implementation and evaluation of activities (Grey, 1989). In the case of the Arctic Borderlands Ecological Knowledge Co-op, the motivation to launch the program was multidimensional. At the national level, a program initiative of Environmental Canada[2] in 1994, called the Ecological Monitoring and Assessment Network (EMAN), was launched as a coordinated Canada-wide response to issues of global change. The top-down directive to regional offices offered initial seed funding to establish "EMAN Sites" to monitor change in ecosystems. Reviewing the directive, Environment Canada Yukon reflected on the situation in its own area, including the recent settlement of Native land claims, and recognized the need to think beyond study sites to see its region as a system that includes human communities. Challenging the EMAN call for "sites," the Yukon Environment Canada staff chose to focus on the range of the Porcupine Caribou Herd and its near-shore environment, building on the region's wealth of past research, monitoring, and co-management processes.

The proposed focus on the range of the Porcupine Herd resonated with local communities for whom caribou is a keystone subsistence species. Yet, local communities' willingness to become involved in the project was not immediate. A long and turbulent set of conflicts growing out of oil and gas development-driven research and marginal involvement of local people in past studies left apprehension and concern (Kofinas 1998). With the recent settlement of land claims for most Canadian communities of the Arctic Borderlands region and establishment of several co-management boards, conditions were right for altering the region's historic disconnect between scientists and local communities.

In 1994, Environment Canada/Yukon Region held its first workshop to introduce the EMAN concept. Native participants stated that in principle, they endorsed the idea, but only if it included a meaningful role for local communities. The sharp exchange between locals and scientists at that meeting reflected the dysfunctional situation of the time, with a university-trained biologist suggesting that local people would require a formal education to be involved, and several local representatives responding that community members are significantly more knowledgeable about their region than scientists. From this tension, the idea of community monitoring emerged as a key component of the program, along with the realization that inclusion of local knowledge and the work of rigorous science in monitoring could be complementary.

In the initial scoping of community concerns, three areas were identified as the focus for monitoring—climate change, regional development, and contaminants, with the understanding that the list could change as new issues emerged and older ones disappeared.

The unique conditions of the ABC enterprise required a set of operating principles that would ensure that the program would be sustained. Clearly, short-term monitoring programs have limited value in advancing knowledge. The absence of a single large source of funding suggested that the program must be economical and reduce its vulnerability to budget cuts by drawing on multiple funding sources. Keeping the work of monitoring relevant to communities would be critical to maintaining local support. The program's success would also require that the organization of monitoring fit into existing resource management activities.

From a governance standpoint, the ABC is a cooperative program of partners. Organizationally, it has been established as a nonprofit organization; its application for charitable status is pending; when approved it will be eligible for grants and donations. It was also decided that all ABC operational decisions be made at "annual gatherings" by all those attending, and the powers of its elected board of directors be limited to implementing the decisions of those participating at the annual gatherings. The organizers decided that the ABC would not make political recommendations, but instead track the status and trends of ecological conditions and assess their implications to sustainability. Environment Canada Yukon assumed coordinating responsibility (not ownership) for the Borderlands Co-op's initial operations, with the understanding that the oversight responsibility could change to another organization when/if the partners so decided.

Selecting Indicators and Building a Database of Information Sources

The selection of indicators in ecological monitoring represents an *a priori* statement about system relationships and dynamics. The appropriateness of an indicator is best assessed by evaluating its sensitivity to change, its critical role as linked to other aspects of the system, the availability of baseline data, and logistics and cost considerations in continuing with its measurement.

To arrive at a set of indicators for the Borderlands Co-op, a special workshop was organized that we later called the First Arctic Borderlands Annual Gathering. The indicator identification exercise involved mixed small groups of managers, community subsistence users, and researchers (ten to eighteen per group) focusing on one of three habitat types—terrestrial, freshwater, and marine systems. Through discussions and evaluation, seventy-one potential indicators were identified, with many overlapping across habitat types. Human dimensions were a part of all discussions, with the greatest number of social and economic indicators (e.g., community population and demographics, level of household income, high-school graduation rate, etc.) identified in the regional development process. An evaluation was conducted to determine which of the seventy-one indicators were currently being measured by agencies, researchers, and communities; which were not being measured; and which were the best candidates for tracking. In short, the ABC would draw on the many data collection efforts of existing programs to summarize and report on indicators in a comprehensive way.

"What About All Those Past Studies?"—Building a Metadatabase of Information Studies

Before embarking on the measurement of those indicators, partner communities directed Borderlands Co-op organizers to review and catalog available data from past studies. To meet this objective, a meta database was created for studies of the region. What could have been an enormous endeavor was made relatively simple with the support of regional researchers who retained actual data, but provided key information on the variables and years covered in their datasets. By year two, the Borderlands Co-op had produced a meta database of information sources of the region, posted it on the web, distributed it as a paper document to communities, and made it available electronically.

Establishing a System for Community Monitoring

A pilot project for community monitoring was launched in 1996 to ascertain the potential contributions of local knowledge to the monitoring program, and has since been repeated each year for five additional years. Interviewing for the community monitoring program is undertaken by locally hired monitoring associates, and involves a limited number (fifteen to twenty) of local experts in each community. Interviews focus on indicators of ecosystem health about which local communities are especially aware.

Local experts are asked to comment on weather, subsistence fish resources, berries, caribou, other animals, and overall conditions of human communities. A semi-structured questionnaire guides a discussion on conditions observed during the past year. Questions are both closed and open-ended, allowing local experts to elaborate where necessary and also facilitating the work of locally hired monitoring associates who have limited to no interviewing skills. Local monitoring associates are selected by local organizations, and participate in a three-day training/orientation program. As a part of the training, local monitoring associates review the questionnaire and make changes to improve its effectiveness and speak to emergent issues.

Selecting and Paying Local Experts

Local experts are community members considered to be among the most active subsistence harvesters and most knowledgeable about land and resources. Since much of the interview is focused on current conditions, elders who no longer spend time on the land are generally not included. Selection of local experts is made by elected officers of the local contact organization—the Inuvialuit Hunters and Trappers Committee and the Ehdiitat Gwich'in Renewable Resource Council of Aklavik, the Tetl'it Gwich'in Renewable Resource Council of Fort McPherson, the North Yukon Renewable Resource Council in Old Crow, and the traditional village councils in Arctic Village and Kaktovik. An effort is made to select a representative group of local experts from the full spectrum of community family groups and where possible, repeat the participation of experts from year to year.

In the terms of social and natural scientists, the method of working with local experts is similar to working with key informants to document "expert knowledge" (see Huntington, 2000, and Ferguson et al., 1997). Interviews last one to three hours. Local experts are mailed a printed summary of interview findings each year and are asked to evaluate the monitoring program as a part of the interview. At the direction of community leaders, each local expert receives an honorarium of five gallons of gas (increased to ten gallons in 2001–2002) in exchange for his or her contribution to the program; a gas payment is preferred to cash because it supports local subsistence harvesters with the means to continue their traditional way of life.

As requested by communities, local experts remain anonymous in the reporting of monitoring results. While people are generally pleased to share their knowledge, they are not comfortable with interview comments being tagged with personal identifiers. All data are retained as property of the contributing community; the dissemination of results is only completed with the approval of local contact organizations.

Keeping Local Knowledge Monitoring Results in Perspective

It is important to remember that the findings of the community interviews should not be generalized to the community level in the same way as the results from survey or random sampling methods. The results do, however, represent the perspectives of those who are considered by community leaders to be their best informants on the topic.

Data of the community monitoring program are entered into a Microsoft Access database. Throughout the six years of the program, the questionnaire and accompanying maps have been modified to allow for improved spatial referencing of observations and explanations. Information of the Local Knowledge Monitoring Database is coded by several analytical categories, including:

- long-term changes,
- recent patterns,

- unusual observations,
- explanations, and
- rules of thumb.

Given the Borderlands Co-op's limited funding, no tape recorders are used to document the narratives of participants. Instead, interviewers capture the ideas of local experts as best they can, and those statements are not reported as quotes.

Creating Communications Tools for Sharing Perspectives

Participating communities want access to the results of the monitoring in a form that is understandable and useful in the process of exchanging knowledge and enhancing learning among and between groups. Three strategies of communication are used by the Borderlands Co-op to meet these goals:

1. Written reports in the form of inexpensively produced reports, documents, and posters
2. A web-based resource that posts current trends and status of select indicators, Borderlands Co-op activities, and the findings of the community-monitoring project (see www.taiga.net/coop)
3. Face-to-face meetings, including annual gatherings of community members, agency managers, and researchers of the region where community monitoring and science-based findings are reported and discussed. In addition to the annual gathering, local monitoring associates and others make presentations to co-management boards and local organizations

Together, these communication tools are intended to achieve a synthesis of understanding that improves our ability to discuss the implications of change and our options for responding to it.

Community Monitoring Results and Their Interface with Science

Over the course of the community monitoring program (1996–present), a tremendous amount of information has been reported and documented. Presented below is a sample of key topic areas of the community monitoring program, which provides an overview of the information being collected. In some cases, we include a discussion of science-based monitoring results to illustrate the interface of the two systems through the Borderlands Co-op experience.

About Local Experts and Time on the Land

We have interviewed an average of eighty-three people annually (twenty-one people/community/year) during the first five years of the Community Monitoring Project, with two thirds of them being over the age of fifty. About 6% of the selected local experts are less than thirty years of age, representing young people of the community who are especially active subsistence users.

During the second year of the community monitoring project, community members asked that a set of questions be added assessing local experts' level of experience. This request was, in part, the result of skepticism by some of the reports from fellow community members and the need to include a level of rigor in assessing the accuracy of findings. In 2001–2002, age and a spatially explicit description of each expert's on-the-land travels were added, as well as a mapping exercise in which the area of an expert's lifetime on-the-land travels are documented.

Community members report about changes in the quality and quantity of their time spent on the land in the past three decades, and the implications of their transition from an era in which children were raised at hunting camps to today's more motorized and village-focused lifestyle. As noted by community members, time spent on the land is an important indicator for assessing the transmission of traditional knowledge to younger generations. A summary of interview findings shows no significant recent change in the time local experts are spending on the land (see Figure 2-5), although reports from community members indicate that the current trend is otherwise. Aklavik's monitoring associate of three consecutive seasons noted that while many may spend half their time on the land, in 2001 the last of community members—elderly people—living full time in the Mackenzie Delta had moved to town because of health issues. Improving job opportunities and changes in lifestyle have also affected younger people's choice to live in town. Several Old Crow residents, however, noted an increase in the time on the land spent by members of their community from 1996, a shift resulting from the implementation of Old Crow's land claim and availability of new funding to support traditional on-the-land pursuits. Passing spring in Crow Flat by family groups is a traditional activity that for years was previously abandoned, but is again being practiced.

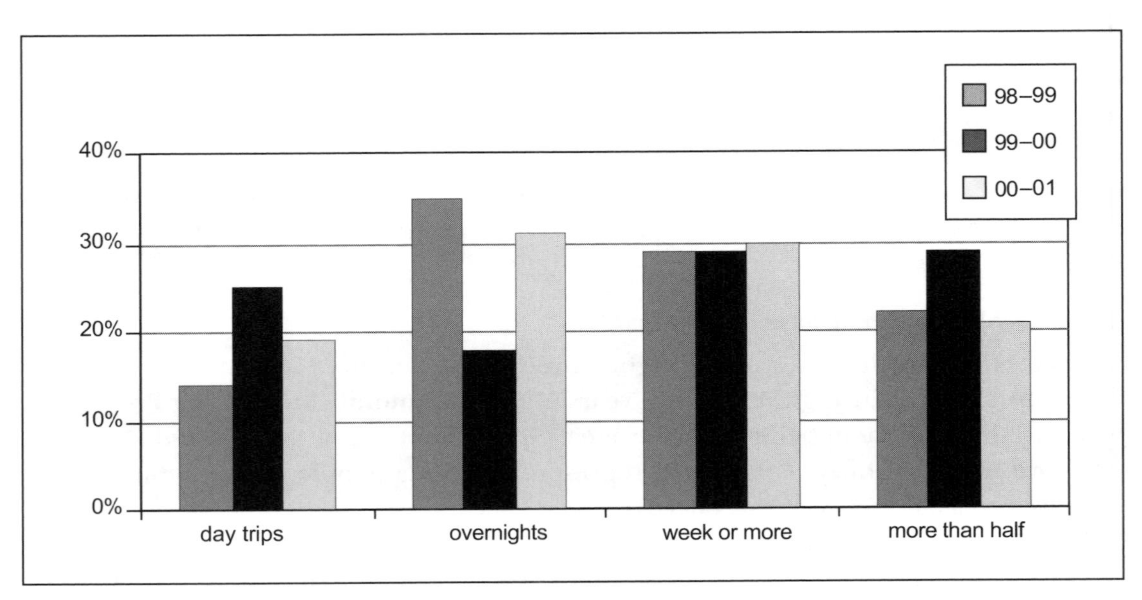

Figure 2-5. Local experts' time on the land, 1998–2000.

Weather, Water Levels, and Seasonal Transitions

Interview questions on weather ask local experts to share their impressions of conditions of storminess, snow levels and snow pack, water levels, freeze-up and break up, and overflow on ice. Local weather observations provide an account of seasonal patterns and local conditions not available from quantitatively based data of weather stations or water survey measures, in some cases providing narratives of the sequence of events and unfolding of seasons. Observations of significant long-term changes in weather are noted by locals of all communities and captured below as a set of generalized conditions.

- *Increased variability in weather events*—Observations of extremes in weather are reported, especially as related to fall storms (e.g., December 2000 rain and ice storm in Old Crow area) and late snow pack melt (e.g., entire region for spring 2000 and 2001). As noted by several elders and similar to reports from Sachs Harbour (see Jolly et al., this volume), "The weather today is harder to know."
- *A general drying of river and lake systems*—A drying of conditions and change in the overall hydrologic system was initially reported by communities as the Borderlands Co-op was launched in 1994. These reports have been repeated in interviews since the community monitoring was launched in 1996. In particular, Old Crow residents report that several lakes in the Crow Flat area, known for their outstanding fishing and muskrat trapping, had drained in the early to mid 1990s. Arctic Village residents share similar reports about lakes of their region. Local experts of the Mackenzie Delta area note a general lowering of water levels with associated increase in the growth of willow.
- *Fewer very cold days in early winter; fewer extended periods of extreme cold*—Old timers from all partner communities commented that periods of extreme cold (i.e., below –40°C), commonly experienced in the past, had not occurred in recent years as in the past. These elders noted that temperatures in past years would drop during the months of December and January and stay below –40°C. While cold temperatures occur on occasion now, they are said not to be sustained for days at a time.
- *Change in pattern and rate of fall-to-winter transition*—Local experts describe a general change in the overall patterns of seasonal transition. This observation is captured in comments on increased frequency of fall storms as well as comments on a general change in the sequencing of events, such as timing and rate of freeze up and snowfall.
- *Increase in the number of summer storms (especially frequency of high wind events and lightning)*—Old Crow residents describe storm events with an unusually high frequency of lightning. Aklavik residents who spend summers at Shingle Point on the Yukon's Beaufort Sea coast tell of an increase in storms that are accompanied with high winds. As well, they report an increased frequency of east winds. Recent tornado events in the Mackenzie Delta during the 2001–2002 season have raised concern among local community members.

Offered as a more richly detailed account, several of these weather conditions are captured in the 2000–2001 summary report of Inuvialuit Aklavik interviews conducted by Annie B. Gordon and presented at the sixth annual gathering (Figure 2-6):

> In August we had a big storm on the [Yukon] coast, with a very strong northwest wind. We even got a warning from the people in Kaktovik, [Alaska], phoning Aklavik to let us know by trappers' radio that a big wind was coming. But luckily, we only got the tail end of the wind. Even that was strong enough to cause a problem for us, like losing our new muktuk stage at Shingle and table to work on if we had gotten a whale. Plus, we lost the old stage from the year before. The wind was very strong so we had big waves, causing a break through between middle camp and the point and leaving us on an island. We had to be taken out by a chopper! It was too rough to even try with a boat. When we flew by chopper, all we could see was water right up to the hill. You could just see patches of land here and there all the way to Bird Camp. Also, for the first time in this region we had a thunderstorm and tornado weather warning. I think we were lucky it didn't reach Shingle Point. We saw the dark clouds and it went towards the east from around Blow River. So we never got that weather.

Local observations on changes in weather, in most respects, mirror the recent findings of science-based research on northern climate change, especially as related to increases in summer storm surges and overall drying of taiga and tundra ecosystems. Prompted by Old Crow residents' reports of the drying of Crow Flat, research was conducted by Environment Canada to quantify the percentage of total net loss of surface water and establish the geographical extent of drying. Findings based on analysis of satellite imagery both confirm the perceptions of local experts, and indicate a 5% net loss of surface area of water since 1973. Follow-up Environment Canada studies show changes in some select areas of Yukon's North Slope, but no overall drying trend (pers. com. Jim Hawkins, Environment Canada). In recent years, government budget cuts have resulted in closing several long-term water survey-gauging sites in northern Yukon, thus adding to the importance of ongoing community reports.

Figure 2-6. Anne B. Gordon shares a report on her findings at the 2002 Arctic Borderlands Annual Gathering in Fort McPherson, Northwest Territories.

Effects of Weather on Access to Subsistence-use Areas

While weather is a key driver in the ecological dynamics of subsistence resources, weather also affects community access to harvesting areas. In several cases, impeded access to hunting grounds has limited communities' ability to meet their subsistence needs. A sample of locals' reports about the weather-specific conditions affecting access to hunting grounds includes:

> More frost than usual on trees and willows make it hard on hunters travelling the land. (Fort McPherson 1997–98)

> Warm weather and lots of snow made for unsafe [travel] ice conditions on lakes and rivers. (Fort McPherson 1997–98)

> Snowed for two months straight. So much water [overflow], so we couldn't use usual trail (Aklavik Inuvialuit 1997–98)

> [Snow is] different this year. It's harder to pull a sled with Skidoo. (Aklavik Inuvialuit 1997–98)

> Wind on coast affected boat travel for fishing and beluga hunting. (Aklavik Inuvialuit 1999–00)

Described in the 2000–2001 Old Crow community monitoring summary report of Shirley Kakfwi,

> Because of too much change in the weather, it makes it hard for people to go out in the bush. There is not much permafrost and the ground is still too soft under the snow. The grounds usually make cracking noises when it gets really cold, but we don't hear that anymore.

The report of "frost on trees" refers to difficulty local hunters and trappers have encountered when trees are laden with ice—a condition described as the result of warmer and more moist winter atmosphere. The absence of overflow ice on rivers and lakes is said to occur when constant low early winter temperatures allow adequate freezing before deep snow falls. However, not all documented years have been problematic with respect to overflow. In the 1999–2000 year, Old Crow residents report that a slow but steady cooling in fall-to-early winter froze the Porcupine River in a way that allowed for unusually smooth and fast travel. Higher early winter temperatures, as noted above, can result in dangerous travel conditions, making some areas inaccessible, especially for older hunters who have trouble with rough, snow-free trails. More variable freeze-thaw conditions of recent springs are attributed to creating greater "candle ice" conditions on lakes, impeding access to some hunting areas and making travel unsafe. In the 1997–98 year, active subsistence hunters of Aklavik dramatically limited their usual spring harvesting activities due to these conditions. The final statement regarding a recent increase in winds and dangerous boating refers specifically to a summer 2000 wind storm that resulted in three fatalities, which has been followed by two subsequent seasons of unusually high summer winds on the coast. Moving towards a more detailed reporting system on access conditions, new questions have been added to the interview to ask explicitly about access problems and their overall effects on people's day-to-day lives. The first year of responses to those questions show mixed reports on the extent to which weather created a "problem" (see Table 2-2).

"Yellowberries," "Knuckleberries," **"Akpiks,"** *Known Outside the Region as Salmonberries* **(Rubus chamaemorus)**

Berries are an important part of communities' wild food diet and are regularly harvested, especially by elderly women. Berry pickers report on the annual quality and quantities of salmonberries, cranberries, and blueberries, comparing their berry crop to the previous year and to "average years." Of these berries, salmonberries or *akpiks* are particularly sensitive to climate conditions. Reports from local experts reflect a good understanding of the sequence of weather conditions necessary to achieve a rich and abundant salmonberry crop. As well, locals' reports also indicate concern for what is perceived by many to be a recent increase in intensity of summer heat.

> Shady and wet areas grow a good berry crop. In open areas [drier] very little crop. . . . Early, hot weather burns the flowers. (Fort McPherson 1997–98)
>
> It was too hot [for salmonberries], and we had no snow from April. The snow melted in April. Usually the snow stays until June and waters the ground. (Aklavik Inuvialuit 1998–99)
>
> Sun seems hotter, cooks the berries. (Aklavik Inuvialuit 1997–98)

Table 2-2: Weather conditions created problems for people on the land

Have weather conditions this year (2001) created any unusual problems for getting out on the land?

community	n	Weather affected effort in unusual way		No unusual problems noted
		hard	easy	
Aklavik Gwich'in	20	35%	0%	65%
Aklavik Inuvialuit	20	85%	0%	5%
Arctic Village	10	60%	20%	20%
Fort McPherson	21	57%	0%	43%
Old Crow	20	75%	0%	25%
all communities	91	63%	2%	33%

> The weather burned the berries sooner than they were picked. They were either small or shriveled by the time we went to pick them. (Old Crow 1999–00)

> The berries really never grew this year due to it being too cold and windy. There were a lot of leaves but no berries. (Aklavik Inuvialuit 1999–00)

> Pretty good year, big and juicy. May be due to lots of moisture, sun, and a nice summer. (Old Crow 1999–00)

Aklavik Inuvialuit also note a locally observed indicator for recognizing good *akpiks* years.

> When the cotton [grass] grows lots ahead of *akpiks,* then the berries will not grow much. When the "cotton" don't grow, there's lots of berries. (Aklavik Inuvialuit 1998–99).

One of the objectives of the monitoring program is to track conditions over time. A summary of five years of salmonberry reports shows there is little unanimity among local experts on the overall berry crop in some years. Community-to-community differences within a single year reflect the regional patchiness of climatic conditions. For example in 1997–98, coastal conditions were warm but not overly sunny, making for an adequate crop, while further inland at Fort McPherson, early rains were reported to be followed by intense summer heat, making for outstanding salmonberries. The years 1998–1999 and 1999–2000 were poor years overall, followed by a good year in 2000–2001. Five years of data are not enough to identify a pattern of change; it will be important to track these patterns through time to produce reports on the trend (see Table 2-3).

Apart from the actual quality of the berry crop, local experts provided other noteworthy comments on berries. Fort McPherson harvesters expressed concern about dust from the Dempster highway having a deleterious effect on berries. Along with the observation that local berry harvesters are not venturing far from the road when picking berries, there is concern among some from that community that berries are affected by highway dust and contaminated with "chemicals." The relationship between poor berry years and problems with animals has also been noted. For example, Old Crow participants reported in 1996–97 that poor blueberries resulted in a higher number of human-bear encounters and an increase in bear damage at bush camps. Another noteworthy concern is that certain harvesting areas are being repeatedly used and "picked out," and there is need for people to give harvesting places an occasional rest. Finally, there was a concern that fewer people are picking berries than did a decade ago.

Porcupine Caribou and Caribou Hunting

Caribou is an important part of the diet and culture of all communities in the Arctic Borderlands Region (Caulfield, 1983; Kofinas 1998; Wein, 1994; Wein and Freeman,

Table 2-3: Reports on quality of salmonberries; 1996 to 2000

	Community			
Year	**Aklavik Gwich'in**	**Aklavik Inuvialuit**	**Fort McPherson**	**Old Crow**
96/97	Mixed reports	Average	Average and above average	Average with some below average
97/98	Good with a few poor	Mixed reports	Good and exceptionally good	Mix reports
98/99	Very poor	Poor	Poor	Mixed reports
99/00	Very poor	Good to poor	Poor	Mixed reports
00/01	Mixed reports	Good	Exceptionally good	Good

1992, 1995), and community concerns regarding caribou health are tied closely to the Borderlands Co-op's three issues of contaminants, regional development, and climate change. As a part of the community monitoring, interviewers asked local hunters a series of questions about the seasonal movements of caribou, size of observed groups, caribou body condition, observed abnormalities, predation, availability of caribou, unusual observations, and other issues of herd health.

Local observations of the movements and distributions of the herd provide an understanding of seasonal conditions not available from conventional tracking of caribou using radio collars. As well, the narrative of a season's migratory patterns provide an explanation of coupled environmental conditions that affect caribou distribution patterns and human harvesting behavior. An excellent example of this contribution is the reports from Aklavik on the fall-to-winter caribou movements of the El Niño year of 1997–1998. During that year, Aklavik hunters observed that several thousand Porcupine caribou overwintered on the Yukon Coast, a distribution pattern that is not unheard of, but is considered unusual. Reports of large numbers of caribou on Yukon's North Slope were a surprise to biologists, whose limited number of satellite radio collars had not detected the anomalous distribution pattern.

Aklavik hunters explained the unusual event by describing that the Beaufort Sea ice pack was more distant from the Yukon North Slope coast than in most years, resulting in warmer than usual temperatures in the coastal region and growth of rich "caribou food." As caribou dispersed east into Canada from their traditional calving grounds on the Coastal Plain of Alaska's Arctic National Wildlife Refuge during the month of July, several large groups were observed to linger on the coast to take advantage of the rich forage, staying into September. As it happened, the first snows of that year fell without an icing event, which typically makes feeding difficult for caribou and drives the herd south to its more common wintering grounds. Without ice, caribou could continue to paw through the soft snow and access the summer's rich forage. At the beginning of the rut in October, a group of several thousand caribou had moved east of Hershel Island and remained on the coast throughout the winter. Hunters who had been tracking the animals through their fall hunts and other on-the-land travels discovered that the coastal caribou were in better condition (fatter) than the group wintering near Aklavik in the northern Richardson Mountains. Consequently, many hunters expended the additional effort to travel to the coast throughout the winter and harvest the better animals.

Hunters have also made long-term observations about changes in herd movements. For example, caribou are considered to be more skittish today than twenty years ago, reacting more to planes and snowmobiles. Local experts from all of the communities also report that migration occurs in smaller, more scattered groups today than it did years ago.

However, not all reported observations and explanations from local hunters are ecologically "rational" in the paradigm of western science. At the 2000 annual gathering, a hunter reported that the Porcupine Caribou Herd had passed by but not overwintered near his community, a movement pattern he explained by the recent death of a local hunter who "slept to the caribou." (Sleeping to the caribou among traditional Gwich'in refers to having a special relationship with the animal that involves intimate communications, sometimes achieved through dream world.) No challenges to the local's claim were made by participating scientists, and the assertion was entered as a part of the formal record. As well, it served as a reminder that the spiritual elements of human-animal relations are an important part of indigenous community knowledge.

Perhaps the most controversial aspect of caribou science for local communities are population estimates derived from aerial surveys and locals' long-standing skepticism of those numbers (Osherenko, 1988; Freeman, 1989; Kofinas, 1998; Kruse et al., 1998; Kendrick, 2000). The Borderlands Co-op's science-based indicator tells us that numbers of Porcupine caribou have declined in recent years from a census peak in 1989 of 189,000 to a current population of 123,000, and that the decline is not correlated with other science-based indicators such as caribou birth rate, calving distribution, or harvest. Caribou biologists have not identified a definitive cause for the decline.

To assess community members' views on caribou population trends, interviewers posed questions about changes in population in 1998–99. More than half of the local experts questioned reported that they believe there has been no decline in the herd (see Figure 2-7). Of those perceiving a herd decline, many attributed the change to higher predation, human harvest levels, and human disturbance. Some of these hunters noted the increase in caribou numbers from the early 1970s to the late 1980s and its correlation with changes in community subsistence patterns during the same period.

> Change in community lifestyle [decrease in total harvest] has resulted in the higher herd population in recent years. (Fort McPherson 1998–99)

Those not perceiving a decline suggested that the aerial census of the Porcupine Herd (considered by caribou biologists as one of the most accurate counts of all North American caribou herds) does not account for all caribou through aerial photographs. Local experts referenced the effects of overpopulation of the herd as a concern equal in importance to that of the agency's decreasing herd numbers, although there are no science-based studies in the region currently assessing a possible decrease in forage quality due to overgrazing. The cause for the decline in Porcupine Ccaribou is still an unanswered question by biologists and community members who share the perception.

Caribou body condition is a useful indicator of caribou health because it integrates many ecological factors that influence caribou productivity and is recognized by biologists and hunters alike as meaningful (Allaye-Chan, 1991; Elkin, 1999; Kofinas et al., 2001). Community hunters who regularly work with caribou are familiar with indicators of caribou body condition and health (see Figure 2-8), both before harvesting animals and while dressing out a carcass (see Table 2-4). To assess the condition of the Porcupine herd, the Yukon Territorial Government has conducted a detailed and quantitatively based body condition monitoring collection of about fifteen to twenty cows biannually for the past ten years. From that limited sample size, the study extrapolates to the health of the herd as a whole. The ABC community monitoring, on the other hand, asks hunters to report on their overall impressions of the condition of the animals they harvest. While not providing the fine detail of the body condition scores of the Yukon study, these reports do provide an assessment based on the observation of sixty hunters for as many as a thousand animals annually. Hunters are asked to provide a general assessment of body condition of harvested caribou with qualitative measures, ranging from "poor" (i.e., low to no back fat and no gut fat) to "very fat"

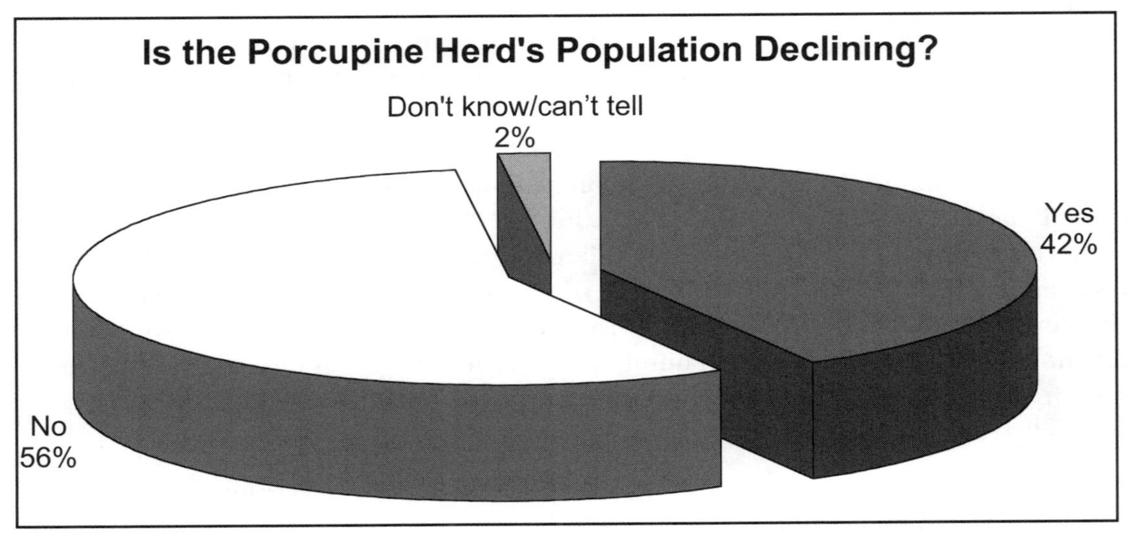

Figure 2-7. Perceptions of caribou herd declining.

Figure 2-8. Georgie Moses, an experienced Vuntut Gwitchin hunter and dog musher of Old Crow, inspects a caribou's stomach on Crow Mountain. Skilled hunters of the region assess the quality of a caribou before and after making each kill.

Table 2-4: Porcupine caribou hunters' indicators of good quality caribou: Body condition and overall health (Kofinas, 1998:166)

Indicators hunters look for when selecting caribou	• Size of rump • Gait or waddle of walk • Whiteness of mane • Size of rack • Symmetry and overall shape of rack • Number of configurations or points on rack • Size and shape of shovel • Grayness of rack • Social role of individual in group • Posture of animals when moving
Post-mortem indicators of caribou health	• Quantity of "backfat" (i.e., rump) • Quantity of stomach fat • Color of marrow • Tone and color of lungs (e.g., lungs stuck to chest indicate poor health) • Color of kidneys and liver • Absence of pus bags on kidneys • Absence of "water" in muscles (water being produced when animals is worked) • Contents of stomach (e.g., grass-filled may indicate sick animals) • Presence of parasitic larvae in kidneys

conditions (>1 inch of back fat). Hunters comment on the seasonal, annual, spatial, and individual variability of body condition (e.g., caribou east of the Dempster Highway are in good shape while those around the Eagle Plains area are "poor"). They are also asked to provide information on harvested animals with abnormalities and generate a ratio of total animals harvested to number of animals observed with abnormalities.

In the 1998–99 interviews, caribou hunters were asked to evaluate the overall health of the Porcupine Caribou herd. The findings show that a surprisingly high number of hunters in each community stated that, overall, the Porcupine herd is not healthy. Many hunters who perceive the herd as unhealthy related its poor health to the herd's exposure to disturbance from human activities (e.g., snowmachine hunting, aircraft overflights, road traffic and hunting on the Dempster Highway), while others pointed to the herd's high population as a cause of poor health. However, hunters have not reported seeing exceptionally high numbers of "sick animals" or animals with parasites (warble flies, bot flies, brucellosis, etc.). Recent interviews indicate a shift in perception: community hunters believe the herd is healthy.

Like caribou body condition, community hunters have noted that the total number of caribou harvested annually is an important indicator for monitoring the herd. There are currently thirteen different systems for collecting harvest data on Porcupine caribou and no comprehensive method of synthesizing those data to arrive at a single estimate of annual harvest levels. For the most part, current harvesting programs involving local communities in Canada are intended to establish need levels as a part of land claims settlement agreements.

A new set of questions was included in 2000–2001, asking hunters to report on overall availability of caribou to a community for a given season and whether local needs were met (see Table 2-5). When caribou are not available or are obtained only through considerable effort, local hunters are asked to explain the conditions creating this situation.[3] We would like to explore if this line of questioning, if asked repeatedly of all primary caribou user communities over time, can be used to more easily estimate annual harvest levels than the currently inconsistent systems of harvest monitoring.

Fish and Local Fishing

Questions on fish and fishing were initially structured by asking each fisher to report on his or her two most important fish resources. The interview has since been broadened to include reports on eleven species of fish (although some species are not harvested by all communities). Fishers' report on coney or inconnu *(Stenodus leucichthys),* crooked back *(Coregonus clupeaformis),* chinook salmon *(Oncorhynchus tshawyscha),* dog or chum salmon *(Oncorhynchus keta),* silver salmon *(Oncorhynchus nerka),* Arctic grayling *(Thymallus arcticus),* herring or Arctic cisco *(Coregonus autumnalis),* jack fish *(Esox lucius),* lake trout *(Salvelinus namaycush),* loche, burbot *(Lota lota),* Arctic char *(Salvelinus alpinus),* and whitefish or broad whitefish *(Coregonus nasus).* Fishers evaluate the overall quality and quantity of fish, timing of run (where appropriate), parasites, and physical abnormalities (Figure 2-9).

Table 2-5: Fall 2001 caribou availability and meeting needs

	How available were caribou to your community for hunting this past fall?			**Were enough harvested last fall to meet needs?**	
Community	**Close by and easily found**	**Not close and required lots of effort**	**Not at all available**	**Yes**	**No**
Aklavik Gwich'in	11%	78%	11%	17%	50%
Aklavik Inuvialuit	40%	50%	10%	27%	53%
Arctic Village	20%	80%	0%	40%	60%
Fort McPherson	100%	0%	0%	74%	0%
Old Crow	44%	56%	0%	38%	23%
All communities	47%	49%	3%	39%	37%

We stress to fishers that the Borderlands Co-op community monitoring program is not interested in the total number of fish taken or the location of private fishing spots. Quantitatively based data on harvest levels are documented with other programs and there is concern that enquiries into the location of good fishing areas will be perceived as intrusive. Instead, fishers report on their general knowledge of fish health and fishing conditions, and where appropriate they are asked for an explanation of their observations.

For the period of 1996 to 2001, reports from local fishers indicate that most species used as a subsistence resource are healthy, although there have been ongoing concerns about observations of unusual loche or burbot *(Lota lota)* livers. It is important to remember that reports of healthy stocks do not negate the need for or importance of community monitoring; good news reports are as valuable as observations of abnormalities.

As a part of hunters' and fishers' traditional knowledge of animals, livers are key indicators for assessing an animal's health and condition. In some communities, there have been taboos for pregnant women eating livers. Community monitoring reports

Figure 2-9. Randall Tetlichi, 2002 ABC monitoring associate for Old Crow, pre-tests the questionnaire by interviewing Edward Lennie of Inuvik at the Borderlands Coop's orientation and training workshop.

from all communities noted that fishers were catching loche with unusual livers (discolored, lumps, spots). Expressing concern about possible contaminants in loche livers and associated human health risks, residents wished to know more definitely the source of the problem.

Our community monitoring program responded with a set of specific questions about loche livers, which indicated that the concern was widespread among local fishers, particularly in the Aklavik area. At the fourth annual gathering of the co-op in 1999, a follow-up study was recommended to investigate more fully livers quality. During the winter of 1999–2000, the co-op initiated a study, funded by Environment Canada, with the goal of working with local fishers to collect what are perceived to be "normal" and "unusual" livers and then to compare them through laboratory analysis of fat levels and the presence of contaminants. The testing results showed that livers classed as "unusual" were darker than the "normal" ones and sometimes mottled or spotty. However, the "normal" and "unusual" livers did not differ in fat content, shape, or contaminants. The concentrations of major contaminants were lower than or comparable to those in most Yukon loche livers (pers. com. Joan Eamer, Environment Canada).

Findings from the loche study stimulated additional discussion among community members, who asked "If not contaminants, then what?" Several elders who are longtime fishers of loche added to the discussion by reporting a decrease in size of this fish since "long ago," with some attributing the problem to drying of creeks and low water levels. Alternative explanations for the unusual loche livers have been suggested, including the presence of parasites, water temperature, and the location and timing of some loche harvesting. Additional follow-up studies have been recommended, while a general report has been released telling local residents the findings of the study.

The community monitoring program also focuses on the quantity and intensity of fish runs. The statements from Old Crow fishers below describe some of the local indicators and explain the environmental conditions that affect salmon.

> Horseflies are a signal that the king salmon will begin to run in four or five days. (Old Crow 1996–97)

> If water is low, then dog salmon [chum] get battered up. When there is high water, fish come quickly. When water is dropping, fish tend to go faster up river, and therefore get more battered. (Old Crow 1998–99)

> If there is lots of water, there are very few fish. If there is low water, there will be a lot of king salmon. (Old Crow 1996–97)

> In high water, fish tend to stay in the middle; fewer than average fish between this year and last due to low water. (Old Crow 1996–97)

> There is lower water so fish had hard time to come by. It is very unusual that king salmon come with dog salmon. Dog salmon usually comes a month later. (Old Crow 1996–97)

Local fishers from Aklavik and Fort McPherson have reported the unusual observation of salmon running on the eastern Beaufort Sea coast and up the Mackenzie River since 2000. Some speculate that the unusual sightings are due to changes in ocean water temperatures, with some also noting that the occasional observation of salmon in locations as far up river as Rogue River (off the Peel River) has occurred in the past.

The limits of community knowledge-generated results were recognized early in the community monitoring program. For example, in 1998 Old Crow fishers of the Porcupine River observed salmon to be in good numbers, contrary to Canada Department of Fisheries and Oceans statistics indicating low escapement up river. The discrepancy is partially explained by fishers' effort, gear, and perceived abundance and partially explained by the difficulties associated with addressing specific questions through the use of open-ended questions. For example, responses such as "lots," "many," and "more" are not necessarily comparable. As a result, the questionnaire has been modified to also include specific descriptors (check boxes) and additional opportunities for more narrative responses.

Local Communities

Local communities of Arctic Borderlands view themselves as integral to their environment. Reflecting this view, social indicators of community well being were included in the list generated at the Borderlands Co-op indicators workshop, but have not been tracked in any detail. This decision is the result of debate among community members on whether social indicators are appropriate for a program with an ecological monitoring focus. As well, some community members expressed a sense of vulnerability about exposing internal aspects of community life that may appear highly dysfunctional to outsiders. Since no consensus could be achieved on the inclusion of social indicators, it was decided that the monitoring program would await further direction from the communities to pursue that topic area. As a result, no questions about community conditions were included in the initial phases of the community monitoring project. Yet, in spite of the absence of such questions, local experts have talked at length about social elements of change, local economic and social conditions, education, and the transmission of traditions to young people. Most recently, a series of open-ended questions has been added, providing an opportunity to comment on cultural, social, and economic conditions. While it is difficult in the five years of the project to discern trends, some are apparent (e.g., perceived changes in employment opportunities that correspond with the current development boom in the Canadian Western Arctic). As illustrated below, the recorded comments of locals do provide a picture into current conditions and concerns.

> People aren't going out (on the land) as much. . . . Probably because of jobs in town. Living is much easier today. (Old Crow 1997–98)

> Lot of our younger people are unemployed; there's no work around here. There's really nothing for them economy wise. (Aklavik Inuvialuit 1999–00)

> Less jobs. If we get jobs it isn't long enough to get unemployment. (Aklavik Inuvialuit 1999–00)

> Everyone stuck in town, inside their warm houses. (Fort McPherson 1999–00)

> Lots of jobs this year for those that want to work. (Aklavik Gwich'in '00–01)

> There is a lot of mining and pipeline preparation. This has brought more education and training opportunities. A lot of local people are employed in these fields now. (Fort McPherson 2000–01)

> Language is being taught in school; the kids are being taught on land activities, which is very good. (Aklavik Inuvialuit 1999–00)

> Language is taught . . . but is not used in the community. (Aklavik Gwich'in 1998–99)

The inclusion of social indicators in the ABC community monitoring program is a topic of discussion currently underway by ABC partner communities, and is especially relevant as regional oil and gas development activities of the Western Canadian Arctic increase.

Reflections on the ABC Community Monitoring Experience

What have we learned from the ABC community monitoring experience? How has the local knowledge of indigenous communities informed our monitoring process? What are the outstanding challenges in the establishment and evolution of the ecological monitoring that support meaningful community involvement?

The reports of local experts on the status, trends, and anomalies of weather, plants, animals, local hunting and fishing, and community demonstrate how local knowledge stands on its own to provide a richly detailed and holistic account of ecological conditions and process. Through the sharing of oral and written summary reports by community associates at the Arctic Borderlands Co-op's annual gatherings, local knowledge broadens beyond the focus of a single community to create a regional image. In this respect, community monitoring contributions through the ABC process are transformed to address multiple scales and to become regional knowledge.

Local knowledge documented through the ABC process and science-based indicators are also complementary. Reports of Old Crow local experts on the drying of lakes in Crow Flat illustrate how community monitoring served as an early warning signal that prompted quantitatively based remote-sensing analysis to determine the extent of change. In the case of the El Niño year of caribou movements on Yukon's North Slope, community hunters supplied an account of caribou behavior undetected by the remote sensing tools of biologists. Going further, those reports provide a rich explanation of the year's anomaly that integrates climate, extent of sea ice, forage quality, caribou body condition, and human economic components of the Arctic system.

Community contributions to ecological monitoring can also challenge the basic assumptions of science-based monitoring. One example noted above is caribou hunters who questioned agency reports of a decline in the Porcupine caribou herd's population. As highlighted by the Gwich'in hunter referencing how caribou avoided his community because of the death of a man who "slept to caribou," an indigenous perspective helps us remember that the Arctic system should also be viewed in the context of human-nature relations and not purely as an object of interest worthy of study. The statement also serves as an important reminder that indigenous communities of the Arctic Borderlands approach the idea of "ecosystem" through a worldview that resembles but also differs from that of the West.

As important as the information gleaned from the community monitoring interviews are, the contributions of indigenous communities shaping the ABC's process of communication and collective learning. Clearly, the informal cooperative structure of the ABC has allowed for a level of experimentation, learning, and adaptation not possible in many other monitoring programs. The Borderland's operating principles—staying relevant to local communities, thinking long-term, economizing, and moving slowly—have also served to keep us focused on our overarching objective, and maintained our momentum. As a result, Arctic Borderlands Co-op annual gatherings have become important events in the Arctic Borderlands Region, gaining the strong support of various communities, First Nations, Inuvialuit organizations, co-management boards, and government agencies. Annual gatherings serve as a time when researchers, managers, and subsistence users alike come together to exchange ideas about ecological conditions and discuss the implications of change in a setting that that is not politically charged. The use of "elders' round tables" and special presentations by visiting scientists are symbolic in recognizing the legitimacy and outstanding contributions of several types of experts. More functionally, the exchanges between indigenous people and scientists have served to identify data gaps, pose new hypotheses, and direct future studies.

The atmosphere of collaboration generated by the ABC is best evidenced with the recent willingness of Arctic Village and Kaktovik to participate in the co-op activities—communities that hold dramatically different political positions on proposed oil development in the Arctic National Wildlife Refuge of Alaska, but which also share a common interest in sustainability. The extent to which Alaska communities' participation in the ABC can flourish within the current U.S. management regime that offers limited rights and financial support to local communities is unknown.

ABC's web-site at www.taiga.net/coop addresses the need for improved information exchange and is a communication tool that is increasingly valued by local organizations as Internet connectivity is improved and made more affordable across the North. Given the recent rate of change and high uncertainty surrounding the question of what is changing and why, the full suite of ABC communication tools increases the likelihood that groups in the region can improve understanding and shape future changes in ways that best meet the needs of communities. Said another way, the ABC provides a means for enhancing the capacity of Arctic communities to respond to change.

In spite of these benefits, challenges associated with community involvement in a regional monitoring program remain and questions are unanswered. Agency managers and community leaders are typically overburdened with the demands of immediate crises, leaving their focus on ecological monitoring on the back burner. Funding cycles tend to reward "flavor-of-the-day" ideas, leaving more established programs struggling to survive. Currently, the majority of the Borderlands Co-op financial support is from the Canadian government, through annual contribution agreements with departments and through special time-limited project funding. Government planning and funding processes require clearly defined measurable results within specified time frames—something that can be difficult to mesh with the needs of the Borderlands Co-op program for continuity and shared ownership. How an informal collaborative alliance like the ABC can avoid the myopic tendencies of society, while remaining attentive to community needs over the long run, is uncertain.

Issues of community capacity and human capital pose another set of challenges in community monitoring. Operating with limited talent pools and organizational infrastructure, local communities of the Arctic Borderlands region are currently engaged by the demands of land claims implementation, federal and territorial government devolution, and community self-governance. Consequently, partner communities of the ABC, while supportive of the monitoring program, have also been unwilling to assume a lead responsibility in its execution. Community interest in building human capital among its membership creates a dilemma for local leaders, who must choose between applicants who have a financial need for employment and demonstrate a readiness to learn on the job and those who currently hold the professional skills and community knowledge necessary to do the work well. The consequence of the choice has far-reaching implications for the quality of the information documented as well as the overall continuity of the program.

Following from the challenge of cultivating community capacity is a set of methodological issues. What is the best method for documenting local knowledge on a regular basis when there are limited resources for transcribing tapes and when local interviewers have adequate-to-limited experience with administering questionnaires? Using a structured interview format has provided local researchers with limited interviewing experience, a clear road-map for asking questions. However, structured interviews are not well suited for interviewing local hunters, whose reports come in the form of narratives. Associated with the challenge of the interview format is the problem of asking local experts to compare current conditions with previous years and "average years." Such measures raise questions about the extent to which memory interacts with emotions such as nostalgia (e.g., "things always used to be better than they are now days"), and the tendency of all people to focus on anomalies and outliers.

To what extent should community monitoring draw on the observations and explanations of traditional local hunters and fishers, and to what extent should it become a process through which local subsistence users learn and employ technical methods to monitor their homelands? Clearly, training local residents to serve as research technicians is not what is implied by community monitoring with local knowledge in the ABC context. As we have demonstrated, it is important not to view the two systems of knowledge as an either-or question. Therein lies a fundamental challenge of any knowledge co-production process—achieving relationships that are respectful of indigenous cultures and sensitive to the dominant and hegemonic forces of western science. In the ABC program, an effort has been made to respect local knowledge as a distinct tradition of knowing. From our own experience with the Arctic Borderlands Co-op, we find that strong links and even a bit of tension between local- and science-based approaches offer interesting opportunities for generating innovation. Working through community associates who understand and undertake technical tasks of local monitoring (e.g., collecting specimens, recording plant phonology) and who also interact with local experts to document community knowledge may help facilitate this tricky interface.

Conclusion: Community Monitoring Beyond the Arctic Borderlands

The Arctic Borderlands Ecological Knowledge Co-op with its community monitoring program is currently being referenced as a model for indigenous community involvement in the development of other programs across northern Canada and Alaska. Nascent monitoring programs with a local knowledge component are currently being explored by the Nunavut Wildlife Management Board, Mackenzie Valley Cumulative Impacts Monitoring Program, the Beverly Qamanirjuak Caribou Management Board, the Bathurst Caribou Planning Group, the Labrador Inuit Association, and U.S. Fish and Wildlife Service, Division of Subsistence (fisheries) in Alaska. As these new programs evolve, we face the challenge of scaling from the community and regional levels to an inter-regional and Pan Arctic scale. How will this increased scope be achieved? Ultimately, the future success of community monitoring in a larger milieu of knowledge co-production will depend on our view of power, not as a finite resource to be controlled but as source of energy that is expandable and shared.

Acknowledgements

Thank you to the many local monitoring associates who have worked with the program as interviewers since its inception in 1996. They are Joe and Glenna Tetlichi, Carol Arey, Myrna Nerysoo, Christine Nukon, Carol Arey, Dolly Peterson, Ellen Blake, Roberta Alexie, Richard Gordon, Norm Snowshoe, Sherri DeBastien, Annie B. Gordon, Susan Ross, Vicki Josie, Roxanne Koe, Danny Greenland, Randall Tetlichi, Connie Stewart, Dolly Peterson, Shirley Kakfwi, Johnny Edwards, and Joel Tritt.

A special thank you is extended to Joan Eamer, Deborah Robinson, Mike Gill, Scott Gilbert, Aileen Horler, and Kent Sinnott for their outstanding assessment of indicators and overall organization of the project and to Joan Eamer, Barney Smith, James Andre, Deborah Robinson, Heather Swystun, and the editors of this volume for their helpful review of this document. A thank you also goes to Johnny Edwards, Joe Tetlichi, Robert Charlie, Billy Archie, Carol Arey, Don Russell, Herbert Felix, and Billy Day for their help.

Funding support for the work of Gary Kofinas with the community monitoring project has been through several funding sources, including Environment Canada and a grant from the National Science Foundation, Office of Polar Programs Award #OPP-9709971, Integrating Traditional and Scientific Knowledge in Large Mammal Research. Time to prepare this paper for publication was provided through the National Science Foundation's Sustainability of Arctic Communities Project Award #OPP-9521459.

The ABC project would not be possible without the contributions of the many local experts who contribute to our project annually. They are Abe Stewart, Abe Stewart Sr., Alan Blake, Alfred and Catherine Semple, Alfred Semple, Alice Vittrekwa, Allen Koe, Allen Koe Sr., Allen Tritt, Amos and Rebecca Francis, Amos Francis, Andrew and Shirley Charlie/Kakfwi, Andrew Gordon Sr., Art Furlong, Beatrice Jerome, Bella Ross, Bertha Francis, Bertha Frost, Betty Frost, Billy Archie, Billy Bruce, Bobby Gilbert, Brian Lord, Carl Charlie, Caroline Kaye, Charlie Stewart, Chris Bonnetplume, Curtis Netro, Dale Semple, Danny Kassi, Danny C. and Annie C. Gordon, Danny A. and Annie B. Gordon, Danny James Gordon Jr., Danny Kassi, Darius Kassi, David Charlie, David Lord, David Smith Sr., Dennis Arey, Dennis Frost, Dick Nukon, Dolly Josie, Donald and Elizabeth Aviugana, Donald Frost, Dorothy Alexie, Dorothy Koe, Dorothy McLeod, Dorothy R. Koe, Dwayne James, Ernest. Vittrekwa, Effie Jane Snowshoe, Eileen Koe, Elizabeth Colin, Elizabeth Kaye, Ellen Vittrekwa, Emma Kaye, Ernest Vittrekwa, Ernestine Elias, Esau Schaffer, Eunice Mitchell, Fanny Charlie, Florence Netro, Florence Thomas, Fred Kendi, Fred Koe, Freddie Kendi, Freddy and Liz Frost, Freddy Frost, Gene Flitt, George Allen, George and Judy Edwards/Selamio, George Koe, George Selamio, Georgie Moses, Gideon James, Harold and Theresa Frost, Harvey Kassi, Henry John, Ida Inglangasuk, Ida Stewart, Issac Thomas, Jacob and Elisabeth Archie, Jacob Archie, Jacob Kowana, James (Jumsey) Edwards, James Itsi, James McDonald Sr., Jim Vittrekwa, Jim and Jane Montgomery, Jimmy Linklater, Joe Tetlichi, Joe A. Vittrekwa, Joe Arey, Joe Benoit, Joe H. Vitterkwa, Joe Ricky, Joel J. Tritt, John Carmicheal, John Joe Kaye, Johnny Charlie Sr., Jonas Meyook, Jonas Meyook, Joseph Bruce, Joseph Vittrekwa, Joseph Vittrekwa, Julieanne Koe, Keith Colin, Kenneth/Patti Tetlichi, Kenny Tetlichi, Kias Peter Sr., Kim and Leonard Nukon, Larry Semmler, Larry Sittichinli, Lawrence Lord, Lee John Meyook, Lincoln Tritt, Liz Colin, Louise Linklater, Lucy Greenland, Lucy Wilson, Lydia Thomas, Mary Teya, Mabel Firth, Mabel Tetlichi, Margaret Blake, Margaret Kendi, Margaret Meyook, Marion

Schaffer, Marvin Frost, Mary Effie Snowshoe, Mary Jane Moses, Mary M. Firth, Moses and Barbara Kayotuk-Allen, Moses Kayotuk, Moses Lord, Nancy Flitt, Neil Colin, Nellie and Larry Arey, Percey Kaye, Peter and Brenda Charlie, Peter Charlie, Peter Elanik Sr., Peter Francis, Peter J Kaye, Peter Josie, Peter Tizya, Phillip Rispin, Rachel Villebrun, Randal Frances, Randall Tetlichi, Rebecca Francis, Renie Arey, Richard Ross, Richard Ross Senior, Robert Alexie Sr., Robert Bruce, Roger Kaye, Roy Moses, Sarah Kaye, Sarah Meyook, Shawn Bruce, Simon Snowshoe, Stanley Njootli and Maureen Vitts, Star Tyrrell, Stephen Frost Sr., Sterling Firth, Steven Bonnetplume, Steven Frost Jr., Steven Tritt, Tabitha Nerysoo, Teresa Kendi, Thomas Gordon, Thomas Koe, Thomas Stewart, Timothy Sam, Tom Elanik, Tracy Kassi, Tremble Gilbert Sr., Vicky Josie, Wally McPherson, Wally Tyrell, Walter Gardlund, Walter McPherson, William Josie, William Peterson, William Ben Vaneltsi, William Teya, Wilson Malegana, Woody Elias, and others.

Notes

1 The Arctic Borderlands Ecological Knowledge Co-op is a program of the Arctic Borderlands Ecological Knowledge Society, a new nonprofit organization. For more detail about the Arctic Borderlands Ecological Knowledge Co-op see www.taiga.net/coop.

2 Environment Canada is Canada's federal department of the environment.

3 Statistically normalized 2001–2002 data show a strong relationship between reported availability and meeting needs ($r^2 = .947$).

Bibliography

Allaye-Chan, A. C. 1991. Physiological and ecological determinants of nutrient partitioning in caribou and reindeer. Ph.D. dissertation. Fairbanks: University of Alaska Fairbanks Institute of Arctic Biology.

Anderson, D. G. 2000. Rangifer and Human Interests. *Rangifer: Keynote Lectures of the Arctic Ungulate Conference, 9–13 August 1999, Tromso, Norway* 20(2–3): 153–174.

Berger, T. 1977. Northern Frontier, Northern Homeland: The Report of the Mackenzie Valley Pipeline Inquiry. Ottawa, Printing and Publishing, Supply and Services Canada.

Berkes, F. 1981. The Role of Self-regulation in Living Resources Management in the North. In: M. Freeman, ed. *Proceedings, First International Symposium on Renewable Resources and the Economy of the North.* Ottawa, Association of Canadian Universities of Northern Studies, Canada Man and the Biosphere Reserve Program: 166–178.

Berkes, F. 1994. "Co-management: Bridging the Two Solitudes." *Northern Perspectives 22*(2–3): 18–20.

Berkes, F. 1999. *Sacred Ecology: Traditional Ecological Knowledge and Resource Management.* London and Philadelphia: Taylor and Francis.

Brydges, T., and A. Lumb. 1998. Canada's Ecological Monitoring and Assessment Network: Where we are at and where we are going. *Environmental Monitoring and Assessment 51*(1–2).

Caulfield, R. 1983. *Subsistence Land Use in Upper Yukon Porcupine Communities, Alaska.* Juneau, Alaska Department of Fish and Game, Division of Subsistence.

Cruikshank, J. 1998. *The Social Life of Stories.* Omaha: University of Nebraska.

Elkin, B. T. 1999. Community-based Monitoring of Abnormalities in Wildlife. J. Jensen and L. A. Walker, Environmental studies—Canada. Dept. of Indian Affairs and Northern Development: 123–125.

Feit, H. 1986. Hunting and the Quest for Power: The James Bay Cree and Whiteman in the Twentieth Century. In: R. B. Morrison and C. R. Wilson, eds. *Native Peoples: The Canadian Experience.* Toronto: McClelland and Stewart. 171–207.

Feit, H. 1988. Self Management and State Management: Forms of Knowing and Managing Northern Wildlife. In: *Traditional Knowledge and Renewable Resource Management in Northern Regions.* Edmonton: Boreal Institute for Northern Studies, University of Alberta.

Feit, H. A. 1973. Ethno-ecology of the Wasanipi Cree; or How Hunters Can Manage Their Resources. In: B. Cox, ed. *Cultural Ecology.* Toronto: McClelland and Stewart. 115–125.

Ferguson, M. A. D., and F. Messier. 1997. Collection and Analysis of Traditional Ecological Knowedge about a Population of Arctic Caribou. *Arctic 50*(1): 17–28.

Fienup-Riordan, A. 1990. *Yup'ik Lives and How We See Them.* New Brunswick: Rutgers University Press.

Ford, J., and D. Martinez, eds. 2000. Special Issue: Traditional Ecological Knowledge and Wisdom. *Ecological Applications 10*(5).

Freeman, M. 1989. Graphs and Gaffs: A Cautionary Tale on the Common Property Resource Debate. In: F. Berkes, ed. *Common Property Resources; Ecology and Community-Based Sustainable Development.* London: Belhaven Press.

Freeman, M., and L. Carbyn, eds. 1988. *Traditional Knowledge and Renewable Resource Management in Northern Regions.* Edmonton, Boreal Institute for Northern Studies, University of Alberta.

Freeman, M. M. R. 1992. The nature and utility of traditional ecological knowledge. *Northern Perspectives 20*(1): 9–12.

Grey, B. 1989. *Collaborating: Finding Common Ground for Multiple Problems.* San Francisco: Jossey-Bass.

Heiman, M. K. 1997. Science by the people: Grassroots environmental monitoring and the debate over scientific expertise. *Journal of Planning Education and Research 16*(4): 291–299.

Huntington, H. P. 2000. Using Traditional Ecological Knowledge in Science: Methods and Applications. *Ecological Applications 10*(5): 1270–1274.

Inglis, J. T., ed. 1993. Traditional Ecological Knowledge: Concepts and Cases. Ottawa, International Program on Traditional Ecological Knowledge, International Development Research Centre.

Irving, L. 1958. Naming Birds as Part of the Intellectual Culture of Indians at Old Crow, Yukon Territory. *Arctic 11*(2): 117–122.

Kendrick, A. 2000. Community Perceptions of the Beverly-Qamanirjuaq Caribou Management Board. *The Canadian Journal of Native Studies 20*(1): 1–33.

Klein, D. 1991. Caribou in the Changing North. *Applied Animal Behavior Science*(29): 279–291.

Kofinas, G., D. Russell, and R. White 2001. Summary of Proceedings of Workshop on Community Monitoring of Caribou Body Condition, Whitehorse, YT, posted at http://www.dartmouth.edu/~arctic/rangifer/wahtek/.

Kofinas, G. P. 1998. The Cost of Power Sharing: Community Involvement in Canadian Porcupine Caribou Co-management. Ph.D. dissertation. Vancouver: Faculty of Graduate Studies, University of British Columbia.

Kruse, J., D. Klein, L. Moorehead, B. Simeone, and S. Braund. 1998. Co-Management of Natural Resource: a comparisons of two caribou management systems. *Human Organization 57*(4): 447–458.

Kruse, J., and The Sustainability Project Research Team. 2000. Sustainability of Arctic Communities: An interdisciplinary Collaboration of Researchers and Local Knowledge Holders. Unpublished manuscript.

Levi-Strauss, C. 1966. *The Savage Mind.* Chicago: University of Chicago Press.

Maxwell, B. 1992. Arctic climate: potential for change under global warming. In: F. S. I. Chapin, R. L. Jefferies, J. F. Reynolds, G. R. Shaver, and J. Svoboda, eds. *Arctic ecosystems in a changing climate: An ecophysiological perspective.* San Diego: Academic Press. 11–34.

Nadasdy, P. 1999. The Politics of TEK: Power and the "Integration" of Knowledge. *Arctic Anthropology 36*(1–2): 1–18.

Nelson, R. K. 1983. *Make Prayers to the Raven: A Koyukon View of the Boreal Forest.* Chicago: University of Chicago Press.

Nelson, R. K., K. H. Mautner, and G. R. Bane. 1982. *Tracks in the Wildland: A Portrayal of Koyukon and Nunamiut Subsistence.* Fairbanks, Alaska: Cooperative Park Studies Unit, University of Alaska.

Osherenko, G. 1988. Can Co-Management Save Arctic Wildlife? *Environment 30*(6): 6–13, 29–34.

Overpeck, J., K. Hughen, D. Hardy, R. Bradely, R. Case, and M. Douglas. 1997. Arctic Environmental Change of the Last Four Centuries. *Science 278:* 1251–1259.

Page, R. 1986. *Northern Development: The Canadian Dilemma.* Toronto: McClelland and Stewart.

Pinkerton, E. W. 1994. Where Do We Go From Here? The Future of Traditional Ecological Knowledge and Resource Management in Native Communities. In: Barry Sadler and Peter Boothroyd, eds. *Traditional Ecological Knowledge and Modern Environmental Assessment.* Vancouver: Canadian Environmental Assessment Agency, International Association for Impact Assessments, and University of British Columbia, Centre for Human Settlements: 51–60.

Rowntree, P. R. 1997. Global and regional patterns of climate change: recent predictions for the Arctic. In: W. C. Oechel, T. Callaghan, T. Gilmanov, et al., eds. Global change and arctic terrestrial ecosystems. New York: Springer-Verlag. 82–112.

Russell, D., G. Kofinas, and B. Griffith. 2000. Need and Opportunity for a North American Caribou Knowledge Cooperative. *Polar Research 19*(1): 117–130.

Russell, D. E., A. M. Martell, and W. A. C. Nixon. 1993. The Range Ecology of the Porcupine Caribou Herd in Canada. *Rangifer* Special Issue 6: 168.

Scott, C. 1996. Science of the West, Myth to the Rest. In: L. Nader, ed. *Naked Science: Anthropological inquiry into boundaries, power, and knowledge.* New York: Roultledge. 69–86.

Slobodin, R. 1962. *Band Organization of the Peel River Kutchin.* Anthropological Series No. 55. Ottawa, National Museum of Canada: 97.

Slobodin, R. 1981. *Kutchin. Handbook of North American Indians: The subarctic.* Washington: Smithsonian Institute 6: 514–532.

Stevenson, M. G. 1996. Indigenous Knowledge in Environmental Assessment. *Arctic 49*(3): 278–291.

Usher, P. 2000. Traditional ecological knowledge in environmental assessment and management. *Arctic 53*(2): 83–94.

Wein, E. E. 1994. Yukon First Nations Food And Nutrition Study. Report to the Champagne and Aishihik First Nations, the Teslin Tlingit Council, the Vuntut Gwich'in First Nation, the Yukon Department of Health, and the National Institute of Nutrition, National Institute of Nutrition Post-doctoral Fellow.

Wein, E. E., and M. M. R. Freeman. 1992. Inuvialuit Food Use and Food Preferences in Aklavik, Northwest Territories, Canada. Canadian Circumpolar Institute, The University of Alberta. *Arctic Medical Science, 51:* 159–172.

Wein, E. E., and M. M. R. Freeman. 1995. Frequency of Traditional Food Use by Three Yukon First Nations Living in Four Communities. *Arctic 48*(2): 161–171.

Zhang, X., L. A. Vincent, W. D. Hogg, and A. Niitsoo. 2000. Temperature and Precipitation Trends in Canada during the Twentieth Century. *Atmosphere-Ocean 38*(3): 395–429.

Figure 3-1. Banks Island has one of the largest populations of musk ox in the world. Local people wonder about the impact that climate change will have on this animal. Photo by Neil Ford.

3

We Can't Predict the Weather Like We Used to:
Inuvialuit Observations of Climate Change, Sachs Harbour, Western Canadian Arctic

Dyanna Jolly,[1] Fikret Berkes,[2] Jennifer Castleden,[3]
Theresa Nichols,[4] and the community of Sachs Harbour[5]

An emerging theme in arctic climate change studies is to communicate climate change from the perspective of local people and to create community-based research partnerships to improve the understanding of impacts. Traditional and local knowledge are important in this regard, and so is local involvement at the community level (Bielawski, 1997, Kassi, 1993; ICC, 2001). However, there are a limited number of studies that have taken arctic climate change research to the community level and involved local people.

One recent project with a community-based approach is the *Inuit Observations on Climate Change (IOCC)* project. The project materialized as a response to Inuvialuit concerns in the Western Arctic community of Sachs Harbour, on Banks Island, about recent changes in the weather, land, and oceans, considered without precedent. In the 1990s, residents from arctic communities such as Sachs Harbour were beginning to talk about changes they observed when hunting, fishing, and travelling. Their concerns were voiced at several northern workshops and meetings, and communicated to the International Institute for Sustainable Development (IISD) in Winnipeg. Observations of environmental changes, and the impacts of such changes on subsistence activities in Sachs Harbour were taken to IISD by Rosemarie Kuptana, a board member and a prominent Inuit leader born in Sachs Harbour. In 1998, the IOCC project was initiated as a collaborative research effort between the IISD and the people of Sachs Harbour, with the goal of raising awareness of how climate change was affecting the lives of Inuvialuit on Banks Island.

---

1. Centre for Maori and Indigenous Planning and Development, Lincoln University, New Zealand
2. Natural Resources Institute, University of Manitoba, Winnipeg, Canada
3. International Institute for Sustainable Development, Winnipeg, Canada
4. Department of Fisheries and Oceans, Winnipeg, Canada
5. c/o the Sachs Harbour Hunter and Trappers Committee, Sachs Harbour, Canada

The IOCC project built on earlier environmental change research conducted under projects such as the Mackenzie Basin Impact Study MBIS (Cohen, 1997) and the Hudson Bay project (McDonald et al., 1997), that documented change as explained through local and traditional knowledge. It also complemented other research occurring at the same time looking at Inuit knowledge of climate change (Thorpe, this volume; Fox, this volume). However, it was unique in making indigenous perspectives of climate change its primary focus, advocating a community-driven approach to research, generating an interdisciplinary effort to bring together scientific researchers and local experts, and using video as a medium to communicate climate change as explained by Inuvialuit through traditional knowledge.

Two primary goals framed the project's design and implementation (Ford, 2000):

- To produce a video that will demonstrate to Canadian audiences, interest groups, and decision-making fora that climate change is making an impact on the traditional lifestyle and livelihood system of Inuit on Banks Island in the Beaufort Sea.
- To evaluate the potential contribution of traditional knowledge, local observations, and adaptive strategies to scientific research on climate change.

To achieve these goals, an interdisciplinary working partnership was created involving the IISD, researchers from five different institutions and local experts. The project had two parts—a video component and a scientific component. The video component was designed to produce a video to raise awareness about climate change as experienced by northern communities. Community members and the film crew worked together to determine the kinds of changes that would be filmed and how the impacts of such changes would be communicated. The scientific component was designed to develop an in-depth discussion of the role of traditional knowledge and community-based assessments in understanding climate change. A group of northern guest scientist researchers joined in the project, including a lead scientist from the Inuvialuit Settlement Region.

This chapter is a review and analysis of the Inuit Observations on Climate Change project. Following an introduction to the community of Sachs Harbour and local observations of change, the chapter describes the participatory research model developed for the project and the study methods used. Next, we discuss some of the main findings of the project regarding the changes observed by the people of Sachs Harbour and their response to change. The analysis section focuses on the two themes that emerge from the project objectives: the documentation and communication of Inuit perspectives to national and international audiences, and the contributions of traditional knowledge to the study of climate change. The review presented here provides a synthesis of materials produced by the project to date, including trip reports, the IOCC Final Report (IISD, 2001), several academic papers (Ford, 2000; Riedlinger and Berkes, 2001; Berkes and Jolly, 2001, Nichols et al., in prep.), and a graduate thesis (Riedlinger, 2001).

Inuvialuit Observations of Climate Change

In recent years, people in Sachs Harbour have observed changes associated with the weather, land, ocean, and wildlife that they consider to be different from normal, or expected variability. More frequent and intense storm events, higher temperatures, increased rainfall, reductions to multiyear sea ice, increasingly visible permafrost thaw and soil erosion, as well as incidences of southern bird and fish species are examples of phenomena identified as indicators of larger scale environmental change in the region. Inuvialuit elders say that the seasons are "getting crazy now" and that the weather has become unpredictable. Such changes are significant enough that people in Sachs feel it is necessary to begin monitoring them, and to raise awareness about the impacts on the Inuvialuit way of life.

> The weather didn't change that much years ago. It was always cold. Not like today. Now, you can't even tell when the weather is going to change. Mild weather—means it is going to be a storm coming, and we get ready for it. But today it changes so much—we are expecting a big storm and next day, clear as can be. I can't predict the weather anymore like we used to years ago.—Peter Esau, Sachs Harbour, 1999.

> Never saw salmon here before. People here have been setting nets for quite a while. That is the first time I ever seen that. Even herring [least Cisco] for that matter. It is kind of changing around here for us. I really find a difference with the fish that they are catching. Chars are getting bigger then we used to catch.—John Lucas Sr. Sachs Harbour, 1999

> Used to get almost no rain on the Island. Used to be snow in the fall. Now rain. This place used to be a dry place.—Andy Carpenter Sr., Sachs Harbour, 1999

Observations of climate change by people in Sachs Harbour are based on their knowledge of the weather, or *sila,* and the interaction between the weather and other environmental phenomena. This knowledge is one aspect of the larger body of traditional and local knowledge relating to the land, seasonal cycles, processes and relationships between the elements of the environment (Figure 3-2). As with other circumpolar communities with land-based livelihoods, there is a rich tradition of understanding, interpreting and predicting the weather and knowing about the weather is an integral part of community life. People watch the weather, because the weather tells when it is a good time to go out on the land, leave for a camp or plan a hunting trip. The weather dictates if a plane will come that day, bringing mail and supplies. It influences when the geese will arrive, when the sea ice will begin to break, if the fish will bite, or a storm will come.

Cumulative experience and observation gives Inuvialuit a sense of expected variability and fluctuation in the arctic environment. Assessments of change are derived from this normal, as it provides a baseline against which to compare change. Thus local residents can identify a window of expected variability relating to such phenomena as the timing of freeze-up and break-up of the sea ice, seasonal temperatures and precipitation, wind strength and direction at a given time of year, or rates of coastal erosion.

> Sila (the weather) has changed alright. It is a really late falltime now, and a really fast and early springtime. Long ago the summer was short, but not anymore.—Mrs. Sarah Kuptana, Sachs Harbour, 1999.

> It used to be really nice weather long ago when I was a kid. Bad weather now. So many mosquitoes. Sometimes it was hot, sometimes cold—not like now. [Things happen at the] wrong time now, it is way different now. August used to be cool off time, now it is hot. It is really short in the winter now.—Edith Haogak, Sachs Harbour, 2000.

The Community of Sachs Harbour

Sachs Harbour *(Ikaahuk)* (71° 59' N, 125° 14' W) is the most westerly island of the Canadian Arctic Archipelago. It is the smallest of six Inuvialuit (western arctic Inuit) communities in the region covered by the comprehensive native land claims agreement, the Inuvialuit Final Agreement of 1984 (Figures 3-3 and 3-4). In 1999, the community population was approximately 120.

Sachs Harbour became a permanent settlement in 1956, when seasonal trappers began to spend their entire year on the Island. As with other arctic communities, Sachs Harbour has undergone significant socioeconomic changes in the last 50 years, from trapping to the DEW lines, oil development and more recently, co-management agreements. The community now has a school from kindergarten to grade eight, a nursing station, arena, guesthouse, and the *Ikaahuk* co-op. There is a small airport, and two flights a week come from the mainland center of Inuvik. Subsistence harvesting and other land-based activities are now part of a larger mixed economy, and re-

Figure 3-2. Sarah Kuptana and her daughter Rosemarie Kuptana in the home of Sarah Kuptana, discussing the changes they have seen on the land. Women in the community are especially knowledgeable about the health of the animals that are harvested. For example, women have observed changes in the fat of animals when the animals are skinned, providing an indication of the relative health of the animal.

main a central component of community life. Following the land claims settlement, management of fish and wildlife resources is a shared responsibility between Inuvialuit and federal and territorial governments.

The history of land use and occupancy of Banks Island is well documented in several studies for the region. Most notable is Usher's (1970a, 1970b, 1971) series on the economy and ecology of Sachs Harbour as a frontier trapping community. Historical information is also found in the Banksland Story (1969) written by Father Lemer, O.M.I., along with community elders; Stephansson's (1923) writings on the Canadian Arctic Expedition; the Aulavik oral history project (Nagy, 1999); Condon et al.'s (1993) book on the Holman area; and Freeman's Inuit Land Use and Occupancy project (1976).

Model of Community Research Partnership

The IOCC project sought to establish a model of community research partnership, recognizing that the meaningfulness and accuracy of the project results were directly related to ensuring the commitment and participation of the community. The project was to benefit the people in Sachs Harbour, as well as providing information for others. This was achieved through using a diverse, participatory set of methodologies that worked to involve the residents in all components of the project, i.e., in project design as well as implementation and maintain flexibility and adaptability. The project developed as a process that was finely tuned to feedback and could adjust when

Figure 3-3. Aerial view of Sachs Harbour, Northwest Territories. June 1999. The ocean surrounding Sachs Harbour is frozen for most of the year. Recently, summer break-up occurs earlier.

necessary. The project team included an IISD facilitator group, film crew, local experts and liaison people from the Inuvialuit Settlement Region, a group of research scientists, and a university group, with the Hunter and Trappers Committee serving as an advisory committee, working alongside a community liaison.

Several key features facilitated planning and implementation of the project. One of the most important was a planning workshop, held in Sachs Harbour in June 1999 (described in more detail below). The idea was to have the project spin-off from an initial workshop, setting the foundations for participation and partnership. The intent was to pass ownership to the community, with the outside team providing support and facilitation (Ford, 2000). Initiating the project through a planning workshop set a precedent for the work to follow. As John Keogak, member of the Hunter and Trappers Committee commented, "I think all projects should be considered this way, where you meet with the people and start from where the knowledge is—and work your way up from there."

A second key design feature was planning the study around four seasonal visits, at times of the year determined by the community, in order to document changes on a seasonal basis. During the initial planning workshop in June 1999, community members were clear that the project had to provide a picture of climate change in all seasons, and within the context of seasonal activities. Thus, three additional seasonal visits

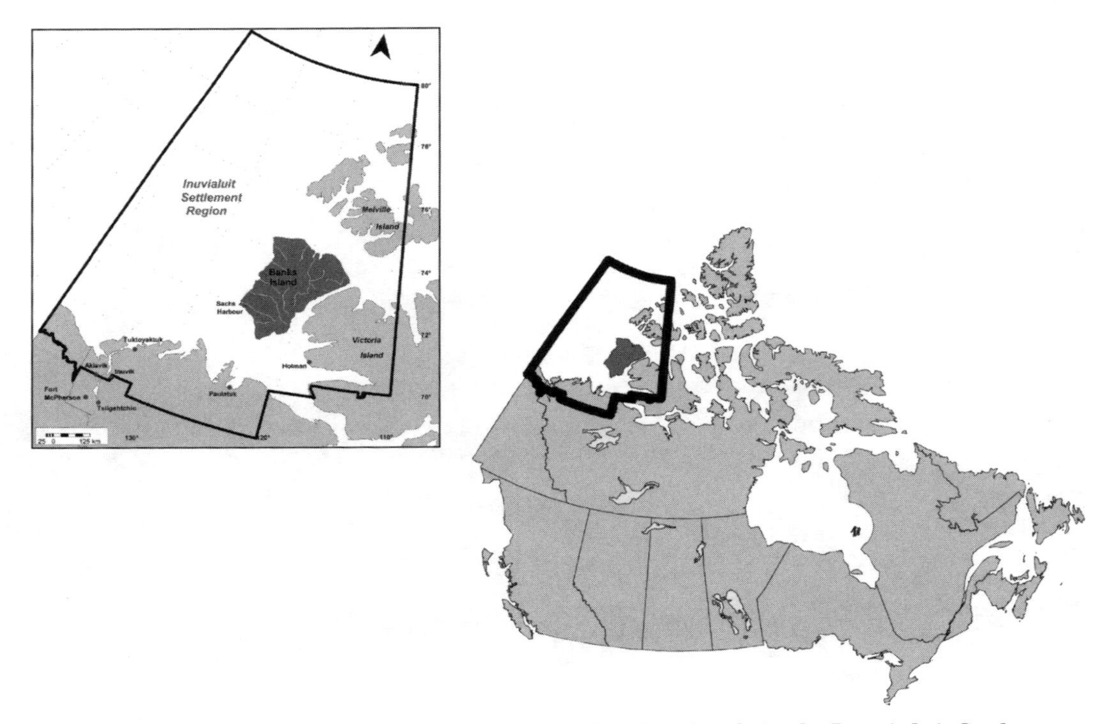

Figure 3-4. Map of the community of Sachs Harbour, Banks Island, in the Inuvialuit Settlement Region, Northwest Territories. Source: Department of Resources, Wildlife and Economic Development, Government of the NWT.

were organized, with each trip organized around a specific theme, or group of observations (Table 3-1). Repeat visits gave the project continuity. For example, they allowed the video team to show preliminary footage that had been taped in previous trips, and provide the opportunity for feedback and checks on how the video was progressing. It allowed the science team to follow up on specific topics. Most importantly, repeat visits encouraged a more relationship-based approach to research (Riedlinger, 2001).

Other key features relate to communication tools and feedback, as well as the organization of a final workshop. These features are described in Table 3-2.

Communication and collaboration within the IOCC project was facilitated by a series of participatory methodologies. These included techniques and tools as part of community-based planning workshops, semidirected interviews, travelling/ collaborating with local experts out on the land (participant observation), and documenting observations and knowledge of change on video. Of primary importance was that the methods facilitated communication of traditional knowledge and allowed the community to present climate change as they saw it. They were utilized in a way that was involving, and that encouraged building relationships between the IISD team and the community. These methods are described below:

Starting Off the Right Way: Holding a Planning Workshop

The first trip to Sachs Harbour was made in June 1999, to hold a planning workshop (Figures 3-5 and 3-6) for the project. This initial two day workshop had three purposes: (i) to propose a project to the people in Sachs Harbour; (ii) to allow residents to describe their livelihood system and discuss the changes they are experiencing; (iii) to work with the community to plan further trips. Using participatory exercises and techniques, the project team worked with community members to design how the project would proceed in terms of presenting Inuit observations and knowledge of climate change through film and scientific interviews.

The focus of the planning workshop was to have Inuvialuit identify and analyze the changes they were experiencing and the impacts on community life. This was accomplished using brainstorming (issue identification), problem trees (cause-effect analysis), timelines (establishing history), ranking exercises (highest priority research items) and seasonal calendars (harvest activity). These methodologies contain elements from Participatory Rural Appraisal (Chambers, 1991), emphasizing direct participation, with community members playing an active role in setting priorities and determining research topics. They also drew on ZOPP (Ziel Orientierte Projekt Planung, or Objectives Oriented Planning), used by GTZ, the German development agency (Ford, 2000). ZOPP is an approach to planning which emphasizes quality and process, intended to be adapted in the planning process to suit the unique goals, partnerships and cultures of those involved. The participatory techniques are designed for use in community-based workshops, to allow participants to identify and analyze issues, problems, and priorities as they see them. Tools such as problem trees focus on encouraging participation by everyone at the workshop, regardless of age or gender. A more detailed description of these methods is found in Table 3-3.

Table 3-1. Summary of topics covered during each trip (from Riedlinger 2001)

Trip 1 (June 1999)	Community workshop, introduction of all issues, brainstorming session, identification of priority concerns
Trip 2 (Aug. 1999)	Sea ice related change: changes related to multiyear and pack ice conditions, changes in the distribution and extent of sea ice, sea mammals (ringed and bearded seal; polar bear), ice thickness, ice safety, ice features such as pressure ridges, indicators of change, freeze-up and break-up. Changes relating to summer seasonal activities such as gill net fishing and seal hunting.
Trip 3 (Feb. 2000)	Wildlife related change: caribou, muskox, wolf populations over time, the relationship between species, animal condition/health and severe weather events, snow depth on winter ranges, occurrence of freezing rain, vegetation changes, changes related to insects, and the impacts on animals. The impacts of change experienced during fall and winter harvesting and community activity.
Trip 4 (May 2000)	Potential impacts of climate change on permafrost: active layer processes, thaw slumps, inland lakes and erosion, coastal erosion, water levels and ice wedges. The impacts of permafrost melt on travel, accessibility, buildings and roads, and changes experienced during spring seasonal activities, such as ice fishing and goose hunting.

Table 3-2. Key features of the IOCC project design

Key Features of Project Design	Description
Planning workshop	Used to collaboratively plan and design how the project would proceed. Established boundaries, objectives, identified key issues and provided a guide for future work.
Repeat Visits	Captured climate change in all seasons, gave the project continuity, and helped to build relationships within the project.
Feedback and communication tools	Used to encourage and recognize participation, and offer means of resolving conflict. The project used a series of newsletters, rough cuts of the video, and trip reports to facilitate feedback.
Final workshop and verification	One held in Ottawa, Canada for funders, participants and interested organizations, and one held in Sachs Harbour. Provided a review of the project, discussion of strategies for communicating and distributing results and future opportunities.

Figure 3-5. The community planning workshop encouraged participants to write down observations of climate-related changes to their local environment. Using flip chart paper and working in groups, participants formed cause and effect chains for each observation.

Figure 3-6. This planning exercise was used to instruct the project team as to which climate-related observations were of greatest importance to the community. Lists of observations were posted on a wall and grouped into categories. Participants were then asked to rank the categories. Participants each voted with five colored dots by placing their dots next to the climate change phenomena that they considered most important.

Table 3-3: Six primary activities took place during the planning workshop in Sachs Harbour, June 1999

Activity	Description
Issue identification	Participants wrote down examples of the environmental changes that they have been experiencing that may indicate that the arctic climate is changing. Each observation or example was written on a separate card. People wrote their own observations, or they were recorded by the project team or other community members.
Cause-effect analysis	Community members arranged the cards into problem trees, with the extent and scope of each phenomena forming the "roots" of the tree, and the effects of each phenomena forming the "branches" of the tree.
Timeline	Community members, particularly elders, traced the changes they have experienced (environmental, wildlife-related, social, economic) back through time in living memory, providing a historical context for recent observations.
Ranking	A set of categories was developed by the participants to organize the observations collected in the issue identification exercise. Each participant was then given three colored dots to vote with, by placing a dot next to the category they considered most important.
Annual calendar	A circular chart of the seasons and seasonal harvest activity was developed by participants, along with the identification of key times of the year that changes were most noticeable.
Trip planning	Based on all of the exercises, residents then decided when the best times of year would be for the project team to return, and what phenomena would be best to focus on in both the video and interviews.

The planning workshop set the groundwork for all further research in three ways. First, it created a collaborative process, whereby the people of Sachs Harbour are active partners in designing and implementing the project. Second, it identified a wide range of environmental changes being observed by the community, observations that would provide the basis for further research. Finally, it identified the times of year that the next three trips should take place, and the phenomena that should be investigated.

Semidirected Interviews

In the three community visits that followed the planning workshop, the science team used semi directed interview techniques to facilitate the process of learning from local experts (Figure 3-7). The method is described as an informal listening technique that uses some predetermined questions and topics, but allows for new topics to be pursued as the interview develops (Chambers, 1991). Interviews were structured as expert with expert discussions, whereby a researcher had an opportunity to learn about climate change from local experts, as well as share information and discuss explanation.

Each trip interviewed many of the same people, but involved a different guest scientist and looked at a different climate change phenomena. Interviews generally began by establishing when the person came to Banks Island and the extent of the person's knowledge of the area, including primary hunting, trapping areas, travelling and camping areas. General questions and topics were prepared ahead of time, to guide the interviews. Observations documented as part of the issue identification exercise during the initial planning workshop were used to initiate discussions of changes. For example, an interview might begin with a precursory statement such as "the last time we were here, people said there was less ice in the summer," or "during the planning workshop, people told us the land was going down in some places." This approach allowed the interviewee to start from a familiar starting point. Subsequent discussion added more detail, helped to set observations in time and place, allowed for the separation of climate related from non-climate related observations of change, provided insight on indicators used to measure change, and helped with species identification.

A total of thirty-six interviews were conducted with seventeen community members over the length of three trips. Interviewing people more than once helped to maintain consistency and continuity in our interviews. When the number of interviews is considered in the context of the population and age structure of Sachs Harbour, the interviews represent the full population (and not a sample) of elders and those families who are active harvesters.

The science team also utilized several additional communication tools and techniques. Topographic and sketch maps, pictures and photographs were used during the interviews to help communicate phenomena that were often difficult to convey verbally. Maps were used to show change over time, indicate locations, place names, regional phenomena, etc. Pictures were used to ensure that we were all talking about the same thing (i.e. thaw slumps), or to provide a starting point for discussion. They were also used for species identification and to clarify Inuvialuktun terminology. Photographs taken on the island of land features, such as thaw slumps, were used to identify place-specific features. During the project, we learned that these tools could have been used more widely. For example, in discussing the project with one university-based climatologist, it was suggested that satellite imagery of sea ice topography might be a useful tool for discussing changes related to the distribution and extent of pressure ridges, leads, and offshore cracks.

Collaborating on the Land: Participant Observation

Participant observation is an established social science methodology considered appropriate for community-based traditional knowledge research (Dene Cultural Institute, 1991). It relies on informal, direct participation and experience with the people you are learning from, and is a central feature of many northern research projects. Participant observation worked to strengthen the relationship between the outside project team and the community, through spending time together on the land—in places important to the people, such as during the spring goose hunting camp. This was a very important part of the project, according to the community, since "going out on the land" is how people are supposed to learn about the environment (Berkes and Jolly, 2001).

Both the video and science components of the project utilized elements of participant observation. Being out with people on a daily basis was the only way to truly capture people's perspectives on climate change. For the video team, communicating climate change meant having people talk about it in the context of their way of life, while they are out on the land engaged in daily activities. For the science team, an interview might occur as an informal, mini-workshops that took place on location. Talking to people about the changes they are seeing in the places they are seeing them.

In addition, Dyanna Jolly (Riedlinger), then a graduate student working with the project, spent additional time in the community after each team trip, living with an elder (Mrs. Sarah Kuptana). Participant observation in her research was a integral component of "relationship-based research"—the creation of a research space guided by respect, sharing, reciprocity, humour and humility (Riedlinger, 2001). Participating in everyday activities (travelling, camping, fishing, plucking geese) as part of the research process lent to a fuller understanding of the changes people were seeing, and how they perceived them in the context of the Inuvialuit way of life. This link to the community gave the project continuity, and provided opportunities for followup research and liaison work.

Figure 3-7. Norm Snow and Theresa Nichols conducting an interview with Roger Kuptana about his observations on climate change around Sachs Harbour (August, 1999). Scientists were invited to Sachs Harbour based on the initial community planning workshop where priority areas were identified. Specifically, scientists with expertise in the areas of wildlife, sea ice, and permafrost participated in the project, corresponding to community concerns.

Telling Our Own Stories: Documenting With Video

The video was designed to show Inuvialuit observations of environmental change in the context of daily life and seasonal harvest activity. As a communication tool, it was produced to demonstrate the close connection between Inuvialuit and the environment, to show outsiders some of the impacts of climate change, and to offer a discussion of climate change in the context of traditional knowledge (Ford, 2000). As with other aspects of the project, careful attention was paid to ensuring that the video would reflect what Inuvialuit, and not the project team, had to say about climate change.

During each trip, the film crew worked alongside community members to determine the kinds of sequences that were taped, and how traditional activities were portrayed (Ford, 2000). Most filming was done outside, during half day or day trips around in the area. In some cases, such as the spring goose hunt, the crew would remain out at hunting and fishing camps. Interviews were also done in peoples' homes, particularly interviews with elders. These interviews often put observations into a historical context, through stories of long ago and the way things used to be. Key to filming was flexibility. Schedules and plans had to accommodate variability associated with the weather, locating a caribou herd, or finding the elder you want to talk to at home. Working in February in the High Arctic did not allow many hours of useable light to get a musk ox hunt on film! (Figures 3-8 and 3-9.)

Rough cuts of the video "Sila Alangotok" were shown in Sachs Harbour on two occasions before the final version was produced. These showings allowed people to critique how community perspectives were communicated. They allowed a forum to discuss key issues, such as the caution advised by some hunters about using harvesting sequences, including footage of a hunter shooting a seal from a boat and fleshing it on shore. Such images were viewed as potentially vulnerable to being taken out of context and used for purposes other than the video, necessitating important discussions such as ownership of images. In this case, a commitment was made that all video production would be approved by an advisory committee (the Sachs Harbour Hunter and Trappers Committee) before it was shown to outsiders, and that no sequences could be taken from the video and used for other purposes (Ford, 2000).

Results and Findings

> I am not worried about it, but our dad when he was ... before he died, he used to talk. He said that around here, our land, it is going to quit freezing later on, from long long time from now he said. Our dad said that. And there is going to be no more winter. Long ago he didn't talk about warm weather, but he used to talk about the North getting warm and the south getting colder than the North. That is what our dad used to talk about.—E. Haogak, Sachs Harbour, 2000

Observations by Sachs Harbour residents were remarkably consistent in indicating tangible evidence of climate change. There is an abundance of local expertise in the community relating to both historical and current sea ice conditions, weather patterns, erosion and permafrost melt, wildlife populations and the linkages and relationships between these phenomena. The extent and scope of Inuvialuit knowledge of climate change is closely connected to the history of land use and occupancy for the region and to current land-based livelihoods. Living memory is extended through oral traditions, and observations are often related through peoples' own life experiences and histories. Thus, the changes people are seeing are tied to how these changes are impacting the way people do things.

In this section we describe climate-related changes that have been observed by Inuvialuit in Sachs Harbour, focusing specifically on the sea ice environment. Next, we describe how such changes are effecting livelihoods and land-based activities, and provide a brief discussion on coping with such changes. Research conducted within the main project found that the question of adaptation is a lens through which to further investigate Inuvialuit understandings of climate change.

Figure 3-8. The permafrost on Banks Island is melting at a faster rate than before. The melting has resulted in the ground collapsing around inland lakes and has caused rapid erosion of seaside cliffs as in this photo. Here, John Keogak shows the film crew coastal slumping and permafrost thaw in August 1999.

A: Community-based Assessments of Climate Change: Climate Change and Sea Ice

Documenting what the community of Sachs Harbour knows about climate change articulates a community-based assessment of climate change for this region, as expressed through traditional knowledge and local expertise. Several prominent themes are evident in this assessment:

- increased *variability;* more sudden and intense changes in the weather;
- changes are most noticeable in the transition months (spring and fall);
- isolated events or anomalies are becoming more frequent; there are more *extreme events;*
- there is increased *unpredictability;* changes are quick, not gradual, and the rate of change appears to be accelerating.

Four categories of changes were used to summarize community-based knowledge of climate change over the course of the IOCC project—changes related to sea ice environment, permafrost, weather/seasons and wildlife (Table 3-4). These categories reflect those used by local residents to organize their observations in the initial planning workshop. Here, we describe recent changes observed by people in Sachs Harbour as they relate to one of the most noticeable groups of observed changes—the sea ice environment. By restricting our discussion to sea ice rather than all four groups, we can provide a more detailed case illustrating the richness of knowledge that exists about climate change and its implications to the arctic ecosystem.

Changes relating to the sea ice represent the largest group of observations and information, and were a recurring theme throughout all four trips during the project. They include changes to the seasonal extent and distribution of ice, timing of freeze-up and break-up, ice thickness, surface topography such as pressure ridges, leads and cracks, and the distribution and abundance of sea mammals. Sea ice conditions are monitored from both the kitchen window and on the ice itself. Knowledge of the sea ice includes a rich understanding of interrelated factors such as temperature, salinity, wind direction and strength, currents, shoreline topography and previous ice conditions. In general, people in Sachs Harbour are seeing less ice overall, both annual and multiyear. In addition, there is a marked increase in variability associated with sea ice conditions in the region. Such changes are viewed as an indicator of larger scale environmental

Figure 3-9. Filming a musk ox hunt in February 2000 on Banks Island with John Lucas Sr. and his family. There is an abundance of knowledge in the community relating to historical and current wildlife populations, behaviour, health, and of the relationship between weather and wildlife.

Table 3-4: Examples of recent environmental change related to climate, as described by Inuvialuit in Sachs Harbour. Source: Jolly and Berkes 2001

Phenomena	Observed change
Sea Ice	Less/no multiyear ice in July and August; more open water (and "rougher" water)
	More ice movement than before
	Not able to see the permanent pack ice to the west
	Ice breaks up earlier and freezes up later
	Rate of ice break-up has increased
	Annual ice in harbour is weaker, thinner (not safe)
	Less and thinner landfast ice (shore ice)
	Changes in distribution and extent of local pressure ridges
	Leads (openings in ice) farther away from shore
	Ice pans do not push up on shore anymore
	Open water in winter is closer than before
	Changes in ice colour and texture
Permafrost	Land subsiding in some places
	Increasing slumping and landslides, both inland and along the coast
	Exposed ground ice (ice lenses) on hillsides
	Increased depth of active layer in spring
	Increased rate of melt in spring
	Water and "ice pebbles" in the ground rather than ice in some places
	Increased puddles, ponds, water in pools on the flat land
	Pingos decreasing in size
	More mud on the land
Changing seasons	Longer, warmer summers
	More rain and wind in the summer
	Melts faster in the spring than it used to
	Spring comes earlier now than it used to
	Shorter, warmer winters
	August is warm month now; used to be "cooling off month"
	Autumn comes later now than before
	Seasonal change is more unpredictable now
Fish and Wildlife	Two species of Pacific salmon caught near the community
	Increased numbers of *Coregonus sardinella* (least cisco)
	Fewer polar bears in area because of less ice
	Increasing occurrence of "skinny" seal pups at spring break-up
	Observations of robins; previously unknown small birds
	Increased forage availability for caribou and muskox
	Changes to timing of intra-island caribou migration
Weather	More rain in the fall; rain that should have been snow
	Different kind of snow, less rough drifts
	Longer duration of "hot" days, now a whole week rather than one or two days
	Thunder and lightning—none seen before
	Weather more unpredictable
	Changes in wind velocity and direction; more intense wind storms

change. Residents also fluidly relate physical changes occurring in the sea ice to changes in the local ecology and adaptive modifications.

Multiyear ice, or ice that has survived for a minimum of one full summer melt, was once considered a common year-long feature offshore of Sachs Harbour. This ice, having purged its brine content, is largely composed of freshwater ice, with a typical thickness greater than 3 metres (WMO, 1971). Residents unanimously agreed that the offshore multiyear ice had decreased in size over recent years, and had begun disappearing from view by July. Several people also commented on a possible connection between changes in water temperature and the loss of the multiyear ice. In addition, several families commented on the traditional use of melted multiyear ice for potable water during travel and hunting, and the negative implications of losing this natural resource.

> A lot less multiyear. Seems like the water in summertime is getting warmer as well.—Roger Kuptana, Sachs Harbour, 1999

> Long ago, there was always ice all summer. You would see the [multiyear ice] all summer. Ice was moving back and forth this time of year. Now, no ice. Should be [multiyear]. You used to see that old ice coming from the west side of Sachs. No more. Now between Victoria Island and Banks Island, there is open water. Shouldn't be that way.—Frank Kudlak, Sachs Harbour, 1999

> Used to be ice close by most of the summer. Now you see ice in June, sometimes part of July. After that no ice. A few miles out from Sachs we used to hit ice in our boats. In June, twenty miles out hunting seals in open leads we used to go.—Andy Carpenter Sr., Sachs Harbour, 1999.

First-year sea ice continues to be a dominant feature over the Western Arctic seascape for much of the year; however, residents reported a number of significant changes to timing, composition and physical properties in recent years. First-year sea ice forms annually, and is comprised of freshwater ice, brine, and air, although the ratio varies over the season largely as a function of temperature. Local experts report that the thickness of the ice has decreased significantly over recent decades. This assessment was most often linked to the elements of ice safety and travel, as well as ringed seal habitat and breathing hole locations. Several hunters explained how ice thickness, ice age and ice strength could be determined from ice texture and colour. In turn, the information determines safe travelling routes as well as identifying good hunting and fishing sites.

> The problem is that the ice is getting thinner—landfast ice has gotten a foot thinner. This is the most obvious in the last few years.—Roger Kuptana, Sachs Harbour, 1999

> It's not as thick as it used to be. I remember going out hunting. We never had to worry about ice. It's not as safe—it moves around alot.—John Keogak, Sachs Harbour, 1999

> There are lots of thin spots where you can see the ice is broken up—the ice is fairly thin. You can tell by the colour.—Peter Esau, Sachs Harbour, 1999

Inuvialuit elders and hunters frequently identified the timing of seasonal changes as a clear indicator of climate change. Earlier break-ups are more noticeable than later freeze-ups (with the exception of 1998–99, where open water extended into late November), but both events are considered to have changed. Since the 1950s, the timing for break-up in Sachs Harbour was reported to take place in late June to early July. All surveyed respondents agreed that the ice break-up season was occurring earlier, and at an accelerated rate, than previous decades. Residents also reported that it had been common to have had sea ice remain in the harbour throughout the summer. That had not been the case in recent years.

> The ice goes away really quickly. That ... has really changed. Really noticeable. [It] melts earlier and goes away.—Joe Apian, Sachs Harbour, 1999

> Years ago it never opened up in the springtime the way it does now. In the springtime ducks used to be in the cracks, small cracks, and they would starve. Long ago just a little crack. Never happens anymore. Water always there now. Ducks have no problem.—Joe Apian, Sachs Harbour, 1999

> Now it is breaking up earlier and earlier. When we first came here in 1955 all the people used to wait for the ice to go so they could go across [to the mainland]. In July we already came from DeSalis Bay, walking over by the land. Fred Carpenter was still waiting for the ice to leave. Finally the ice took off. We were walking from that way [northeast] and Fred Carpenter was on top of the hill looking that way for open water with binoculars. We met him on top of the hill when we were coming in.—Lena Wolki, Sachs Harbour, 2000

> There is a lot of difference if there is not ice out there in the falltime. It doesn't freeze up for a long time because you always have wind smashing up the ice and taking it out. When there used to be ice quite a few years back it used to freeze up right away. But now there is no ice out there, nothing to hold when the ice is formed. It just keeps breaking up with the wind.—John Lucas Sr. Sachs Harbour, 2000.

Residents did not limit their observations or knowledge to the changes in the sea ice, but linked those modifications to other changes in the landscape and surrounding ecosystem. In a number of cases, they described the relationship between the presence of ice, wind, wave energy, ice safety, the abundance and health of local marine species and implication to the community's hunting success. Among the dominant themes that emerged was the aspect of wind. As sea ice does not become "safe" for travel until later in the freeze-up, boats are a mainstay in seal hunting and fishing. The greater the amount of ice (and related floes), the calmer the open water, and the safer conditions for boating. In contrast, less ice and more open water was linked to stronger winds, more violent wave motion, less safe boating conditions, and a drop in hunting and fishing opportunities. The increased wave action was further associated with the coastal erosion, which was also considered by all those interviewed to be accelerating in rate.

> When there is lots of ice, you don't worry too much about storms. You get out there and travel in between the ice [floes]. But last few years there has been no ice. So if it storms you can't get out. People used to go by Cape Kellet to hunt seals. Last few years they haven't been able to. Early July they go out there for a while, but after the ice is gone, end of July, there is no more travelling out on the ocean.—Andy Carpenter, Sachs Harbour, 1999

> Long ago, (we) used to have ice floes all the time out there. We used to go hunting. You know, it's not rough like that when there is ice.—Lena Wolki, Sachs Harbour, 1999

The effect of early melt and the reduced amount of sea ice were linked to the biological responses of polar bears and ringed seal. The presence of thick landfast ice in the area was related to higher numbers of local seals, and a higher number of healthy pups born in the spring. High concentrations of seals in the area were met with higher numbers of polar bears. From the hunter's perspective, the combination of higher numbers of seals, calmer open waters (in leads) and safer boating or snow machine travel conditions resulted in more successful seal hunts. In addition, the increased amount of polar bears lead to successful polar bear hunts and tourism.

The lack of sea ice in the area was identified as having the opposite effect—fewer seals in the area, rougher waters, dangerous boating conditions, and increased risk to the hunters. Under these conditions, hunters are limited to the nearshore area, effectively limiting the hunt. Polar bears are also forced onto land much earlier, resulting in a longer seasonal fasting and poorer health.

> Even in the summertime, the bears stayed on the ice. Never came to land. But now (you) never see the ice (during the summer). ... Lucky if you see ice now from a distance in the summertime. It used to come in—ice used to come close to the settlement. That was a good time. Lots of seals.—John Keogak

> In the 1970s, you could go out thirty or forty miles in winter hunting polar bear. Then only twenty miles, then ten. Last year only six miles out and you reach ice you have to worry about.—Roger Kuptana, Sachs Harbour, 1999.

> When there is no old ice, there is not very much seal too. They travel on the big old ice. Long ago when the old ice came in, we would head for the big ice out there. Just stop and the seals start coming in. Nowadays it is not like that anymore. Those small ice [ice floes], hardly any seals on them. They are too small. It's warmer summers, I think, that melt those ice right down. That is why there is hardly anymore ice. Used to be old ice coming from the North all the time. And when it freeze-up with old ice, seems to be more bears all the time. When it has leads, around March, boy lots of bears when you go out! I used to feel safe when I would camp on the old ice. Just like islands that old ice. Good for tea! That old ice always open up too, all kinds of seals. Because that old ice is smooth, good for seal holes. Geddes Wolki, Sachs Harbour, 1999.

B: Coping with Climate Change

Climate-related changes experienced in Sachs Harbour are relatively recent, and they are affecting subsistence activities. However, many of the impacts have been absorbed through the flexibility of the seasonal harvesting cycle and the Inuvialuit way of life. There are two components of the adaptive capacity of the community of Sachs Harbour to deal with climate change. One component is the actual response to change, which may be called coping strategies. A second component is related to Inuit adaptations for life in a highly variable and uncertain environment; these may be called long-term adaptive strategies (Berkes and Riedlinger, 2001). In this chapter, we deal mainly with coping strategies.

For the most part, Inuvialuit coping strategies relate to adjusting or modifying subsistence activity patterns—changing when, where or how hunting and fishing occur. The observed responses in Sachs Harbour to environmental change and fluctuation can be considered under five categories: modifying timing of harvest activity; modifying location of harvest activity; modifying method of harvest activity; adjusting the species harvested; and minimizing risk and uncertainty.

Modifying the timing of harvest activity. Of the five categories of coping strategies identified, changes to when harvesting activities take place are most noticeable. Increased seasonal variability is forcing people to adjust the timing of their seasonal calendar. For example, warmer temperatures and unpredictable ice conditions have resulted in hunters going out earlier for polar bear. In spring, people describe not going out on the land for as long, in response to shorter, warmer springs and increased rates of snow and ice melt. They return to the community after the goose hunt, rather than proceeding to the lakes to ice fish. Waiting is a coping strategy; people wait for the geese to arrive, for the land to dry, for the weather to improve, or for the rain to end (Figure 3-10).

Modifying the location of harvest activity. Erosion and slumping at one fishing lake has necessitated that people fish at other lakes instead. More bare ground and unreliable snow conditions mean families are travelling along the coastal sea ice rather than along inland routes. Recent changes to the sea ice leave hunters staying close to the community because of safety concerns, while the animals they seek remain farther out. Permafrost thaw in many areas has left travelers making new trails and routes to avoid slumps, mudslides and erosion.

Adjusting how harvesting is done. Community members describe using all terrain vehicles instead of snowmobiles to travel to spring camps when there is not enough snow. They also describe hunting seals from boats in the open water, an adjustment necessitated by the lack of ice floes where the seals normally are in the summer months (Figure 3-11).

Adjusting the mix of species harvested. Inuvialuit are reporting catching more *qaaqtaq* (least cisco) in nets at the mouth of the Sachs River and two species of Pacific salmon were also caught in the summer of 1993 (Figure 3-12). In the spring of 2000, the lack of open water and bare ground resulted in the geese arriving late and fewer eggs being laid, and people collected almost no eggs. Pintail *(Anas acuta)* and mallard ducks *(A. platyrhynchos)*, considered mainland ducks, have been observed, and there are

Figure 3-10. John Keogak with his son and daughter, ice-fishing at Middle Lake goose hunting camp in May 2000. People ice fish while they wait for the geese to arrive.

higher numbers of white-fronted goose or yellow legs *(Anser albifrons)* and tundra swans *(Cygnus columbianus)*, birds that were historically not abundant on Banks Island.

Minimizing risk and uncertainty. In response to increased variability and unpredictability associated with the weather and other environmental phenomena, the Inuvialuit describe the need to monitor conditions more closely, such as the rivers in the spring. There is a heightened risk of getting caught on the far side of the river because it is more difficult to tell when the ice will break. People indicate that "you really need to have experience to travel on the sea ice now," and describe being more careful when they travel.

The Inuvialuit of Sachs Harbour draw on accumulated knowledge and experience to come up with these coping strategies. They make it clear that they have always adjusted and adapted to change—social, political, and economic change, as well as to environmental change. The ability to deal with change is part of Inuvialuit adaptations. People may now use all-terrain vehicles more than dog teams, but as one man described, "it is pretty well the same, how we do things" (J. Lucas Sr., Sachs Harbour, 2000). When asked about the impact of changes on hunting, trapping or fishing, most people were quick to point out that "we always find some way of getting something."

Figure 3-11. Lena Wolki, Geddes Wolki, and Terry Woolf on the shores of Banks Island. The ocean in the background is completely free of ice. Lena and Geddes Wolki have said that for a few years now there have not been ice floes in summer, and therefore they are unable to hunt for seals, which normally would be basking on top of the ice.

Some people described how, in one sense, it is easier to cope with environmental change now than in the past, since the community does not rely exclusively on country foods.

Different kinds of changes have various impacts, and require different kinds of responses. Although this chapter focuses on impacts and adaptations associated with harvest and subsistence activity, climate change also has other economic and cultural impacts. For example, the lack of sea ice makes some people "lonely for the ice." Other environmental changes may not pose a direct threat to subsistence activities but have direct impacts on various aspects of community life, such as the maintenance of buildings and roads in the case of permafrost thawing, slumps, and erosion.

Analysis

A: Documenting and Communicating Inuvialuit Perspectives

The first objective of the IOCC project was to produce a video that would raise awareness and demonstrate to Canadian audiences, interest groups and decision-makers that climate change is impacting northern communities such as Sachs Harbour. This goal was achieved through the production of "Sila Alangotok: Inuit Observations on Climate Change." A full-length version of the video is forty-two minutes in length (a

Figure 3-12. Roger Kuptana removes an arctic char from his net in Sachs River. While arctic Char is a common fish caught in local waters, residents have recently caught salmon. The presence of salmon in this region is unusual and has been linked to climate change.

television hour) and a fourteen minute summary version of the video was produced for screening to decision-makers and the media.

Once the video was near completion, the IISD project team developed a communication and engagement strategy to complement the video. Specifically, the key audiences targeted in the strategy included media, decision-makers, civil society, scientists, and students. The strategy was designed to deliver the message that climate change is having a direct, concrete impact on people in the arctic. In order to maximize media attention during the release of the video, the communications team sought to time the launch with a high-profile international news event which expected to draw considerable media attention to the issue of climate change. Accordingly, the video was simultaneously launched in the Hague (at the United Nations Climate Change Sixth Conference of the Parties [COP 6], Figure 3-13), in Ottawa at a media conference with the community representatives from Sachs Harbour, the David Suzuki Foundation, and the Sierra Club of Canada, and in the community of Sachs Harbour itself in November 2000.

As it happened, the embargo on the story was broken a day before the launch, resulting in substantial media coverage both before and during the launch. The story was covered by twelve newswire services, twenty-four U.S. newspapers (with a cumulative circulation of over four million), twenty Canadian newspapers (with a cumulative circulation of almost three million), *Maclean's, Panorama* and *Outside* magazines (cumulative circulation of two million), and at least twenty-two major online sources including National Geographic.com, @Discovery.ca, One World Net, ABC News.com and CNN.com. The project team took part in nine major radio interviews including ABC Radio, BBC, Radio Netherlands, CBC *As It Happens,* CBC Syndication (thirteen separate city interviews), CBC Radio International, KWAB, Great Lakes Radio Consortium (140 stations in ten U.S. states) and CFAX Radio. The project was also covered on fourteen major television networks including Associated Press Television Network (330 broadcasters worldwide), BBC, France 1, France 2, ARD Television Swiss Romande, National Geographic channel, CBC (the National), CTV, Global, CBC *Canada Now,* CBC Newsworld, Discovery Channel, and the Aboriginal Peoples' Television Network. There were an additional twenty-eight print, radio and Internet stories on the project prior to the launch. Subsequent to the launch, the community's observations have remained profiled through various independent stories through a journal series in the *Globe and Mail,* inclusion in the CBC's Nature of Things' special on Climate Change, and in a production on climate change by Swiss Television. As well, the project team has provided project summaries for conference proceedings and general audience publications such as the World Wildlife Fund's *Arctic Bulletin.*

The video was screened or distributed at a number of key decision-making forums to enhance the ability of civil servants and political figures to reference human impacts associated with climate change. In addition to COP6 noted above, the video was screened at such events as the Canadian Joint Ministerial Meeting of energy and environment ministers, the Arctic Council Ministerial meeting, and a U.S. Congressional Field Hearing on climate change in Alaska.

Beyond the video, Inuvialuit perspectives on climate change and the project results were published in peer-reviewed journals and presented at several conferences and workshops. The project team achieved remarkable success in producing and publishing journal papers. In total, seven papers have been published in science or development journals. An additional document was produced specifically on local adaptation to climate change for the Government of Canada's Climate Change Action Fund. A further three papers are in preparation. The publishing of science papers has added rigour to the analysis of Inuit observations and shared the findings with other researchers.

To communicate and engage scientists and other researchers visually, the video was shown at twelve strategic conferences and scientific events between June 2000 (draft version) and July 2001. Major events ranged from a Traditional Knowledge Summit in Honolulu to the Global Change Open Science Conference in Amsterdam.

The value in engaging the scientific and decision-making audiences in the Inuvialuit perspective on climate change is already evident. Following the launch of the video and implementation of the communication strategy, resources have been committed and new partnerships have developed between governments, northern communities, researchers/research institutes and aboriginal organizations to continue documenting climate change impacts in the Arctic and to strengthen research between indigenous experts and western scientists. Notably, a pan-Canadian initiative

Figure 3-13. Rosemarie Kuptana at a press conference to launch the Sachs Harbour video at the Sixth Conference of the Parties to the United Nations Framework Convention on Climate Change in The Hague, The Netherlands, November 2000.

has begun with climate change monitoring and reporting at the community level and an indigenous perspectives on climate change project for the circumpolar region is being led by the Tampere Polytechnic in Finland.

Communicating and engaging from an educational perspective, the Manitoba Department of Education has included the Sachs Harbour video in its high school curriculum on earth sciences and has developed a teacher's guide to traditional knowledge and climate change. Within Canada, venues such as the Royal Saskatchewan Museum and Science North in Ontario have profiled the Sachs Harbour observations of climate change in their educational displays. The video has been included in several undergraduate courses in departments such as environmental studies, anthropology, development studies, and geography.

The project's Internet site (http://www.iisd.org/casl/projects/inuitobs.htm) has proven to be a useful mode of communication. As of the end of 2001, over 5,000 visits have been registered on the main page and the summary version of the video has been screened from the Internet site over 2,500 times—well in excess of the 435 physical copies that have been distributed by IISD. The project's four trip reports, which are also available online, have been downloaded over 2,400 times cumulatively.

As a medium for communicating climate change from the perspective of local people, video proved to be a tangible and accessible output for reaching key audiences. Employing video as a significant mode of communication provided a catalyst for reflection by those involved in the documentation of climate change impacts in Sachs Harbour, in addition to communicating those impacts to outside audiences. The project's inclusion of a community advisory committee and open meetings for frequent reviews of the draft video—with opportunities to edit—prevented inappropriate/incorrect presentation of knowledge. The video director chose to subtitle any dialogue in Inuvialuktun, rather than do a voice-over in English to ensure people had the fullest opportunity to express themselves and their knowledge. The full version of the video is narrated by Rosemarie Kuptana, a resident of Sachs Harbour and an IISD Board Member.

The dual project goals of science and public awareness were mutually reinforcing and may have strengthened the policy and knowledge interface. The video conveyed urgency regarding the issue of climate change—a topic frequently confusing to the layperson—by documenting the human element of climate change. The visual images and descriptions of climate change impacts in Sachs Harbour provided a means for communicating climate change to a nontechnical audience. The science papers added rigour and analysis to an audience—scientists and decision-makers—who principally rely on western methods of documentation to ensure credibility.

B. Sharing Information Between Western Science and Traditional Knowledge

The second objective of the project was to evaluate the potential contribution of traditional knowledge, local observations, and adaptive strategies to scientific research on climate change. Traditional ecological knowledge, once considered esoteric, is increasingly accepted as a source of legitimate knowledge to be used alongside Western scientific knowledge. In the Canadian Arctic, for example, traditional and local knowledge is being used in participatory environmental research and management in at least seven areas: fish and wildlife co-management; protected area co-management; marine coastal and integrated management; ecosystem health monitoring; contaminants research; environmental assessment; and climate change research (Berkes et al., 2001).

The use of traditional knowledge in climate change is relatively new, as this volume documents. However, given the extent and scope of climate-related traditional knowledge in communities such as Sachs Harbour, the potential contribution of that knowledge to understanding climate change in the Arctic is considerable. As in many arctic communities, the extensive use of the land and the sea ice in harvesting food resources provides the basis for local environmental expertise, guided by generations of experience. The knowledge held by indigenous experts enables local scale understandings of impacts, and can be used as a guide for future research.

Documenting this knowledge serves to illustrate the richness of Inuvialuit knowledge of climate change, and its potential for understanding climate change. However, documentation is only the first step; recognizing and including local expertise requires building relationships between scientists and communities—those that are studying change and those that are experiencing it. The project started with the assumption that this kind of collaboration was both desirable and achievable. Much of the work of the science team was directed at developing a framework for linking Western science with Inuvialuit expertise.

One outcome of the project was the development of a conceptual framework for linking science-based research with local knowledge (Riedlinger and Berkes, 2001). The framework was articulated through five interrelated convergence areas, that is, research areas that could facilitate collaboration and communication between scientists and local experts (Table 3-5). These are the use of traditional knowledge (a) as local scale expertise; (b) as a source of climate history and baseline data; (c) in formulating research questions and hypotheses; (d) as insight into impacts and adaptation in arctic communities; and (e) for long term, community-based monitoring. The five areas highlight the ability of local experts to address the complexities of arctic climate change at spatial and temporal scales often underrepresented in climate change research.

Conclusions

The Inuit Observations on Climate Change (IOCC) project started with the practical objective of producing a video to document the impact of climate change in Sachs Harbour. In the process of meeting that objective, the project developed a model of community research partnership to facilitate the involvement of the community in all phases of the research, from problem formulation and research design to dissemination. The project was designed around a diverse set of participatory methodologies to

Table 3-5. Five convergence areas that may facilitate the use of traditional knowledge and Western science, in the context of arctic climate change research (Riedlinger and Berkes, 2001)

Local scale expertise	The importance of local-scale expertise has been emphasized in discussions of traditional knowledge in the North. Climate change will be first noticeable through biophysical changes in sea ice, wildlife, permafrost, and weather. These changes will not go unnoticed at the local scale in Inuit communities.
Climate history	Traditional knowledge can provide insight into past climate variability, providing an essential baseline against which to compare change. Climate history is embedded in Inuit history of wildlife populations, travels, extreme events, and harvesting records.
Research hypotheses	Traditional knowledge can contribute to the process of formulating scientific hypotheses as another way of knowing and understanding the environment. Collaboration at the initial stage of research expands the scope of inquiry and establishes a role for communities in research planning.
Community Adaptation	Traditional knowledge lends insight into adaptations to changes, explaining them in the context of livelihoods and community life. How are communities responding to change? What are the social, economic, and cultural limits to adaptation in northern communities?
Community-based Monitoring	Traditional knowledge reflects a cumulative system of environmental monitoring and observation. Monitoring projects have the potential to bridge the gap between science and traditional knowledge by providing a collaborative process.

facilitate communication and collaboration to ensure accurate representations of Inuvialuit observations, knowledge and traditions. These methods included community-based planning workshops; semidirected interviews; use of participant observation techniques by travelling with local experts on the land and learning from them; and documenting local observations by video. Other approaches and features of the project included the use of repeat visits around a seasonal cycle; followup and supplementary research; and extensive community feedback (project newsletters, trip reports, the rough-cut and final versions of the video, community workshops) and verification.

The results of a project evaluation, undertaken in the final stages of the project, strongly indicate that the project accurately reflected the local observations and traditional knowledge of the people of Sachs Harbour with respect to climate change. The evaluation also showed that the planning workshop, the video and the science interviews were effective in meeting the project goals, and that the project was inclusive, participatory and responsive to community concerns. The IOCC project also achieved its goal of communicating Inuit observations of climate change to the outside world. Presentations at seventeen conferences and workshops, such as the Beaufort Sea 2000 Conference, International Climate Change Conference, Fifty-first AAAS Arctic Science Conference, Twelfth Annual Inuit Studies Conference, and the United Nations Conference on Climate Change, communicated the project's findings to other researchers, government decision-makers, and international organizations.

Three main factors lend to the success of the project in describing Inuit knowledge and observations of climate change. First, IOCC was a community-initiated project, and the community fully supported the idea and nature of the project. Second, it was a collaborative research effort. Third, the project took place over a yearly cycle, with four return visits, covering all seasons. Much of the success can be attributed to the time taken to organize and carry out the initial planning workshop in June 1999, setting the groundwork for the project as collaborative research.

The evaluation survey indicated that the planning workshop, repeat visits focusing on activities appropriate for that season, and the reporting of the results back to residents in the form of video, trip reports, and newsletters were all important in meeting community expectations of the project. Evaluation followup discussions indicated some additional reasons for the success of the project: the key role of the science team coordinator, who is well known in the area and has working relationships with the people of Sachs Harbour; the participation of the wildlife scientist whom the community had requested as the scientific expert, a person who spends extensive periods on Banks Island and works cooperatively with local hunters; the participation of a graduate researcher who stayed in Sachs Harbour after the rest of the project team had departed, lived in the community with an elder and helped "humanize" the research component of the project; and the continuity and consistency provided by the science team coordinator and the graduate researcher who took part in all four trips.

The second objective of the IOCC project (evaluating the potential contribution of traditional knowledge to climate change research) built on the robust documentation of Inuvialuit observations of climate change and the substantial body of local environmental knowledge and expertise. The results of the project helped identify a number of ways in which traditional knowledge may complement science. These are: the use of traditional knowledge as local-scale expertise; as a source of climate history and baseline information; in formulating research hypotheses; as insight into impacts and adaptation; and for long-term, community-based monitoring.

Neither Western science nor traditional knowledge is sufficient in isolation to address all the complexities of global climate change (Riedlinger and Berkes, 2001). The use of the two kinds of knowledge together improves the capability to understand the phenomena of change and to address it. The Inuvialuit of Sachs Harbour are dealing with changes by switching species and adjusting the "where, when, and how" of hunting. Berkes and Jolly (2001) identify these responses as coping mechanisms or short-term responses. In addition to these responses, there exist adaptive strategies or long-term responses, such as culturally ingrained mechanisms to provide flexibility in seasonal hunting patterns; extensive traditional environmental knowledge that allows a diversity of hunting strategies; networks for sharing food and other resources; and intercommunity trade to provide for larger-scale sharing and reciprocity.

For many groups living in the Arctic, climate change is not an abstract concept. Communities are already experiencing the early signs of change, and are addressing them. The rich set of coping strategies identified in the IOCC study documents this. However, the Arctic is characterized by unpredictability and change. The existing set of culturally ingrained adaptive strategies, documented by arctic scholars such as Krupnik (1993), shows that the Inuit are experts at adapting to change, as they themselves often point out.

Thus, the big question is not environmental change, but the limits to adaptability in an increasingly unpredictable world in which increased variability and extreme events challenge the system's ability to cope with change (Smithers and Smit, 1997). Although the amount of climate change observed in Sachs Harbour in the 1990s was without precedent and beyond the range of expected variation, it was not beyond the range of the community's ability to respond to it and deal with it. However, the observed changes are relatively recent, and the community's response so far is not necessarily an indication of their ability to adapt in the future. In the final IOCC meetings at Sachs Harbour, community representatives indicated the need to follow up on the results of the project. Since changes were still occurring, the Inuvialuit pointed out, climate change research could not be a one-shot effort; climate change requires monitoring.

Community-based projects working to create partnerships between researchers and communities for investigating climate change are important in this respect (McDonald et al., 1997; Thorpe et al., 2001; Fox, 2000). Climate change research does not end with the documentation of local knowledge or of local impacts and adaptations. Each new project will take the research a step further, finding new ways to bridge the gap between communities and scientists, helping to establish a system of community-based long-term environmental monitoring, enhancing the capacity of communities to deal adaptively with change, and helping them to transmit their concerns to regional, national and international levels.

Acknowledgements

This chapter, and the Inuit Observations on Climate Change project, are a reflection of the commitment, generosity and support of a great many people who have shared their knowledge and experience. Most of all, quyanainni ("thank you") to the people of Sachs Harbour, whose support and initiative made this project happen. To all those and their families that took part—especially Roger and Jackie Kuptana, Sarah Kuptana, Andy and Winnie Carpenter, Lena and Geddes Wolki, John and Samantha Lucas, John Lucas Jr., Trevor Lucas, Frank and Martha Kudlak, Joe Kudlak, Peter and Shirley Esau, John and Donna Keogak, Edith Haogak, Larry and Yvonne Carpenter, Margaret Elanik and Joe Apian—thank you for making this project happen. Our appreciation also for the generous support and untiring efforts of the Sachs Harbour HTC, especially Florence Elanik for her time given to coordinating correspondence and review of materials, and to the project's community researchers—Kimberly Lucas and Lucy Kudlak, for their time, humour and patience. Thanks also to the other half of the research partnership—Neil Ford, Graham Ashford and Jennifer Castleden, Deborah Lehmann and Dennis Cunningham (IISD, Winnipeg), Rosemarie Kuptana (Sachs Harbour), Norm Snow (Joint Secretariat, Inuvik), John Nagy (RWED, Inuvik), Theresa Nichols (DFO, Winnipeg), Stephen Robinson (GSC, Ottawa), Fikret Berkes and Dyanna Jolly (Riedlinger) (Natural Resources Institute, University of Manitoba), Bonnie Dickie (Chestnut Productions, Winnipeg), Terry Woolf (Lone Woolf Film and Television Production Services, Yellowknife), Lawrence Rogers and Stan Ruben (Inuvialuit Communications Society, Inuvik). The IOCC project was made possible through financial contributions from: the Climate Change Action Fund (Public Education and Outreach); the Walter & Duncan Gordon Foundation; the Climate Change Action Fund (Science, Impacts and Adaptation); Indian and Northern Affairs Canada; and the Government of the Northwest Territories. Generous in-kind support was given by the Hunter and Trappers Committee of Sachs Harbour, the Inuvialuit Game Council; the Inuvialuit Joint Secretariat; the Inuvialuit Game Council; the Inuvialuit Joint Secretariat the Inuvialuit Communications Society; the Natural Resources Institute, University of Manitoba; the Department of Fisheries and Oceans; the Government of the Northwest Territories; and the Geological Survey of Canada. Thank you to Graham Ashford and Neil Ford of IISD for the photos, and to Pat Hardy for Figure 3-13. Dyanna Jolly would also like to thank the Walter & Duncan Gordon Foundation, Arctic Institute of North America, and the Northern Scientific and Training Program for additional support. Quanna.

References

Banksland Story. 1969. *Sachs Harbour school yearbook 1968–1969.* Researched and written by Father Lemer, O.M.I.

Berkes, F., Mathias, J., Kislalioglu, M. and Fast, H. 2001. The Canadian Arctic and the Oceans Act: the development of participatory environmental research and management. *Ocean & Coastal Management 44:* 451–469.

Berkes, F. and Jolly, D. 2001. Adapting to climate change: Social-ecological resilience in a Canadian Western Arctic community. *Conservation Ecology* 5(2): 18. [online] URL: http://www.consecol.org/vol5/iss2/art18

Bielawski, E. 1997. Aboriginal participation in global change research in the Northwest Territories of Canada. In: W. C. Oechel, T. Callaghan, T. Gilmanov, J. I. Holten, B. Maxwell, U. Molau and B. Sveinbjornsson, eds. *Global change and Arctic terrestrial ecosystems,* pp. 475–483. New York: Springer-Verlag.

Chambers, R. 1991. Shortcut and participatory methods for gaining social information for projects. In *Putting people first: sociological variables in rural development.* (M. Cernea, ed.) Oxford: Oxford university press, 515–537.

Cohen, S. J. 1997. What if and so what in northwest Canada: could climate change make a difference to the future of the Mackenzie basin? *Arctic 50* (4): 293–307.

Condon, R., and Ogina, J., and the Elders of Holman Island. 1993. *The northern Copper Inuit: A history.* Civilization of the American Indian, University of Oklahoma Press.

Dene Cultural Institute. 1991. *Guidelines for the conduct of participatory community research to document traditional ecological knowledge for the purpose of environmental assessment and environmental management.* Hull, Quebec: Canadian Environmental Assessment Research Council.

Ford, N. 2000. Communicating climate change from the perspective of local people: A case study from Arctic Canada. *Journal of Development Communication 1* (11): 93–108.

Fox, S. L. 2000. Arctic climate change: observations of Inuit in the Eastern Canadian Arctic. In *Arctic climatology project, environmental working group arctic meteorology and climate atlas.* (F. Fetterer and V. Radionov, eds.) Boulder, CO: National snow and ice date centre. CD-ROM.

Freeman, M. M. R. (ed.). 1976. *Report of the Inuit land use and occupancy project.* 3 volumes. Ottawa: Department of Indian and Northern Affairs.

International Institute for Sustainable Development. 2001. *Final Report: The Inuit Observations on climate change project.* IISD: Winnipeg, Canada.

Inuit Circumpolar Conference. 2001. Statement by Violet Ford of the ICC. From consultation to partnership: Engaging Inuit on climate change. In *Silarjualiriniq: Inuit in global issues* 7 (January to March).

Kassi, N. 1993. Native perspectives on climate change. In: G. Wall, ed. *Impacts of climate change on resource management of the North,* pp. 43–49. Department of Geography publication series. Occasional paper no. 16. Waterloo, Ontario: Department of Geography.

Krupnik, I. 1993. *Arctic adaptations: Native whalers and reindeer herders of northern Eurasia.* Hanover and London: University press of New England.

McDonald, M. , Arragutainaq, L., and Novalinga, Z. 1997. (compilers). *Voices from the Bay: traditional ecological knowledge of Inuit and Cree in the Hudson Bay Bioregion.* Ottawa: Canadian Arctic Resources Committee and Sanikiluaq, NWT.

Nagy, M. 1999. *Aulavik oral history project on Banks Island, NWT: Final report.* Inuvik, NWT: Inuvialuit Social Development Program.

Nichols, T., Snow, N., Berkes, F., Ashford, G., and Jolly, D. 2002. Climate change and sea ice: Local observations from the western Arctic. *In prep.*

Riedlinger, D., and Berkes, F. 2001. Contributions of traditional knowledge to understanding climate change in the Canadian Arctic. *Polar Record, 37* (203): 315–328.

Riedlinger, D. 2001. *Community-based assessments of change: Contributions of Inuvialuit knowledge to understanding climate change in the Canadian Arctic.* Unpublished graduate thesis. Natural Resources Institute, University of Manitoba.

Smithers, J., and Smit, B. 1997. Human adaptability to climatic variability and change. *Global environmental change* 7 (2): 129–146.

Stefansson, V. 1923. *My life with the Eskimo.* New York: Macmillan.

Thorpe, N. L., Hakongak, N., Eyegetok, S., and Kitikmeot Elders. 2001. *Thunder on the tundra: Inuit qaujimajatuqangit of the Bathrust caribou.* Vancouver: generation Printing.

Usher, P. J. 1970a. *The Bankslanders: Economy and ecology of a frontier trapping community.* Volume 1: History. Ottawa: Department of Indian Affairs and Northern Development.

Usher, P. J. 1970b. *The Bankslanders: Economy and ecology of a frontier trapping community.* Volume 2: Economy and ecology. Ottawa: Department of Indian Affairs and Northern Development.

Usher, P. J. 1971. *The Bankslanders: Economy and ecology of a frontier trapping community.* Volume 3: The community. Ottawa: Department of Indian Affairs and Northern Development.

World Meteorological Organization. 1970. WMO Sea-Ice Nomenclature, Secretariat of the World Meteorological Organization. Geneva Switzerland. No. 259.

Figure 4-1. Pete Sovalik, Iñupiaq naturalist, on the ice at Barrow in May 1950, working on a project for George MacGinitie at the Naval Arctic Research Laboratory. Photo by Howard Feder.

4

Coastal Sea Ice Watch:
Private Confessions of a Convert to Indigenous Knowledge

David W. Norton
Arctic Rim Research, Fairbanks, Alaska

Having to recount truthfully a research effort's earliest stirrings is probably good medicine. A dose of humility helps me admit that luck and instinct (rather than thoughtful planning, for example) gave early form to the investigations described below. Comparing this retrospective to those of other investigators may contribute to understanding the intersection of two currently prominent themes in polar research: environmental change and indigenous knowledge. Who knows? If these two research pursuits mature and flourish, early steps in their conceptual development may amuse future practitioners.

The following is one duly humble perspective on recent experiences working with members of both the subsistence and research communities. A combination of need and opportunity stimulated formation of an interdisciplinary and intercultural team to design and execute a demonstration project, the subject of which is nearshore sea ice in the western Arctic. Success of this project will be measured in terms of improved capability to predict changes in stability of nearshore ice as a platform for subsistence activities and in terms of methodological contributions.

Our research project took shape in early 1999, when colleagues in Barrow, Alaska, were especially concerned with public safety for the hundreds of people who would occupy landfast ice during the annual spring subsistence hunt for the bowhead whale *(Balaena mysticetus)* in April–June of each year. In mid-May of 1997, a large slab of nearshore sea ice bearing dozens of subsistence whalers had broken loose from landfast ice near Barrow and floated north into the Beaufort Sea. By braving hazardous fog for a night and a day, North Slope Borough Search and Rescue helicopter pilots plucked all 154 marooned crewmembers from twelve camps on the drifting ice (Figure 4-2). Barrow residents became eager to avoid a repeat of that brush with disaster. No community member was more eager than J. C. ("Craig") George, field supervisor of bowhead whale census teams scheduled to camp on, and count passing whales from, the nearshore ice throughout Barrow's spring 2000 whaling season. He reviewed several earlier spring breakoffs of landfast ice, in hopes of learning to predict them.

Experienced whalers and supporting crews shared with Craig George their experiences, including interpretations of what triggered past breakoffs of landfast ice (*"uisauniq"* = ice breakoff that takes people with it). By the spring of 1999, the North Slope Borough's Department of Wildlife Management had acquired a pressure sensor ("tide gauge") with which to record changes in sea level during spring whaling. This investment reflected both the linkage in whalers' minds between sea level changes and past breakoff events and the high regard in which researchers held whalers' interpretations. Archived satellite imagery proved useful in reconstructing breakoff events (Figures 4-3 and 4-4) in 1993 and 1997. During the 1997 emergency, however, scientists found that it was logistically impossible for people in the Arctic to obtain the best remote sensing imagery of sea ice fast enough (real-time or near-real time) for it to help them respond to an emergency as it developed.

While some of us worked on streamlining delivery of satellite imagery to ice-dependent subsistence hunters, the National Science Foundation's (NSF) Office of Polar Programs announced its Humans in the Arctic System (HARC) initiative in 1998–99. HARC solicited research proposals in which the payoffs would include facilitating human residents' adaptive responses to changing arctic conditions. We Barrow-resident scientists instantly imagined NSF supporting enhancement of the small-scale ice research that Craig George had initiated with local Inuit collaborators. From the outset, our idea was to expand the exchange of techniques between two complementary teams of ice observers—the local experts at fine-scale ground truth and the community of scientists who have access to remote sensing through both satellite imagery and instrument installations.

To make our image a reality, we had to craft a research proposal to NSF for financial support. Readers accustomed to proposal-writing as a prerequisite to conducting

Figure 4-2. Whaling crew members at camp on floating ice, watching rescue operation by North Slope Borough Search and Rescue Helicopter. (Photo: J. C. George).

research may not give this step a second thought, but there are two reasons for emphasizing the step here. First is to acknowledge the energy used within a proposal on justifying requests for research support, especially for a new project. Second is to remind ourselves that the proposal-writing process, like the rest of science, is anything but "natural." Wolpert (1993: ix) observes that "the ideas of science are alien to most people's thoughts." He extends alienation to scientific practices: "the way in which science is carried out [is] entirely counterintuitive and against common sense" (Wolpert,

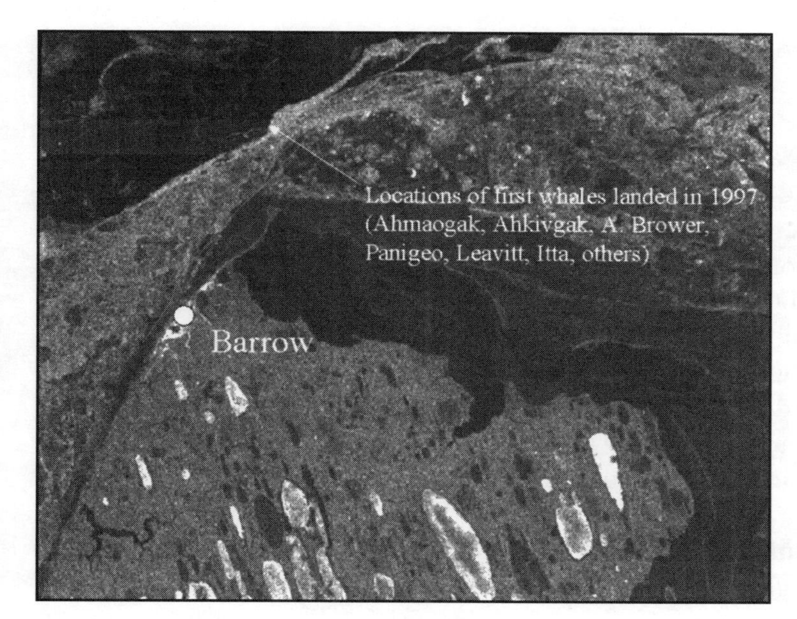

Figure 4-3. RADARSAT image of shorefast ice along the Chukchi Sea coast off Barrow, mid-May 1997, before breakoff (courtesy of J. C. George).

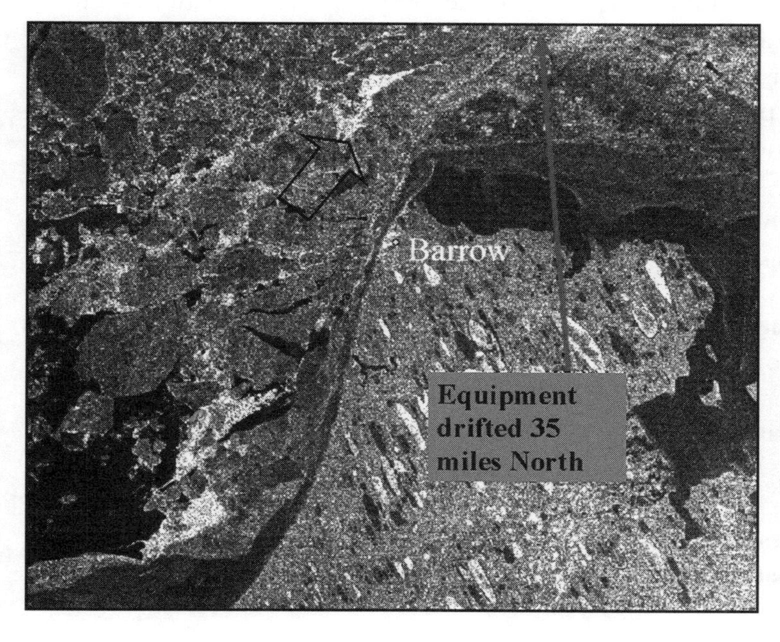

Figure 4-4. RADARSAT image of shorefast and pack ice along the Chukchi Sea coast off Barrow , May 19, 1997, after shorefast ice breakoff and the rescue of 154 whaling crew (courtesy of J. C. George).

1993: 1). Although indigenous collaborators are now essential to some research proposals, the process of seeking peer review, approval, and award of funds to conduct research is no more natural to these naturalists than it was to practicing scientists before the mid-twentieth century. Writing and evaluating proposals are conventions learned and executed in semiexclusive circles, so nonscientist colleagues in the Arctic can be forgiven for believing that scientists have (or should have) more autonomy than we actually do.

Tailoring our appetites for support to the HARC initiative's intentions involved assessing what would appeal to NSF's proposal reviewers. That required little guesswork. Language describing the HARC initiative left little doubt that proposers were assumed to be convinced of the reality of ongoing climate change, and many were expected to propose tapping into indigenous knowledge (also known as traditional ecological knowledge or TEK), two ideas enjoying widespread subscription by the scientific community. In illustration of the scientific favor these two articles of commitment enjoy, consider the fifty-first Arctic Science Conference conducted by the Arctic Division of the American Association for the Advancement of Science (AAAS) in September 2000. The proceedings volume of that annual conference consisted of abstracts for 129 oral and poster presentations, of which forty (31%) dealt with climate change and forty-seven (36.4%) involved indigenous knowledge (AAAS, 2000). Because some presentations involved both TEK and climate change (prescient of the chapters in this volume?) we cannot add 31% and 36% to say that 67% (two-thirds) of arctic research projects today involve one or the other. Still, over one-half do involve TEK, climate change, or both.

A heretical streak made me uneasy about the sheer momentum behind these two canons. Agnosticism about global warming, especially its local manifestations and ultimate causes, seemed a wise attitude. Craig George's inquiry into late twentieth-century breakoffs of landfast sea ice at Barrow did not link them to warming or other environmental trends. The other canon, TEK, seemed to represent an overdue and over-celebrated infatuation with practices that had united scientists and community members in Barrow for years without fanfare (Brewer and Schindler, 2001; Brewster, 2001; Feder, 2001).

Barrow's legacies to science generally, and the community's connection to this research initiative specifically, deserve a brief digression. When Professor Laurence N. Irving became founding director of the Naval Arctic Research Laboratory (NARL) at Barrow in 1947, he established styles of research collaboration that persist to the present (Elsner, 2001; Norton and Weller, 2001). A distinguishing collaborative style was Professor Irving's reliance on and learning from Iñupiat observers, whom he knew to be the best natural historians available. From general natural history studies in the early years of the laboratory, to investigations of contemporary applied problems such as population estimation, behavior, and patterns of migration by the bowhead whale (Albert, 2001) field researchers have continued to regard the Iñupiat of the North Slope of Alaska as indispensible experts. By 1999, locally directed analyses of the nearshore physical and biotic conditions that allow whalers to live on and hunt whales from the outer edge of shorefast ice had entered their third decade as a joint enterprise combining indigenous expertise with scientific approaches to research. An essential ingredient in this successful collaboration has been key scientists earning the respect of whalers. In turn, this respect was built around scientists learning to live and conduct field work safely on the ice alongside whalers for two months each spring.

To expand our inquiry from ice breakoff events at Barrow, we proposed to accomplish four things:

- furnish some outer boundaries to the types of anomalous conditions experienced by traditional users of sea ice, thereby to ensure reality in predicting ice responses to climatic variations in the future;
- gauge the effectiveness of our reliance on Barrow-style TEK in a league where other research projects would also use TEK;
- make experimental use of specialized satellite remote sensing as a stimulus for discussions across disciplines and cultures and as an analytic tool for evaluating changing sea ice conditions;
- include a geographic range of representative case studies from sites and subsistence communities outside Barrow's familiar environs, where nearshore ice dynamics are diverse and well known but not necessarily representative of all or even most other communities in the western Arctic (Figures 4-5 and 4-6).

Methodologically, the proposed approach emphasized that:

- we would demonstrate methods to span the gap separating the hemispheric scales used in general circulation models (GCM) from the fine scales of traditional expertise ("ground truth") embodied in subsistence hunters' experience with nearshore sea ice;
- our case study approach had already stimulated analytic discussions of breakoff events;
- researchers and whalers at Barrow are mutually cordial, having worked together on the ice for many seasons;
- we intended to obtain archived satellite imagery from the Alaska SAR (Synthetic Aperture Radar) Facility (ASF) in Fairbanks;
- we would adopt geographic inclusiveness from U.S., Canadian, and Russian components of a preexisting Barrow-based and National Oceanographic and Atmospheric Administration-funded research team studying marine contaminants from Chukotka to Amundsen Gulf.

After submitting the proposal to NSF in April 1999, the authors had more time on their hands (while the proposal was being evaluated by external reviewers) to agonize over two factors that could affect the chances of an award: possible competitors for the funds and the funding agency's preparedness to support unconventional research. Competitors, if any, were keeping a low profile. Although we saw significant site advantages in basing HARC investigations at Barrow, we knew of no other proposals vying to use these advantages. We assumed that research involving TEK could be proposed only by proven veterans of collaborative fieldwork uniting scientists and indigenous experts. How many other communities in the Arctic rivaled Barrow in offering field experiences? If forty-seven current research projects in the western Arctic relied upon TEK (AAAS, 2000) could we hope that TEK-experienced investigators would be too busy already to compete with us for HARC support? Despite confidence in the scientific and TEK qualifications of the diverse team of investigators we had assembled, we worried about the administrative novelty in our proposal. NSF was being asked to fund a municipality, Alaska's North Slope Borough (NSB), to conduct research. Unlike many universities, the NSB lacked a track record in the performance of basic research for the NSF.

Executing the Project: Experiences and Surprises

We learned soon enough how naive our ideas in the proposal had been, when NSF granted the support we requested, with a startup date of October 1, 1999. To be sure, some aspects of the project have worked well, either illustrating blind luck or validating some of our untutored intuitions and prejudices. To understand how this investi-

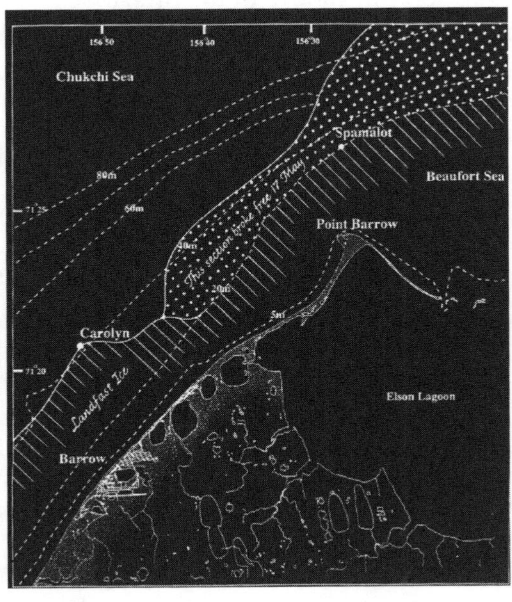

Figure 4-5. Schematic view of ice zones off Barrow and Point Barrow, May 1993 (graphics by Michelle Johnson).

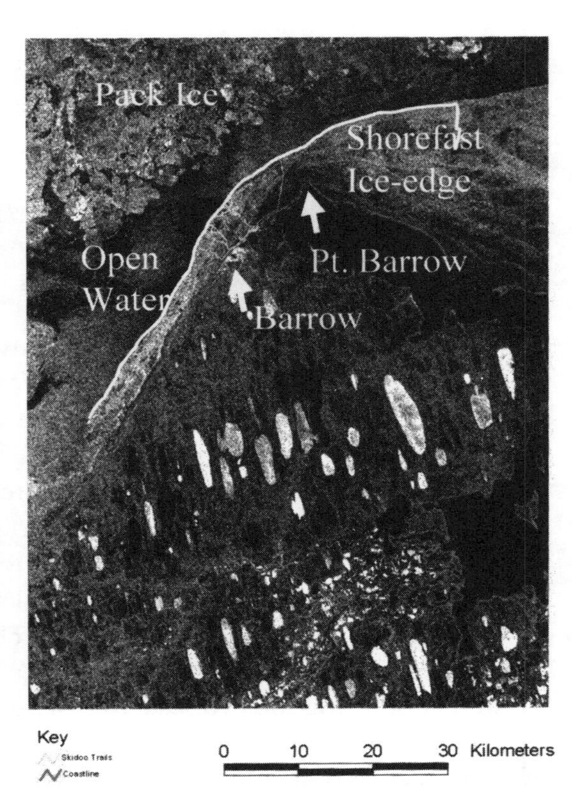

Figure 4-6. RADARSAT image of ice off Barrow, May 15, 2001, showing landfast ice, open lead, and pack ice to the northwest.

gation unfolded, it helps to be aware of fortunate legacies and choices that defined the starting configuration for the project.

Coastal Sea Ice

We inherited (rather than chose) the topic of people's understanding and use of nearshore sea ice. It is hard now to imagine choosing a subject that so deeply captivates both indigenous subsistence whalers and sealers, and the arctic research community. By contrast with sea ice in general, nearshore sea ice was appealing for being a circumscribed environmental topic. Ice in coastal zones, however, has been poorly differentiated from polar pack ice and from terrestrial freshwater ice in the minds of people lacking arctic experience. Cold War priorities may have contributed to this disparity, because studies of coastal ice were overshadowed by research conducted from drifting ice stations during the second half of the twentieth century. In the mid-1970s, planners for the Outer Continental Shelf Environmental Assessment Program (OCSEAP) saw that the unique geophysical and engineering features of nearshore ice (Weeks, 2001: 181) severely challenged arctic petroleum development. During the several years that its environmental studies emphasized coastal ice, OCSEAP encouraged research and contributed refinement to conceptual distinctions among arctic coastal features. OCSEAP investigators managed to enlist indigenous knowledge of coastal ice (e.g., Shapiro and Metzner, 1979) before TEK became a fashionable resource in support of scientific investigations.

U.S. analyses of nearshore sea ice were soon orphaned again (or at least put up for adoption by the petroleum industry). Support for ice studies faded at about the time (1980) that the U.S. Navy vacated its Arctic Research Laboratory (NARL) at Barrow although federal withdrawal from coastal studies was attributed to fiscal austerity and noninterference in affairs within the State of Alaska's three-mile limit. As things worked out, even polar pack ice received only fleeting mention in considerations of

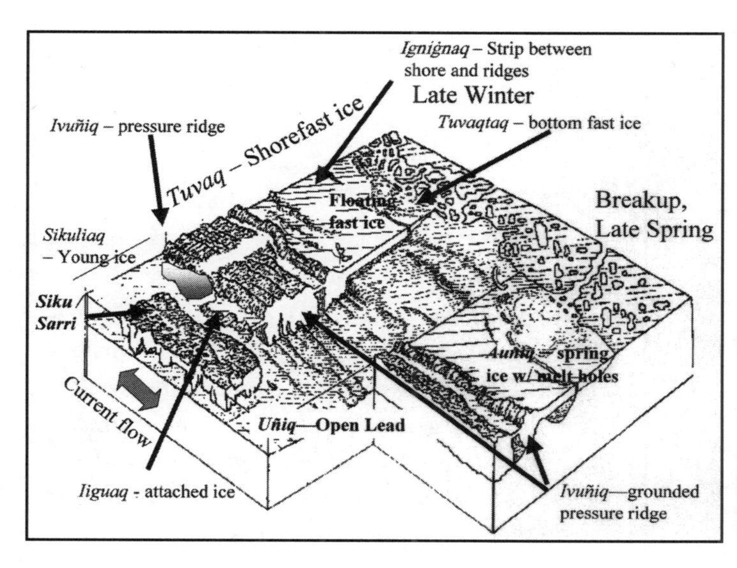

Figure 4-7. Schematic representation of coastal ice, with key Iñupiaq names for structures. Based on various sources, with local names contributed by J. C. George.

northern human ecology and climate change as recently as the mid-1990s (e.g., Peterson and Johnson, 1995). Global climate models and results from the NSF-supported program Surface Heat and Energy Budget of the Arctic (SHEBA) have now sensitized analysts to the pivotal role of perennial arctic pack ice in large-scale climate dynamics.

Coastal ice constitutes a distinct complex and dynamic environment, influenced by both the terrestrial and the marine environments that bracket this nearshore zone (Figure 4-7). By contrast with marine and terrestrial compartments of the northern circumpolar system, coastal sea ice is not extensive enough to affect any but local climate regimes. Although the zone is narrow enough in many places to be obscured by a line penciled onto hemispheric scale maps (Figure 4-8), it is precisely within this boundary ("where the rubber meets the pavement") that traditional resource use and expertise are concentrated. Depending on location and year, nearshore ice consists of varying proportions of annual shorefast ice, annual pack ice, and multiyear pack ice. Each ice-year renews the challenges facing people who must understand and use the coastal zone. Nelson (1969) demonstrated how thoroughly the community of subsistence sealers and beluga whalers at Wainwright, in northwest Alaska, relied on understanding nearshore sea ice in the days before snowmachines. If we were writing a proposal today, we would stress the importance of nearshore ice as a system separate from, and not directly predictable by, dynamics in pack ice systems or in terrestrial systems. Coastal sea ice, we would now argue, needs to be recognized as either an especially contradictory or an especially sensitive indicator of global environmental changes.

Public Safety

It was another good fortune to inherit a focus on public safety from Craig George's pre-HARC interest in breakoff events *(uisauniq)*. Public safety connotes consciously

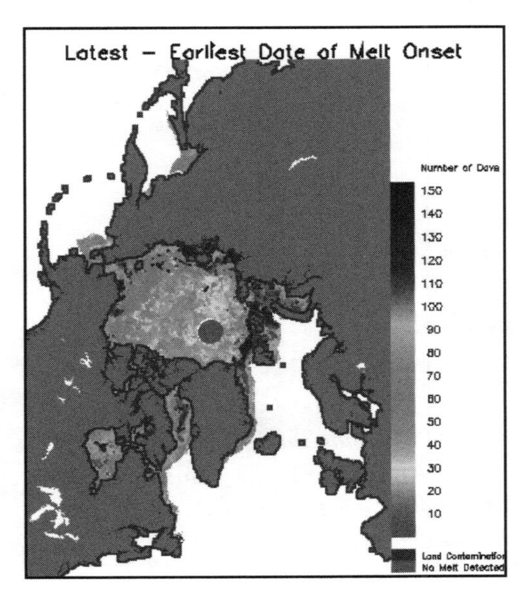

Figure 4-8: Hemispheric map showing ranges in pack ice surface melt dates, 1979–1998. From NSIDC, Sheldon Drobot and Mark Anderson 2002, http://nsidc.org/data/nsidc-0105.html. Darker areas of arctic pack ice indicate greatest variability in onset of summer melt. Coastal zones are too narrow to show at this scale.

avoiding risks of having to escape or be rescued from life-threatening ice conditions. Nelson (1969: 98–132) shows the durability of dominating concern for public safety in the minds and practices of subsistence hunters. Along with falling through thin ice, breakoff *(uisauniq)* has been one of the consistent concerns. Compared to Barrow whalers today, Wainwright hunters were admittedly more patient and better prepared for surviving for days after being cast adrift on floating sea ice during the period of Nelson's studies (1960s), but hunters' vigilance for environmental clues to help avoid risky situations seems not to have changed substantially.

Arctic hunters and fishers seem to build successful careers around revising guidelines for making "correct" (life-sparing) risk-versus-reward decisions. A number of environmental determinants can prompt these revisions, one of which is long-term climate change. Concern for safety framed discussions of how to predict ice instability and helped set a minimum standard for seriousness of abnormal events nominated as cases for consideration by the Barrow Symposium on Sea Ice (BSSI— see below and also Huntington et al., 2001). Whalers' perceptions that our research was motivated by our interest in their safety effectively neutralized the topic of climate change. Allusions to and discussions of climate change arose in Barrow without scientists' prompting. By contrast, our research was perceived by Inuit collaborators in Wainright, another Alaska community some 100 km of Barrow, in 2000 (after the BSSI had taken place) as a study of climate change. That perception of our research injected an emphasis on global change into Wainwright-based discussions of nearshore ice (see Billy Blair Patkotak anecdote, below).

Two Methods Chosen by Intuition

Intuition explains our focus on unusual ice events, rather than on "norms" from which these cases deviated. Although it intuitively satisfied us, our main defense of this choice in the 1999 proposal was that actual events are easily remembered by whalers and easily documented by researchers. Hindsight now provides a stronger rationale for avoiding the temptation to explore and describe "normal" (mean or average) ice conditions. Such average ice conditions are rarely observed. To remind ourselves why this is so, we can imagine two opposite stable environmental conditions ("steady states") such as 0% and 100% snow-covered landscape. If snow-covered and snow-free conditions each last almost half the year, intermediate conditions (between 0% and 100% snow cover) are unstable. The arithmetic mean of two extremes (50% snow-covered) is a momentary condition hardly ever observed. A system passing through that "mean" condition is figuratively hurrying toward one of the stable extremes. The scientific habit of expressing a statistical "central tendency," as if it were inherent in environmental variables, would be misleading in this illustration. As in other comparable cases, the habit should be strenuously resisted (cf. Neill and Gallaway, 1989). We also relied on intuition to choose a public symposium as our primary forum for reviewing case studies and articulating their interpretations from diverse perspectives. Huntington et al. (2001) evaluate the advantages and drawbacks of that choice and of the symposium as it took place.

Preparing for the Symposium on Sea Ice

Our proposal indicated that scientists and indigenous subsistence hunters would convene separately in the fall of 1999 (within less than two months of the award date) to generate lists and brief characterizations of up to a total of ten to twelve anomalous ice events. Whalers and researchers were then to select by consensus six case studies for thorough collaborative analysis at the Barrow Symposium on Sea Ice (BSSI), originally scheduled for March 2000. Project scheduling had been tuned to optimum times of year for whalers' participation in conferences. Midwinter and midsummer are reserved for various other pursuits, while April through June and August through mid-October were dedicated to spring and fall whaling activities. Other than researchers and whalers interacting on the ice itself, March and October-November were the two periods annually during which whalers could devote undivided attention to collaborating with our project in its ice analysis.

There were several instructive surprises in the transition from ten to twelve candidate ice events to the case studies eventually selected consensually by whalers and scientists. The process did not go as planned and took twice as long as expected. Here I was even more at fault than usual. I could have heeded examples set in similar situations by more disciplined and formal Canadian colleagues like Professor Karim-Aly Kassam. Signed agreements between the researchers and the community for participatory action research would have left none of these schedules to informality or chance. To expect our Barrow-resident colleagues to ask whalers to sign an agreement was off-base because whalers might have regarded the request as an affront, a sign of mistrust among old friends. As a semioutsider, though, I should have carried out the task. As it turned out, subsistence whalers had already scheduled business for the fall of 1999 that did not include discussions of past ice events for us scientists. The March 2000 meeting for all participants became a planning session for the BSSI, which in turn was delayed to the next optimum period, November 2000. The delay allowed more time for access to satellite imagery and meteorological data from archives in preparing examples of anomalous ice events. Nine candidate events had been prepared as potential case studies for action at the March 2000 planning session. Each was introduced by a sponsor and supported with documentation and samples of remote sensing imagery from archives.

Satellite Remote Sensing. We knew how eagerly whaling crews pored over any satellite imagery in the visual bands that provided current views of sea ice beyond the local horizon. Single images (e.g., AVHRR, NOAA, or NWS) would be passed from hand to hand and would elicit comments on the extent or width of the alongshore lead off Barrow (in which the whales are hunted) and the configuration of ice as far south as Bering Strait and into the Beaufort Sea to the north and east of Barrow. In our proposal, the improved reliability and timeliness in the delivery of satellite remote sensing images to Barrow were held up as tangible rewards for whalers' giving their own time to this project.

By contrast with aerial photography or satellite imagery in the visible bands of the electromagnetic spectrum, synthetic aperture radar (SAR) imagery affords views of sea ice that are unobscured by either darkness or cloud cover throughout the year. SAR imagery is also captured in fine scale and is especially sensitive to ice surface characteristics (e.g., roughness and smoothness). SAR is currently available from satellite passes at intervals no more than three days apart in the region around Barrow. These SAR features were close to ideal for our goal of bridging the gap between global and hemispheric scale imagery and modeling on one hand and the fine-scale expertise encompassed by indigenous knowledge on the other. The only weakness of SAR imagery is that reflectance from certain surface features or textures can be confused with visible light and dark qualities. We knew from accounts of the 1997 breakoff event and from subsequent interactions with the Alaska SAR Facility, however, that acquiring SAR images in Barrow from digital satellite data downloaded in Fairbanks is a technically specialized and tedious task. We arranged in fall 1999 for this project to acquire digital imagery for retrospective analyses.

In fact, the half-year delay in schedule strengthened the project. More thorough preparation for it meant that the March 2000 planning session at Barrow became a more informative event than would have been possible for meetings three or four months earlier (e.g., for one or more "scoping meetings" originally scheduled for late

Figure 4-9. Kenneth Toovak, Sr., reviewing ice features observed by members of the audience during the helicopter overflight of sea ice on March 8, 2000 (photo by Allson Graves).

1999, just weeks after the funds for the project became available). One experiment was especially noteworthy. We arranged to download near-real-time SAR images of the northern Chukchi Sea in early March. This demonstration was meant to allow collaborators to experience sea ice visually up close and to contrast its visual appearance simultaneously with its appearance to radar sensors. A helicopter took us all on an overflight of landfast ice along the Barrow Peninsula's Chukchi Sea shoreline from the southern to the northern limits of whalers' use of the ice for spring whaling encampments and travel (Figure 4-9). The large-format RADARSAT image of the vicinity of Barrow was downloaded from ASF in Fairbanks and plotted in time to accompany the overflight. Because ENE winds had been blowing steadily for many days, the configurations of the alongshore lead and the shorefast ice were stable, so the SAR image from a satellite pass 60 hours before our helicopter overflight still accurately depicted the ice we observed (Figures 4-10 and 4-11).

Case Study Selections. Formal business at the March session was to select the final case studies for the symposium scheduled for November 2000. To our surprise, participants in March introduced candidate ice events numbers 10 and 11. Both were adopted as case studies without being developed as one of the nine candidates by smaller working groups. The thinking behind adoption of these two case studies was especially instructive.

A whaling disaster at the outer limits of survivors' memories became our earliest case study, KB 57 (Table 4-1). Senior Barrow whalers recall running for their lives as young men, while moving ice destroyed about half the community's gear for spring whaling (boats, sleds, dogteams, firearms, tents, tackle, etc.). The lore associated with

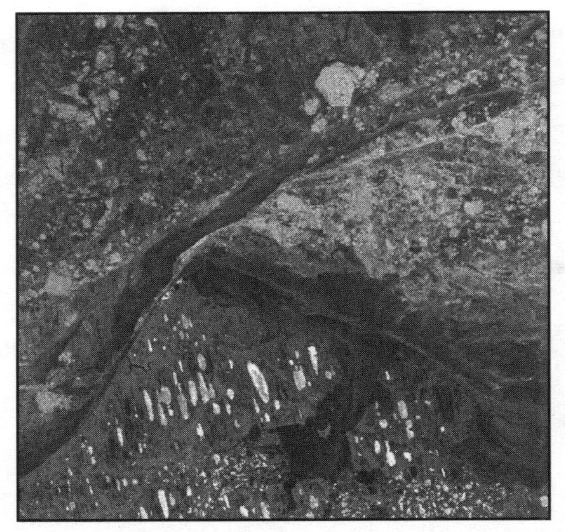

Figure 4-10. RADARSAT image of ice near Barrow, Alaska, March 6, 2000. Note narrow waist in shorefast ice apron opposite NARL.

Figure 4-11. Sea ice from helicopter over Barrow, looking at narrow waist of shorefast ice opposite NARL (photo by Allison Graves).

Table 4-1: Synopsis of five case studies adopted by the project, as addressed by the Barrow Symposium on Sea Ice in November 2000, arranged by year of primary occurrence

Year(s), Designation, First Author(s)	**Nature of the Event**	**Description**	**Notes**
1957, (1) KB57, Karen Brewster	High-energy and high-speed ice override (*ivu*) out on landfast ice, Barrow spring whaling season	Whalers lost a major proportion of gear and dog teams to a surge of sea level and ice; had to "run for their lives."	May 6–7, 1957, depression tracked N over the Siberian mainland, SW winds in so. Chukchi Sea
1975 (1998), (2) KK75-98, Karim-Aly Kassam	Heavy ice year (compared with light ice year); shipping view vs. mammal hunters' views	Sealift blocked at Wainwright en route to Prudhoe Bay by persistent pack ice, Aug–Oct.	Case study extends from Wainwright, AK, to Holman, NWT, highlights local contrasts
1980 (1994), (3) JB80-94, John Burns	Late winter pack ice blockage of Bering Strait, such that marine mammals could not migrate north until mid-May (compared with an early year)	First whales arrived at Barrow on May 23, a month later than normal; possible migration of whales in Siberian waters, far west of normal route.	Ice "arches" formed across Bering Strait, blocking southward extrusion of ice and choking normal lead formation in the Chukchi Sea
1993, 1997, (4) CG93-97, Craig George	Break-off of shorefast ice along Chukchi Sea during spring subsistence season; Safety? Prediction?	Instances since 1980 of whalers being set adrift by calving events, having to evacuate or be rescued.	Hindcasting suggests several triggering events, e.g., shift ENE to SW winds, rapid sea level changes.
2000, (5) RP2000, Russ Page, Dave Norton	Any events such as those in first four case studies that might affect safety or success of subsistence and other activities connected with sea ice	Alongshore lead closed; break-off event avoided by May 30 suspension of whaling; two or three *ivu* events; Aug. storm, effects of which varied locally.	Predictive capabilities partially evaluated and vindicated; lead closure meant that RP2000 required repeating in 2001.

that event established it as a teaching example within the whaling community (Brewster, 1998: 197). Despite agreeing that the disaster occurred in the early days of May, retired whalers could not agree upon the year of this dramatic ice override (known in local Iñupiaq as *ivu*). Their suggested dates ranged from the early 1950s to the late 1960s. As a condition of adopting it as a case study, however, we needed to ascertain the year of the event so that its correlates could be explored.

By poring over Barrow's archived National Weather Service (NWS) records early one morning during the March planning session, Ted Fathauer of the NWS Fairbanks Forecast Office found records of a singularly violent Bering Sea storm on May 4–6, 1957. The low pressure system was strong, and its course over several days ideally positioned it to push a surge of water north through Bering Strait into the Chukchi Sea (Figure 4-12). Confirmation came later from a narrative account that recorded the disruption of Barrow's whaling season and loss of gear in May of 1957 (Reed and Ronhovde, 1971: 384). Although the 1957 *ivu* occurred before any satellite imagery was obtained (Sputnik I was launched in October of that year), its prominence in whalers' memory won us over to using it as a case study. For a while it puzzled the rest of us that senior whalers should be so nonchalant about pinpointing the year in which so many of them nearly died. In time it dawned on us that illustrating a destructive ice event that might recur in the future does not require remembering the year of a past occurrence (a datum that researchers consider essential to evaluating causes). By contrast with the unimportance of a year, local environmental cues that younger whalers might someday use to save their own lives are hugely important.

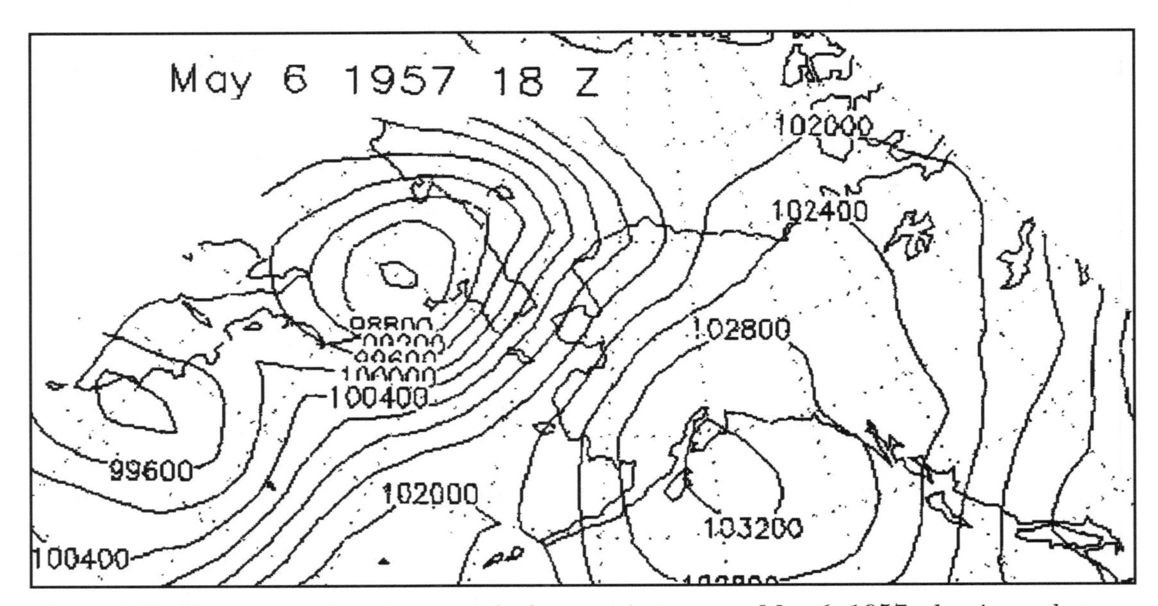

Figure 4-12. Reconstructed surface map for barometric pressure, May 6, 1957, showing a deep depression over eastern Siberia that was linked to high winds driving water through the Bering Strait, and contributing to ice override observed at Barrow that day or the next (KB 57, Table 4-1).

Of the five case studies adopted for the Barrow Symposium, four (KB 57, KK 75, JB 80, and CG 93-97—see descriptions in Table 4-1) were retrospectives on past events, as we had envisioned. The fifth case, however, surprised us by departing from that pattern. Participants at our March 2000 planning session nominated what became RP 2000 (Table 4-1) to focus on the spring whaling season at Barrow, due to begin six weeks after the session. Making future events into a case study signified that researchers and whalers were equally eager to test various signals used to gauge the stability of nearshore ice and the effectiveness of forecast and warning systems in the event of ice instability. In effect, the two groups challenged one another to engage in real-time field analyses of ice dynamics and their public safety implications during the forthcoming hunting season. (As it turned out, the 2000 season itself was full of surprises, to be recounted fully elsewhere.)

Adoption of the 2000 whaling season as a real-time case study was ambitious because it involved fieldwork that we had not proposed to undertake. We decided to forego the planned sixth case study in favor of allocating to RP 2000 the equivalent of two case studies' worth of effort and resources. For RP 2000, we hoped to place reasonably current ("near-real-time") images in the hands of whaling crews and scientists at Barrow. In that hope we were largely disappointed. Bottlenecks in the technology for communicating large files imposed an irreducible minimum delay of two days between receipt of digital data in Fairbanks and their availability as images on paper to people on the ice. Acquisition of images was slowed by inertia at several levels of the system. At administrative levels, questions of international and agency rights to SAR imagery involved a permit review lasting weeks before our NSF funding was accepted as qualifying the project to receive images. At the technical level, procedures for requesting images from specific satellite passes are obscure to nonspecialists. Once requested, images moved over slow transmission links to Barrow. A single large image is a data file that took three to six hours of uninterrupted transmission to download in Barrow. Printing the image on a large-format plotter typically required another ninety minutes. Carrying the print by snowmachine to each of a dozen whaling crews camped over a fifty-kilometer stretch of rough ice would have taken an entire day under the best of conditions. Thus, although systems for delivering SAR images can fulfill requests for archived views as far back as 1993, costs and delays deny arctic coastal residents and users timely views of sea ice. Arctic residents, however, are not being singled out: forecasters with the National Weather Service in Fairbanks and Anchorage have been similarly disappointed in seeking access to SAR imagery.

Indigenous Knowledge. While not attempting to repeat Nelson's (1969) encyclopedic review of how sea ice influences the lives of subsistence hunters, our less inclusive findings have been consistent with Nelson's extensive documentation of healthy indigenous knowledge regarding sea ice. Bridging between subsistence observers and scientific perspectives is an overwhelming undertaking, in terms of volume of information and the time required. Indigenous ice experts in time proved remarkably generous toward our project, by sharing both information and wisdom. Treating sources fairly, however, is a weak link in transferring ideas from oral tradition to written form. Ideal as it would be to connect each idea below to quotations from senior whaler-colleagues, several reasons make it impossible to credit the source(s) of each insight accurately. We planned to forge formal collaboration between the research community (supported by technology) and the whaling guild (using traditional knowledge) but our most informative exchanges tended to take place in the least planned or expected settings. Relaxation perhaps stimulates insights unpredictably and from unexpected sources. We have also learned that some concepts in indigenous knowledge fail to register at first encounter but must be articulated on several occasions, or by several people, before a non-Iñupiaq (*Tanik* is the term used by Iñupiat Eskimos) investigator can fully grasp them (cf. Albert, 2001: 268). Good ideas inevitably travel faster than does credit for them. An idea might be a second- or third-party quote before a *Tanik* thinks to write it down for attribution.

Collectively, our whaling associates take most environmental perturbations in stride. They have witnessed or heard about instances of change in enough varied time scales to expect constantly to match wits against vagaries in their surroundings. Visitors to communities in the North may hear indignation directed at the industrial South for some environmental changes in the Arctic. Chemicals in the environment such as petroleum, persistent organic pollutants, and radionuclides, for example, routinely generate lively finger-pointing. Climate change seems not to be connected in whalers' minds with pollutant transport. Although some whalers perceive that long-term warming is contributing to changes they have observed in spring shorefast ice, their concern is not to identify distant causes or fix blame as much as it is to be prepared for change. Unidirectional change does concern whalers:

> We don't have multiyear ice structures like we used to have. Ice is not as thick as it used to be. We have lots that is two to three feet thick. There's an absence of pressure ridges. We're floating out there a lot of the time and it can shear off at any time. (Eugene Brower, 2000, commenting on break-off risks in recent years on nearshore ice during spring whaling activities. Case Study CG '93-97, Table 4-1)

Concern over long-term changes inspired one of my favorite gems of indigenous wisdom. This gem illustrates several points. Billy Blair Patkotak accosted three of us *Tanik* investigators one morning in June 2000 as we walked along the road in his home community of Wainwright, Alaska. Mr. Patkotak apologized for forgetting a comment that he had meant to contribute during a formal interview about nearshore sea ice conditions the day before (Figure 4-13). Below is my paraphrase of the comment that he meant to contribute formally:

> The rules for interpreting ice [and other variables in the Wainwright nearshore environment] changed after 1975–76. Before that I could tell

> younger whalers what they were seeing and what to expect. After the rules changed, I tried, but I was often wrong. Pretty soon younger whalers stopped listening to me. I tell you, that made me very sad.

Mr. Patkotak's perception that the "rules changed" after 1975–76 (he was born in the mid-1920s, so would have been about fifty years old at that time) coincides with oceanographers' detections of a Bering Sea ecological regime shift dating from 1976. It also distills succinctly the traditional application of senior whalers' wisdom. As noted above, our project's introduction to whalers in Wainwright focused on climate change, and our colleagues in that community emphasized that theme in what they shared with us. Whether or not Mr. Patkotak regrets the environmental change he noted beginning in 1975—76, there can be no doubt that he laments the failure of a social covenant following loss of reliance upon previously useful wisdom. The unstructured setting in which Billy Blair Patkotak shared his insight illustrates why recording and giving due credit to those who share indigenous knowledge sometimes fails. Contrasted with the specific and narrow focus of his interview the day before (Kassam and Wainwright, 2001: 9). Mr. Patkotak's synthesis of information a day later is impressive. Other investigators have treated similar insights from traditional knowledge as evidence for increasing unpredictability of weather and climate in other communities (e.g., Riedlinger, 2001: 97; Jolly et al., this volume; Fox, this volume).

Several senior whalers at Barrow talked of rotations by large pieces of moving ice at or just beyond the outer edge of shorefast ice. These allusions were meant to explain some violent changes occurring at the outer edge of shorefast ice. Speakers sometimes used arms and hands to mimic pans measuring hundreds of meters long swinging in or out like a gate, or careering along the coastal ice margin. I confess to nodding politely and thinking, "why not?" Rotating ice pans made sense intellectually as explanations for some observations at the shorefast ice edge. But it strained my imagination to think that whalers would deduce from indirect clues that big pans may spin during their stately progress in alongshore currents. After all, whalers standing on the surface look down upon very little of their flattish world from above. Translating from their tiny sample of surface-visible environment to inferred motions taking place beyond their horizon requires uncommon skill

Figure 4-13. Billy Blair Patkotak of Wainwright, Alaska, emphasizing a point in discussions of sea ice dynamics, July 2001. (Photo courtesy of Karim-Aly Kassam.)

in spatial relations. Then one day, Lew Shapiro (sea ice specialist at the Geophysical Institute, University of Alaska Fairbanks) treated a small audience of researchers to a narrated showing of his time-lapse film of radar targets in nearshore ice at Barrow. Shapiro's OCSEAP-era film captured images of some ice pans swinging as if hinged, and of others wheeling into and out of radar range. The Shapiro film vindicated traditional ice observers and was one of several epiphanies in my growing admiration for traditional knowledge. Shapiro kindly showed his film again at the November 2000 Barrow Symposium on Sea Ice. SAR images that look down from sequential satellite passes also validate the ice rotation long ago deduced by our Iñupiat colleagues (Figures 4-14 and 4-15).

Some months after the Barrow Symposium, I found myself trying to describe our transdisciplinary and transcultural experiences with nearshore ice environments. A physical scientist and I were trying to capture in writing how much traditional knowledge had already contributed to our understanding of nearshore sea ice (he had participated in the Barrow Symposium). By reviewing various specifics concerning ice that we had learned from our Iñupiat colleagues, we sought to generalize about the inter-

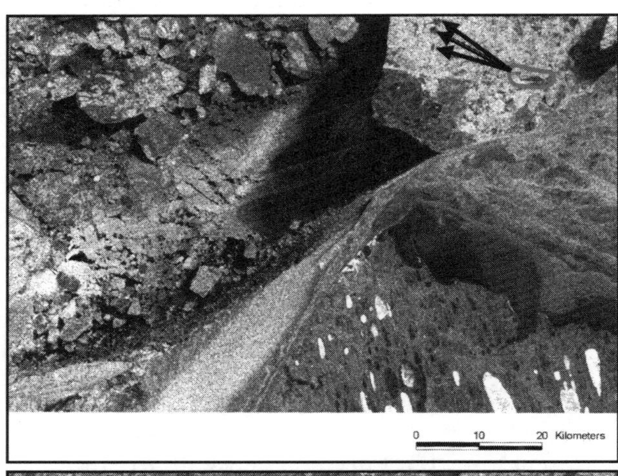

Figure 4-14. RADARSAT image of sea ice, Barrow region, March 28, 2001. Circles and lines highlight orientation of features on a large pan of pack ice. (See Fig. 4-15.)

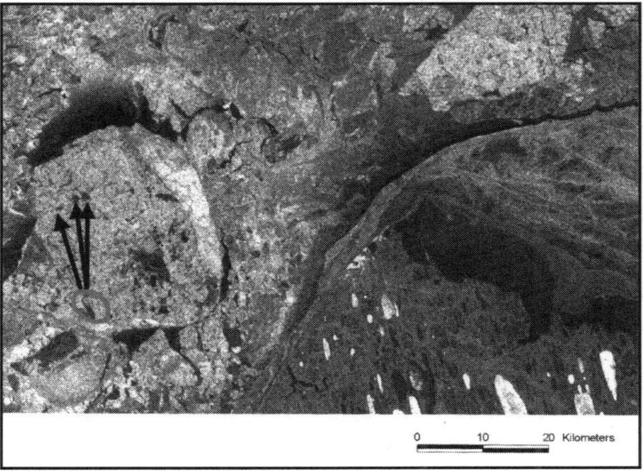

Figure 4-15. RADARSAT image of sea ice, Barrow region, March 31, 2001. Circles and arrows highlight orientation of features on a large pan of pack ice. Note the nearly 90-degree rotation clockwise of this ice pan in the three-day interval since Fig. 4-14.

pretations of phenomena that they had shared, but which—even if presented with the same phenomenon—we researchers might not grasp without their help. Slowly the generalization came to both of us: Whalers observe and think about nearshore ice by thinking in two dimensions besides the ones that nonwhalers primarily use. Everyone can visualize ice in two dimensions. X- and Y-axes define extent of ice or its percent coverage of the surface. Near-universal comfort with Cartesian coordinates makes mapping and interpreting satellite imagery productive for arctic residents and scientists (Robinson and Kassam, 1998: 30; Kassam and Wainwright, 2001: 3). If nearshore and landfast ice responded to a single environmental variable, investigators might expect to learn that the areal extent of nearshore sea ice available to spring whalers has changed as a direct and unambiguous function of climate change. Satellite imagery in all forms affords users essentially a two-dimensional view. On a fine scale, whalers' two-dimensional assessment of nearshore ice still tends to outperform satellite imagery. Their descriptions of surface conditions, for example, are adept at distinguishing multiyear from first-year ice. In turn, knowing the age composition of the surface of nearshore ice helps whalers gauge its strength vs. brittleness (Huntington et al., 2001: 203; Harry Brower, Jr., 2000, quoted below).

In addition to this two-dimensional analysis, subsistence hunters appraise an expanse of nearshore ice by visualizing its third (vertical) dimension. That visualization goes further, and is more active, than passive intellectual awareness that ice does indeed vary in the vertical dimension (thickness). Full appraisal of the third dimension takes into account the energy with which ice collides or is deformed, its fracturing and folding, and the resultant underwater configurations of ice in shallow water. Picturing ice in the vertical dimension allows traditional naturalists to estimate how currents (and sometimes whales) move around features of the underside of the ice. In evaluating the stability of nearshore ice, whalers especially look for surface clues that indicate whether the ice "keels" that are thrust downward beneath fractured and deformed ice in pressure ridges reach deep enough to anchor a section of shore ice to the bottom. It becomes clear from their illustrated explanations that whalers' third dimension includes changes in the atmosphere above the ice, the ice itself, the behavior of the water column on which the ice floats, and the bottom topography of the seafloor. Local naturalists' expertise thus keeps track of a whole array of variables.

Barrow's spring whalers draw on any useful sources of information they can to help with tracking variables. They traditionally accumulate and hone their three-dimensional views of nearshore ice by keeping track of its annual development from freezeup through the winter and into the whaling season at the end of winter. Longitudinal observation of ice development represents a fourth dimension, the historical perspective, or investing the ice with a seasonal "system memory" (Huntington et al., 2001: 204). It is easy to imagine observers who build and share this chronology of nearshore ice being respected experts attentively listened to within subsistence communities (cf. Krupnik, this volume).

The social standing of ice experts undoubtedly integrates a great deal of information and indicates where subsistence communities place their values. Mr. Patkotak's comment on the failure of old rules, noted above, made sense as a social lament. J. C. "Craig" George added his biological perspective on how ice expertise matures. One day early in our research planning, he expressed exasperation with himself for worrying constantly about landfast ice stability and the safety of whalers and researchers. "You know, a few years back, an ice breakoff would have struck me as a lark, a good adventure for everyone. Now, I guess, being older, I've grown timid. It worries me that young risk-takers out on the ice might be as foolish as I used to be." Caution is perhaps a commodity that develops with age in any human society.

Senior whalers' recollections of the 1957 *ivu* at Barrow illustrate one of the few environmental situations that they were unable to take in stride. Like the breakdown in environmental "rules" that Billy Blair Patkotak reported from 1975–76 onward, the 1957 ice override forced Warren Matumeak to forget whalers' rules of behavior on the ice:

> I was in the late 1950s event. Everybody just wanted to stay alive, that's a big story. Wesley Aiken lost everything, I was in his crew. We put the boat on top of the sled, but we should have taken it off the sled and it would have ridden the *ivu*. It would have been okay. Some of us were resting and we sat down and ate and smoked. It didn't take long, it started cracking below us, had to jump and cracks were widening and closing up. We just took off. Didn't even try to get the boats. If it was moving slow, we would have hung on to the boat, but it was moving too fast. It was every man for himself. We were going over the ridges when it stopped. Just hoped those ridges didn't fall over. Can ride slow *ivus*, but that was so fast. The wind wasn't blowing real hard, twenty miles per hour. When the plane went down to the coast and dropped notes to people. The whole town was watching us trying to stay alive. I was running as fast as that *ivu* was moving. I was young then and could run fast. It would stop moving and we'd jump to the next piece. But, it would stop in one place and then be breaking up someplace else. (Warren Matumeak, recounting in 2000 the fast-paced *ivu* event that overtook crews, dogs, and whaling gear in May 1957.)

Harry Brower, Sr., chronicled the 1957 *ivu* before his death in 1992 (and before our project had reestablished the year of its occurrence). The following represents his perspective as a young whaling captain having to urge his crew to do terrible things:

> We got caught on the moving ice, the piling ice. Each man had to try to survive for himself. I told them just before we started leaving the boats and everything, I said, "Don't look backwards, just go ahead and try to save yourself. Don't look at the others. If they get caught, don't go help 'em, just keep on going. Let's see how many of us can make it." Boy, when the ice is piling up like that you have to travel up on top of it and jump around. Scary! That's why I told them, I said, "You kind of have to take to yourself. Don't look backwards. Don't help

nobody." We just threw away all the gear, 'cause we weren't going to be able to save it anyway. We were right in the middle of that moving ice. Most of the dogs were all caught in the ice, too. They were let loose to run around and try to make it through on their own, but we still lost quite a bit of our dogs. (Brewster, 1998: 197)

Teamwork and constant vigilance over the safety of one's crewmates no doubt contribute to the remarkable safety record compiled by subsistence whalers of northern Alaska over the years of recent record. Therefore, being told to ignore one's crewmates, or "don't look backwards," was such an extreme departure from norms of behavior that it indelibly impressed all who experienced the 1957 *ivu* (cf. Nelson, 1969: 379).

Barrow whalers shared proposed mechanisms to explain events and anomalies in local nearshore sea ice. The level of detail presented before and during the BSSI makes it possible to recreate here only a small sample of subsistence naturalists' analytic approach to sea ice. For their part, whalers also absorbed and criticized much scientific information. At times, whalers complained that numeric data presented by scientists were hard to grasp. Scientific exposition is notoriously hard for the layman to grasp (Bielawski, 1995). We tried to speak and present material in plain language at the BSSI (Huntington et al., 2001). Despite those efforts, graphic representation of changes in wind direction during the passage of a storm proved hard to understand, so after the BSSI, arrows depicting directions were added to figures relating numeric values for compass bearings during changes in wind velocities (Figure 4-16).

Ideally, the information and insights flowed both ways between whalers and researchers. Whalers were fascinated by satellite imagery, as always, especially when it was recent. Similarly, when data were retrieved from the pressure gauges beneath

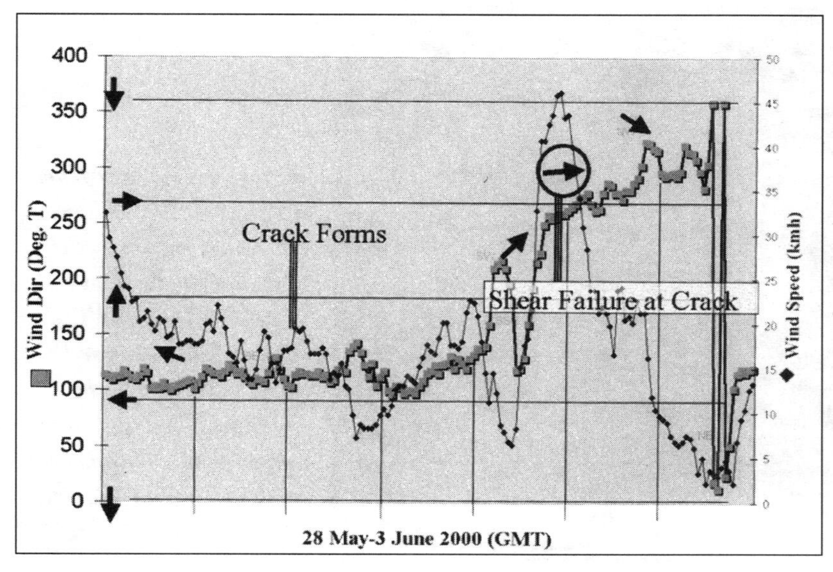

Figure 4-16. Whalers to researchers: make your graphic displays more understandable. Meteorological data May 28 to June 3, 2000, reflecting the passage of a windstorm, shift of winds from easterly through westerly, in relation to a crack and small breakoff event in shorefast ice.

shorefast ice, whalers inspected the rises and falls in sea level to compare those with their own estimates and recollections. The other item that generated special interest when researchers brought it into discussions was the bathymetry of the Chukchi Sea in the vicinity of Barrow's whaling activities. When satellite imagery was georeferenced, isobath contours from files maintained by the North Slope Borough's GIS Division could be displayed with ice features. In those discussions where bathymetry and ice distribution were shown simultaneously, whalers' description of ice and whale movements seemed to reflect the influence of the Barrow Canyon on deep currents (Figure 4-17).

The locations of weakness and the ice types in which fractures occur are a constant concern to spring whalers using nearshore ice:

> One time when out with my dad, around April 26 or 27, there was young ice for miles. I saw waves ripple through the ice going down the width of the lead. Dad warned me about the wave coming back. To watch out for that. It did, and it hit the glacial ice (*piqaaluyak* [multiyear pack ice]) we were camped on. It cracked up all over, shattered. (Harry Brower, Jr., 2000, recalling detailed guidance he received from his father on concerns for spring ice safety in general.)

The attempt to find causes for *uisauniq* (breakoff) events that first inspired our research led to various hypotheses invoking wind, currents, waves, and sea level fluctuations.

> Another time, I was out near the lead edge. It was calm and clear weather. I saw a "bend" in sea level in glassy clear water before the ice broke off. (Warren Matumeak, 1996, as told to J. C. George)

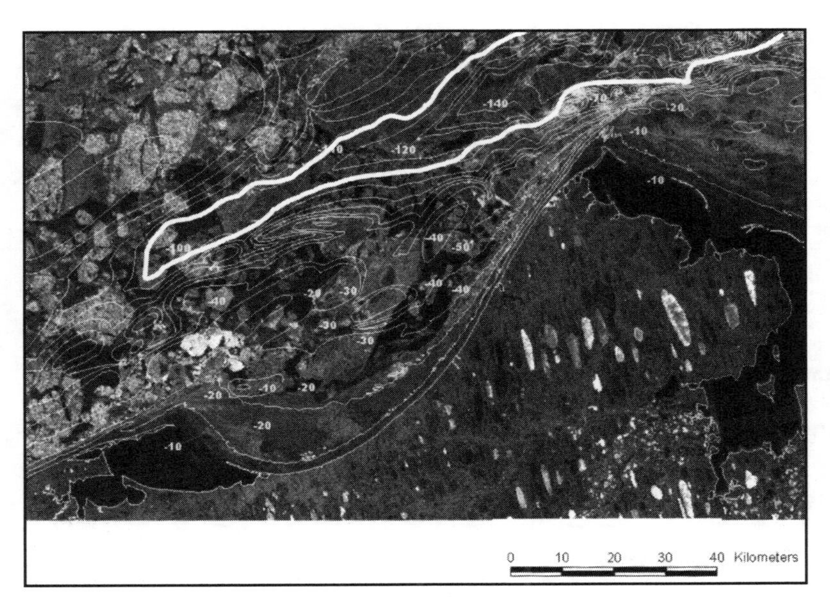

Figure 4-17. Researchers to whalers: here's how Barrow Canyon lies in relation to observed ice types. (RADARSAT, May 15, 2001, overlain with NSB GIS bathymetry data, georeferenced showing the 100-meter isobath in bold. Courtesy Allison Graves).

This was one of a number of unsolicited stories that implicated sea level as the one observable factor that in isolation from others could trigger serious enough ice cracking to turn into breakoffs. The process described by Warren Matumeak is known as the *katak* theory (*katak* = fall, as in ice falling with sea level or "tide"). Later, when asked for more detail, Mr. Matumeak generalized about the process and illustrated the theory:

> When sea level drops due to the change in wind direction to the east, you get a drop in sea level and the ice "falls" or "*katak*." As it does so, the ice on the seaward side of the last grounded pressure ride cracks and releases the floating ungrounded ice. The heavy ice fails when the water level drops. (Warren Matumeak, March 2000 BSSI Planning Meeting Report)

The *katak* theory (Figure 4-18) proved concise and susceptible to testing. Testing the theory justified acquiring and deploying the first of several pressure sensors ("tide gauges") for the project. Subsequent refinements to this and other theories from within indigenous knowledge have led to provisions for additional instrumentation. Our Barrow collaborators now realize that they can profoundly influence the direction of research on nearshore sea ice.

Conclusions

Researchers proposing to draw upon TEK should carefully avoid investigator effects. When climate change was inadvertently represented in the summer of 2000 to indigenous collaborators in Wainwright as our primary research objective, they tended to perceive manifestations of global warming in any local anomalies. They may have prescreened examples of ice anomalies after learning of researchers' interest in global warming. Whether or not deliberate agnosticism toward climate change should be prescribed for projects dealing with the subject, I suspect that our project's focus on public safety in Barrow was more effective at encouraging objective participation. Safety issues stir local interest, whereas global climate change may seem remote and immune to local intervention.

Although comparative Wainwright experiences caused us to wish we had not connected climate change to this project, that lesson was offset by two others for which we can thank Wainwright participants. First, we learned that ice conditions only 100 km

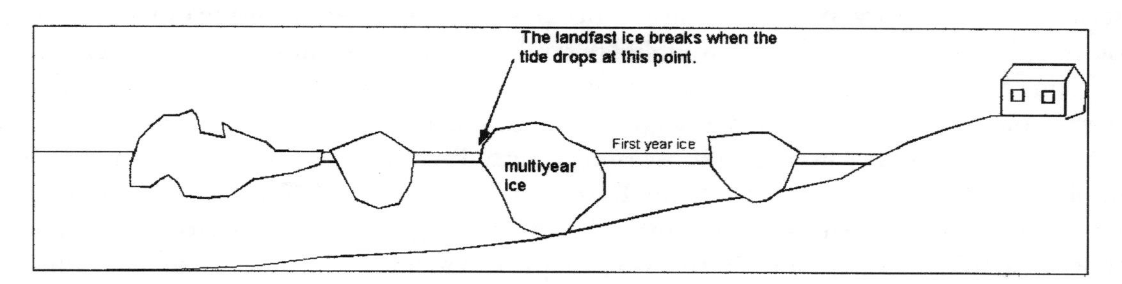

Figure 4-18. Diagrammatic representation of the katak *theory described by Warren Matumeak.*

southwest of Barrow have produced contrasting hunting success in several apparently similar years. The years 1975 and 2000 are examples of seasons in which Barrow's hunters experienced ice choking the alongshore lead and preventing success, whereas abundant ice in those same years contributed to unusually successful hunts for bowhead whales and other marine mammals at Wainwright. Coupled with that contrast, Wainwright's senior subsistence hunters taught us to question the perspective of anyone who labels a given year a "bad" or a "good" ice year. Barge operators bound for Prudhoe Bay in 1975 labeled it a "bad" year because they were stuck in nearshore ice near Wainwright. To hunters in the community, however, the same year was "good" because bearded seals *(Erignathus barbatus)* were especially abundant and accessible under Wainwright's conditions of persistent and heavy nearshore ice. Taken together, these examples of contrasting observations are reminders of the extraordinary richness of detail encompassed by traditional knowledge (cf. Krupnik and Bogoslovskaya, 1999).

Having participated in a working association between technology-rich scientific researchers and field-experienced naturalist-hunters, I am intrigued by what has made this association productive. These two guilds share a long history of transcultural symbiosis (living together) at Barrow, perhaps the longest of its kind in the western Arctic. If we can use the biological metaphor, this symbiosis has ideally evolved into mutualism, in which both parties benefit equally from functioning together. Less desirable would be commensalism, in which one guild gets all the benefits without benefiting or harming the other. Furthest from equitable would be parasitism, in which one lives entirely at the expense of the other (its involuntary host). Judging among these three alternatives is beyond any single observer's abilities, although it remains a thought-provoking question for groups of observers. One trend is nevertheless evident in Barrow's long history of relating to scientists at work in the Arctic. Beginning before the mid-twentieth-century career scientists everywhere have tended to spend a shrinking proportion of their time outdoors studying environmental realities (Sturm, 1999). In Professor Irving's day, scientists and Iñupiat were linked by a camaraderie of being outdoor field naturalists of one specialty or another. Pressures on investigators to reduce time spent on fieldwork have taken a toll. Today, few professional naturalists are left and fewer are being trained for scientific careers. Natural history is instead left to amateurs, and in the Arctic, left to "indigenous knowledge." The end point of surrendering natural history to experts outside formal science is to reduce some environmental researchers in the Arctic to obligate symbiosis, with specialists wholly dependent on sources of TEK for information on arctic natural history. Such a development might work, but its implications deserve serious discussion by the guilds involved.

Specializations narrower than those of technical disciplines tended to develop among members of our research team. Up to a point these are defensible for increasing the team's effectiveness. On the other hand, dividing responsibilities too finely can deprive people of general familiarity with the project as a whole. Generalism and holistic thinking are at a premium in exchanges with indigenous knowledge. For making inroads on participants' versatile capabilities, I have regretted my own and others' entrapment by division of labor. As an example of the first regret, becoming the team's designated proposal-writer saddled me with an awkward burden. The assignment keeps me from an equal turn at fieldwork on the ice; worse yet, it perpetuates a myth that only one or two team members can speak and write some semisecret language understood by the priesthood at the funding agency. As an example of the second, it troubled me that we relied on one or two team members to be the principal daily contacts with whalers and their traditional knowledge. The holistic discourse during the Barrow Symposium eased some of the concerns over these instances of specialization.

Nesbitt (2000: 43–44) likened collaboration between researchers and indigenous naturalists to a "three-legged race." His light-hearted metaphor does not try to analyze which party gets a greater share of benefits from collaborating. It evokes images of partners stumbling around for a while before adapting to one another's strides, and of everyone merrily galloping along once they get the hang of the race.

For years, my exposure to traditional knowledge was largely secondhand. By chance early arctic research interests in the 1960s involved me with small birds nesting on tundra inland of Barrow. Those birds become active in the season when people are just finishing spring whaling and preparing for summer fishing and hunting. Iñupiaq natural history concerning the birds that I was investigating was less extensive than for species large enough to be worth hunting. A number of scientific colleagues in the 1960s worked on subjects that made them heavily reliant on Iñupiat naturalists and grateful for their availability at Barrow. It was thus natural to admire and respect the body of observation, lore, custom, and teaching that permitted people to live from the arctic land and sea. That passive admiration was tempered in the 1970s and '80s. Environmental investigations in those years seemed to be targeted by a rash of noisy folks—mostly outsiders as opposed to practitioners of the quiet competencies that constitute genuine indigenous know-how—who tended to romanticize indigenous practices. Their apparently political antics polarized people rather than drawing them together. Moving to Barrow for the 1990s gave me a second chance at direct involvement with traditional knowledge. By the time NSF's HARC initiative was announced, my admiration for genuine indigenous expertise had been restored. During the project described here, the proficiencies of Iñupiat naturalists converted me from passive admirer to active advocate.

This project may have made believers of others, too. The BSSI enabled us to poll participants on their appraisal of indigenous knowledge. All had been exposed to the resources invested in sea ice analysis by agencies of the federal government and to hopes that reliable ice forecasting reflects the investment. The poll asked respondents to choose the alternative that would make them feel safer out on nearshore sea ice: (a) web connections to satellite imagery and Weather Service ice forecasts, or (a) an Iñupiaq companion traditionally knowledgeable of local ice conditions. Preference for an Iñupiaq guide was unanimous.

What lies in the future for the nexus between traditional knowledge and formal science? If our experience with a topic as challenging as anomalous behavior of nearshore sea ice is representative, TEK is richer than most newcomers can grasp. Traditional know-how necessarily differs markedly from site to site, even between nearest-neighbor communities (e.g., Barrow and Wainwright). Hunters in the two locations apply site-specific predictive understanding to their evaluation of risks. Any current environmental research program confronts a tug-of-war between peripheral and central, specific and general. On the one hand, each community seeks to impress on program managers the uniqueness of its stakeholders' concerns and experiences accumulated as TEK. On the other hand, the predictable similarity of that very litany suggests that a central code of TEK-related ethics and a universal set of TEK methodologies for research should exist. Whether, how fast, and how far convergence in TEK methodologies will proceed remains to be learned. Since we are still learning how richly detailed site-specific expertise can be, a single instruction manual for collaborative investigations seems far off.

Is indigenous knowledge likely to be corrupted by scientific collaboration? (External reviewers of our project recently introduced us to this arresting thought.) Carelessness in science can lead to "investigator effect," an example of which was cited above. But unless research in the North becomes consistently careless and discredited, collaboration with scientists seems to pose little threat of corrupting TEK. Rather, I suspect that technologies will continue to prove more contagious and addictive than will scientists' "unnatural" patterns of thinking. A case in point is Global Positioning Systems (GPS). Years before handheld GPS units became user-friendly, active hunters mastered the user-hostile models of the early 1990s to navigate through darkness and fog. Their proficiency even then with GPS looked so effortless that an observer might have confused it with skills handed down through generations. Parallels apply to outboard engines, firearms, and snowmachines. Snowmachines may have displaced dogmushing technologies and know-how, but arguably did not affect indigenous knowledge at its core.

Richard K. Nelson's experiences over time, primarily with the hunters of Wainwright, are worth highlighting in relation to this question of corruption or resiliency of traditional knowledge. In the mid-1960s, before snowmachines appeared, Nelson (1969: 213) considered bowhead hunting a dying art. When invited back by the North Slope Borough, Nelson found himself revisiting the question of bowhead hunting, which by 1980 was flourishing again at Wainwright and elsewhere (Nelson, 1982: 111). Nelson was also privileged to see Wainwright before and after the advent of snowmachines in 1965, about which he makes a number of perceptive comments (Nelson, 1982: 106).

With respect to understanding nearshore sea ice, fraternizing with scientists shows no sign of displacing traditional knowledge. Nor do scientists' technological shortcuts of satellite imagery, tide gauge data, and measurements of under-ice currents seem as likely to corrupt traditional knowledge as to encourage it. The health of traditional knowledge depends upon continuity, or continued receptivity by coming generations, who must master TEK and pass it on. Cable TV and fast food far outweigh scientific collaboration in any equation or formula predicting the future of indigenous knowledge. Perhaps the health of TEK boils down to whether people can learn not to be in too much of a hurry, so that they can live outside the expectations of instant gratification.

Acknowledgments

National Science Foundation awards #OPP-9908682 and #OPP-0117288 funded our project entitled "Synthesis approach, to link remote sensing information with natural history information and traditional knowledge, through case studies of unusual sea ice conditions." For tolerating my pick-pocketing their ideas, I thank team members and associates Karen N. Brewster, John J. Burns, Arnold Brower Sr., Harry Brower Jr., Hajo Eicken, Ted Fathauer, J. C. "Craig" George, Allison Graves, Henry P. Huntington, Karim-Aly S. Kassam, Warren Matumeak, Russell Page, Billy Blair Patkotak, Peter P. Schweitzer, Greg Tagarook, and Kenneth Toovak Sr.

References

American Association for the Advancement of Science (AAAS). 2000. [Proceedings,] Fifty-first Arctic Science Conference, Whitehorse, YT, Canada. Fairbanks, AK: AAAS, Arctic Division.

Albert, Thomas F. 2001. The influence of Harry Brower, Sr., an Iñupiaq Eskimo hunter, on the Bowhead Whale Research Program conducted at the UIC-NARL facility by the North Slope Borough. In: Norton, ed. *Fifty more years below zero: Tributes and meditations for the Naval Arctic Research Laboratory's first half century at Barrow, Alaska,* pp. 265–278. Fairbanks and Calgary: Arctic Institute of North America.

Bielawski, Ellen. 1995. Inuit indigenous knowledge and science in the Arctic. In: Peterson and Johnson, eds. *Human ecology and climate change: People and resources in the Far North,* pp. 219-227. Bristol, PA: Bristol & Francis.

Brewer, Max. C. and John. F. Schindler. 2001. Introduction to Alaska's original naturalists. In: Norton, ed. *Fifty more years below zero: Tributes and meditations for the Naval Arctic Research Laboratory's first half century at Barrow, Alaska,* pp. 9–10. Fairbanks and Calgary: Arctic Institute of North America.

Brewster, Karen N. 1998. An Umialik's life: Conversations with Harry Brower, Sr. Master of arts thesis, Fairbanks, AK: University of Alaska Fairbanks, Rasmuson Library.

Brewster, Karen N. 2001. Historical perspectives on Iñupiat contributions to arctic science at NARL. In: Norton, ed. *Fifty more years below zero: Tributes and meditations for the Naval Arctic Research Laboratory's first half century at Barrow, Alaska,* pp. 23–26. Fairbanks and Calgary: Arctic Institute of North America.

Elsner, Robert. 2001. Cold adaptations and fossil atmospheres: Polar legacies of Irving and Sholander. In: Norton, ed. *Fifty more years below zero: Tributes and meditations for the Naval Arctic Research Laboratory's first half century at Barrow, Alaska,* pp. 77–80. Fairbanks and Calgary: Arctic Institute of North America.

Feder, Howard M. 2001. A year at NARL: Experiences of a young biologist in the Laboratory's early days. In: Norton, ed. *Fifty more years below zero: Tributes and meditations for the Naval Arctic Research Laboratory's first half century at Barrow, Alaska,* pp. 33–60. Fairbanks and Calgary: Arctic Institute of North America.

Huntington, Henry P., Harry Brower, Jr., and David W. Norton. 2001. The Barrow Symposium on Sea Ice, 2000: Evaluation of one means of exchanging information between subsistence whalers and scientists. *Arctic 54*(2): 201–204.

Kassam, Karim-Aly S., and Wainwright Tribal Council. 2001. *Passing on the knowledge: Mapping human ecology in Wainwright, Alaska.* Calgary: Arctic Institute of North America.

Krupnik, Igor, and Lyudmila Bogoslovskaya. 1999. Old Records, New Stories. Ecosystem Variability and Subsistence Hunting Pressure in the Bering Strait Area. *Arctic Research of the United States, 13* (spring-summer): 15–24.

Neill, William H., and Benny J. Gallaway. 1989. "Noise" in the distributional responses of fish to environment: An exercise in deterministic modeling motivated by the Beaufort Sea experience. In: Norton, D. (ed.) *Research advances on anadromous fish in arctic Alaska and Canada: Nine papers contributing to an ecological synthesis,* pp. 123–130. Biol. Pap.Univ. Alaska No. 24.

Nelson, Richard K. 1969. *Hunters of the northern ice.* Chicago, IL: University of Chicago Press.

Nelson, R. K. 1982. *Harvest of the sea: Coastal subsistence in modern Wainwright.* Barrow AK: North Slope Borough.

Nesbitt, Thomas. 2000. Facilitating the work of northern collaborative research boards: the culture of research. In: *American Association for the Advancement of Science. Proceedings, Fifty-first Arctic Science Conference, Whitehorse, Yukon, Canada,* pp. 43–44. Fairbanks AK: AAAS Arctic Division.

Norton, David W., and Gunter Weller. 2001. NARL's legacy to Outer Continental Shelf studies. In: D. Norton, ed. *Fifty more years below zero: Tributes and meditations for the Naval Arctic Research Laboratory's first half century at Barrow, Alaska,* pp. 233–236. Fairbanks and Calgary: Arctic Institute of North America.

Peterson, David L., and Darryll R. Johnson (eds.). 1995. *Human ecology and climate change: People and resources in the Far North.* Bristol, PA: Bristol & Francis.

Reed, John C., and A. G. Ronhovde. 1971. *Arctic Laboratory: A history (1947–1966) of the Naval Arctic Research Laboratory at Point Barrow, Alaska.* Washington DC: Arctic Institute of North America.

Riedlinger, Dyanna. 2001. Responding to climate change in northern communities: impacts and adaptations. *Arctic 54*(1): 96-98.

Robinson, Michael, and Karim-Aly S. Kassam. 1998. *Sami potatoes: Living with reindeer and Perestroika.* Calgary: Bayeux Arts.

Shapiro, Lew H., and Ronald. C. Metzner. 1979. *Historical references to ice conditions along the Beaufort Sea Coast of Alaska.* Fairbanks AK: Geophysical Institute, University of Alaska Fairbanks. UAG R-268 (processed).

Sturm, Matthew. 1999. Who will be the arctic researchers of the next fifty years? In: *AAAS, Fiftieth Arctic Research Conference Proceedings, 19–22 Sept. Denali Park Alaska,* p. 225. Fairbanks, AK: AAAS Arctic Division.

Weeks, Willie F. 2001. NARL and research on sea ice and lake ice. In: Norton, ed. *Fifty more years below zero: Tributes and meditations for the Naval Arctic Research Laboratory's first half century at Barrow, Alaska,* pp. 177–186. Fairbanks and Calgary: Arctic Institute of North America.

Wolpert, Lewis. 1993. *The unnatural nature of science.* Cambridge MA: Harvard University Press.

Figure 5-1. A Yupik whaling crew cruises in a skin boat at Pugughileq, Southwest Cape. (Photo by Chester Noongwook, 1998).

5

Watching Ice and Weather Our Way:

Some Lessons from Yupik Observations of Sea Ice and Weather on St. Lawrence Island, Alaska

Igor Krupnik
Arctic Studies Center, Smithsonian Institution

This paper discusses the organization and preliminary results of a joint effort in locally based ice and weather documentation, initiated by scholars and Native residents in two Yupik villages on St. Lawrence Island, in the northern Bering Sea.[1] There is a growing interest across the science community in the ways indigenous arctic residents observe and document climate, sea ice, and weather phenomena in their daily life. There is also a mounting pressure on polar researchers to incorporate data and observations from northern residents into scientific models of global warming and arctic environmental change. Such local observations and monitoring practices are gradually gaining recognition as a valuable source of data for studies of arctic climate and ecosystem fluctuations. More efforts are needed, however, to ensure the next transition: from simple recognition to substantive partnerships built on data sharing and resource exchange.

Many efforts to promote such partnership will remain less than effective until the nature of the "other" knowledge, and the ways it is built and transmitted, is properly understood. A productive exchange between the two sets of observations can be feasible only when and if scholars learn more about how northern people watch their environment. Therefore, the collection and thorough examination of local observations, done by Native people themselves, and from their cultural perspective, is crucial for further progress. More and more researchers argue that only by this approach we may learn about the procedures, "matching" techniques, and limitations in efforts to bridge local expertise with the data used in scientific models of arctic climate change (Bielawski, 1996; Fox, 2000; Huntington, 2000a; 2000b; Huntington et al., 2001; McDonald et al., 1997; Riedlinger, 2001).

To test such a transition in a practical way, the U.S. Marine Mammal Commission (MMC) and the Yupik communities of Savoonga and Gambell on St. Lawrence Island, Alaska, agreed to run a pilot project to document local observations of sea ice conditions off St. Lawrence Island "the Yupik way." This was a one-year effort (from fall 2000 to summer 2001) to document local knowledge about arctic weather and sea ice, following Native observations practices and making records in the Yupik language, with subsequent English translation. The project was named *Watching Ice and Weather Our Way* and it was supported by a small contract from the Marine Mammal Commission.

Although the main task of our joint venture was to create such a bilingual (Yupik-English) record of sea ice and weather monitoring (see earlier reports in Krupnik, 2000a; Krupnik and Huntington, 2001), many more materials have been produced during 2000–2001. These include a bilingual illustrated dictionary of Yupik sea ice terms; transcripts of interviews with St. Lawrence Island elders; historical accounts of former ice conditions off the island; comparative reviews of Native observation records and the data collected via satellite sea ice monitoring, etc. Most of these data are now assembled in an illustrated bilingual project report under preparation (*Yupigestun Kellengake,* 2002). This paper reviews the history of the project and its major activities. It also offers an anthropologist's view on the relationships between scientists and Native experts, and between what are commonly called "academic" and "local" science. I believe that the lessons we learned are worth sharing as those relationships continued well beyond the original observation period.

How This Project Originated

The project *Watching Ice and Weather Our Way* originated as a joint initiative in addressing the issues of arctic climate change and in response to concerns that are now shared widely among polar scientists and arctic residents (cf. Ford, 2001). Both arctic scholars and Native subsistence users have, for years, been documenting changes to the arctic climate and natural environment. The key task of the project was to research the ways these two sets of records can be culturally "translated" and mutually "calibrated," so that they become comparable sources of information to both constituencies.

The project originated from a recent interdisciplinary workshop, one of many similar initiatives aimed at bridging the scientific and Native perspectives on arctic climate change (see reviews in Huntington, 2000a; 2000b; Krupnik, 2000a). In November 1998, Caleb Pungowiyi, then the Head of the Eskimo Walrus Commission in Nome, wrote a letter to the U.S. Marine Mammal Commission (MMC) in Washington. The MMC, established under the Marine Mammal Protection Act of 1972, acts as an independent source for policy and program guidance to Congress and the Executive Branch on all issues affecting marine mammals. Pungowiyi's letter pointed out that, despite all the attention being given to climate change studies, few scientists were taking seriously the observations of hunters and elders from Native communities. Pungowiyi, who was himself born in the Yupik village of Savoonga on St. Lawrence Island, argued that such observations clearly indicated significant recent shifts in the characteristics of sea ice, marine mammals, and other aspects of the arctic environment (see Huntington, 2000a:1; Pungowiyi, 2000).

In response to this letter, the commission organized a special workshop on scientific and indigenous observations of change in sea ice and the arctic environment, which took place in Girdwood, Alaska in February 2000. The three-day workshop brought together some fifty scholars and Native experts, including local hunters and environmental specialists from several Alaska communities. The ratio of scientists and indigenous experts was almost 50:50, and it represented a broad spectrum of northern communities, probably the largest ever at so many recent "arctic climate change" meetings.[2] The background conference report (see Huntington, 2000b) included several papers authored by arctic scientists and Native experts, and featured their views on the ongoing shifts in polar sea ice ecosystems as the result of global warming.

During the workshop, Native elders and hunters shared their observations of the sightings of new or unusual wildlife species, of physical changes in the habitat, unusual timing of animal migration, and patterns of animal behavior due to changing ice and weather regimes (see summary in Krupnik, 2000b). In addition, a few focused and more substantial interviews were recorded after daily sessions (Krupnik, 2000b). From those statements, comments, and interviews it became obvious that people in many Alaskan communities clearly saw changes in weather, ice, and marine biota taking place, particularly in recent years.

It also became clear to many participants that bridging local and scientific observations of sea ice and environmental change requires special and diligent efforts. It cannot be achieved via usual round-table discussions that hardly go beyond general summaries, occasional statements by Native participants, or anecdotal "dipping" into Native expertise. Long-term documentation of arctic climate and sea ice change by Native residents themselves should be created that can be shared and analyzed on a comparative basis. The workshop endorsed activities to assist Native people in building such records It also encouraged the participants to seek projects aimed specifically at documenting the ways Alaska Natives observe and communicate sea ice and weather conditions in terms and language(s) of their own (Huntington, 2000b: 7–8).

Two workshop participants from St. Lawrence Island, Alaska, Conrad Oozeva from Gambell and George Noongwook from Savoonga, agreed to explore the possibility of launching observations of sea ice and weather conditions in their respective communities. It was agreed that such observations should be carried out by local monitors and recorded in the Yupik language. This would follow a more traditional way of ice and weather monitoring and would use all the terms and realities that are important in Yupik subsistence activities and worldview. The Yupik text would then be accompanied by a parallel English translation, so that scientists and other Alaska Natives (particularly, younger students) could use observations from two villages for a broader comparative perspective on changes in the Bering Sea/Western Arctic region.

The MMC agreed to support such a pilot documentation project on St. Lawrence Island, from the arrival of new sea ice in the fall of 2000 until break-up in early summer of 2001. By the summer of 2000, a blueprint was prepared for a much larger program in sea ice and weather observations in several Alaska Native communities in the Northern Bering Sea region. It was argued that small teams of local observers that included one or two senior hunters and elders, with younger assistants, should be set up to record their observations during the winter season of 2000–2001 or in 2001–2002. Each village team would decide for itself how to record its ice and weather monitoring, based upon local conditions, cultural tradition, and community's annual subsistence cycle. Preference should be given to recording in Native languages, either Yupik or Iñupiaq (with the subsequent English translation), so that the observers could use all the terms and words important in Native culture. When observations were recorded in English, special efforts would be made to include as many Native terms as possible—terms used by hunters and elders as they monitor the ice and weather in the course of daily subsistence activities.

Thus, the effort was deliberately set up to avoid any preliminary scientific "framing" of Native observation. We never expected people to check their home thermometers and barometers or to produce their village "weather reports." Too often, efforts to record indigenous ecological knowledge eventually devolve into a process in which Native participants are pressed to follow standard formats for local ice and weather monitoring, to use daily observation sheets prepared by scientists, to draw charts, maps, and other forms typical of scholarly research. In our project, the goal was to create a record that the observers could share easily with their parents or with other village elders. We wanted it to be done in the way that "one used to report ice and weather as a young person in the old days," in order to document the cultural specifics of Native observation, to the greatest extent possible.

We sought, however, to get a comparable result for each community: a body of written data organized as a village journal made of short daily (or weekly) entries. Such bilingual village ice and weather journals would be illustrated by drawings and sketches produced by community members to show specific local ice and weather phenomena. Younger people, particularly future Native students, could then use such bilingual village journals as educational materials, in order to learn or remember how elders watched the ice and weather as their fathers once taught them to do. Having elders and experienced hunters as observers would guarantee that the comparable sets of village data and age-based knowledge of ice and weather would be recorded.

We viewed our project *Watching Ice and Weather Our Way* as contributing to broader discussions of arctic climate and ice change through properly collected local observations done by the northern residents themselves and from their cultural perspective. In considering such materials as the project's critical input, we hoped to frame our study as much as a heritage and cultural preservation effort on behalf of participating communities as an exercise in scholarly data collection for further research. This approach helped us recruit many highly dedicated team members and generated enthusiastic local support. Our project participants expressed it strongly in their own words:

> This project is so important, because we have thought many times about documenting our knowledge about ice and sea ourselves but we did not know how to get it started. This is very important for our children and grandchildren, and for all our next generations, so that they would know what we used to do from our fathers and forefathers. I got all my knowledge from watching after my grandfather Nunguk and after many other experiences with people of my early days. So, when these words are written down, they will keep on going for a long time for younger people, as long as this book will last. (Chester Noongwook, Savoonga, 2001)

> All I wanted was to pass to our young people what I learned from way back to the present time and what were some changes in our way of life. As I see it, we haven't changed spiritually and we still respect our ancestral people. It would be good to have to pass our culture from their time to our time and to feel the same, like they had. Our way of life is something that has to be learned specially, like nowadays people get their knowledge in writing and reading at school. (Conrad Oozeva, Gambell, 2001)

> When I met with Igor Krupnik in December 2000 and he invited me to joint the Yupik sea-ice observation project, I told him that I am busy and that I have to take advice from my father Nelson Alowa (Qagaqu), who was sick at that time. When I told my father about this project, he said that many young people of today do not know some ice and weather conditions. Now it may be helpful to put it in written form, so they would be able to read it. "Think about your nephew and your sons, your future grandchildren", my late father said. I said to him that women were not allowed to be involved in men's gathering of data and in hunting trips. He just laughed and said, "I am glad you remember that but also remember, how many times I took you out hunting. This sure will help others if it is put in written words, so more young adult hunters will know what terms were used in the earlier years." (Christina Alowa, Savoonga, 2001)

In October 2000, Conrad Oozeva started his weather and ice documentation "the Yupik way" in the village of Gambell on St. Lawrence Island. In December 2000, I traveled to the second island community, Savoonga, where George Noongwook—another Girdwood workshop participant and the Vice-President of the Alaskan Eskimo Whaling Commission—agreed to supervise local ice and weather documentation. After reviewing the proposal, the Savoonga Whaling Captains Association endorsed the project (Figure 5-2). Chester Noongwook, one of Savoonga's most respected whaling captains and weather experts, volunteered to do the observations and to make daily documentation of ice and weather conditions in his native Yupik language. On December 8, 2001, I put down Chester's first daily weather story in English (see

Appendix 1); shortly thereafter, Christina Alowa, a former Yupik language schoolteacher, agreed to assist in recording Chester's notes in Yupik and to help translate them into English.

Setting the Framework

To fulfill its tasks, the project was set up to document ice and weather conditions for at least one full "sea ice year," from the fall arrival of slush ice until the next summer breakup. The observation record was to be organized as a series of regular entries so that it could be later matched with daily and weekly weather reports, sea ice satellite images and charts, and other forms of scientific data. The project also sought to document local terminology for specific ice and weather phenomena used in each community and to explore local information about past ice and weather events that is stored in language and cultural materials, proceedings of elders' conferences, and other sources. Unfortunately, we did not succeed in generating funds for the bigger effort aimed at covering several communities; so our study was eventually limited to the two villages of Gambell and Savoonga, both on St. Lawrence Island. This pilot project in 2000–2001 was funded by the MMC, with additional assistance offered by some other agencies (see Acknowledgements).

Figure 5-2. Members of the Savoonga Association of Whaling Captains discuss the proposal of sea ice observations "the Yupik way" (photo by Igor Krupnik, December 2000).

St. Lawrence Island is located in the northern Bering Sea, less than 160 miles (250 km) south of the Bering Strait. It is the largest island in the Bering Sea and the one with the highest Native population, both historically and at present (see Hughes, 1984). Despite its position about two hundred miles (320 kilometers) below the Arctic Circle, between 63°00' and 63°38' N, this is fully an arctic environment of the treeless tundra, with the sea covered by sea ice for several months of the year. The island officially became a part of the U.S. territory in 1867, although both of its present-day communities—Gambell at the northwestern tip (Figure 5-3) and Savoonga near its northern tip (Figure 5-4)—are situated closer to the Asian than to the North American shore. Barely forty miles of water separate Gambell from the nearby Chukchi Peninsula, Russia, and the tops of the Asian mainland mountains are visible from the village on clear day. The two existing communities have an almost equal population of about 650. Local residents, with the exception of a handful of contract schoolteachers, are mostly Native Yupik Eskimo. They call their villages Sivuqaq and Sivungaq respectively, and name themselves Sivuqaghhmiit, when speaking in their Yupik language. The official town names, Gambell (given in 1899 to honor its first white missionary-teacher who drowned in 1898) and Savoonga, are commonly preferred when talking and writing in English.

Both towns have similar economic profiles. Most local residents depend upon year-round subsistence hunting (primarily for walruses, whales, and seals), fishing and plant gathering in summer, and various outside sources of cash, including sales of carved and excavated prehistoric ivory. The two communities are closely related by blood, common history, shared language, and extensive family ties.

Despite their geographic proximity, winter conditions in Gambell and Savoonga are quite different. Savoonga is wide open to the north, from where winds and currents bring floating ice that eventually consolidates into a solid ice cover that commonly stays from December until June. Gambell is located at a rocky cape that is exposed to winds from several directions and that produces turbulent conditions in the sea as the currents mix off the cape. Although sea ice arrives here roughly at the same time as in Savoonga (and often earlier), it never becomes as solid and stable because of strong currents and winds. Breakups, ice leads, and patches of open water are the most common phenomena in Gambell throughout the winter; and spring breakup takes place here at least a month or two earlier than in Savoonga. Native elders can talk for hours about the differences in sea ice conditions in two communities located barely 40 miles (66 km) apart:

> Gambell currents, to my thinking, are way much stronger than our currents here, because they go from both sides. Their cape is so different than the place we have here. Here, our currents are much weaker and the wind pushes the ice much stronger to the land, pressures it. Gambell ice is smoother, because there is always some open water there to the west and southwest. Between Gambell and all along northern shore it is very heavy packed ice in winter. So, our ice is

Figure 5-3. Gambell, St. Lawrence Island (May 1999; Photo by Igor Krupnik).

Figure 5-4. Savoonga, St. Lawrence Island (May 1999; Photo by Igor Krupnik).

indeed very different than in Gambell. We do not have that many terms and our terms are often different. This ice here is piled up real high; it makes high ridges along the shore. By the time this ice is chipped out from here in spring, it always moves eastward from here, *uusqayaak* (Chester Noongwook, October 2001).

The ice and weather observations in Gambell were started, as mentioned, in October 2000 by Conrad Ooozeva (his personal Yupik name is Akulki). Born in 1925, he is deeply respected for his expertise in local ice conditions and in traditional Yupik ice terminology. In the 1980s, he compiled a list of some ninety Yupik terms for the various types of sea ice and ice formations, which is included as a part of Yupik language cultural curriculum at both Gambell and Savoonga village high schools (Yupik Language, 1989). He also was a whaling captain of some twenty-five years, and he hunted on sea ice for more than sixty years of his life. Conrad's wife, Elinor Oozeva (Miqaghaq), came as his natural partner in documenting his observations in both Yupik and English. Jennifer Apatiki of Gambell typed both the Yupik and English texts. Conrad's observations were done as weekly, or were bimonthly "block-stories" that he taped in Yupik. Unfortunately, the recordings after late February 2001 have been lost or, at least, were not transcribed.

In Savoonga, the team of Chester Noongwook (Tapghaghmii, born 1933) and Christina Alowa (Sunqaanga), chose to make their observation records on a regular, almost daily basis. The observations were started on December 9, 2000 and ended with the total disintegration of sea ice off Savoonga on June 12, 2001. The six-month record has two substantial gaps: from mid-December 2000 to early January 2001 (because of the passing away of Nelson Alowa, Christina's father) and from mid-April to early-May 2001, when Chester, together with other adult men of Savoonga, was at the village spring whaling camp at Pugughileq, near Southwest Cape. Despite these two gaps, the Savoonga record of some seventy pages may be one of the largest original handwritten texts in the St. Lawrence Island Yupik language. During the later portion of spring observations, Christina started to make pencil sketches and drawings to illustrate Chester's daily ice and weather stories. The Savoonga team was later joined by a local ivory carver and artist, Vadin Yenan (Saywenga, born in Chaplino, Siberia), who made several dozen pencil drawings to illustrate various forms of sea ice. George Noongwook (Mangtaaquli) did style editing and translated the Yupik record into English during the fall and winter of 2001 and 2002, while Valerie Noongwook (Kisighmii) typed the Yupik and English versions of the Savoonga observations. Further editing was done by Igor Krupnik in October 2001 and by Henry Huntington, in November 2001.

Since the Savoonga and Gambell teams followed different patterns of documentation, their records of ice and weather conditions during winter 2000–2001 are quite distinct and cannot be compared on a daily basis. However, both Chester Noongwook and Conrad Oozeva prepared an extensive summary of their observation records—a general review of the unique winter of 2000–2001. I recorded Chester's overview in

English, whereas Conrad put his summary in writing, also in English, himself. Several months of observations produced an unprecedented set of data: literally, dozens of pages of records in Yupik (with English translation). These daily or weekly entries are full of Native terms, detailed explanations of ice patterns, references to rapid shifts in ice and weather conditions, migrations of marine mammals, and local hunting activities. These records by two experienced local elders were later reviewed by Lewis Shapiro, a Bering Sea ice specialist from the University of Alaska Fairbanks (see below).

Winter of 2000–2001: What Did We Learn from One-Year Observations

The winter of 2000–2001 was unusual on St. Lawrence Island and in the northern Bering Sea area in general. Local people have long reported a delay in fall sea ice formation, which now commonly occurs in early December—instead of late October or November as in the "old days." This past winter, however, was indeed very special. The sea ice was not firmly established until mid-late December; then it broke up in late January and February. In the 1940s and 1950s, solid winter ice covered the entire northern Bering Sea for months, from St. Lawrence Island all the way to Nome and the Bering Strait. In recent years, this solid ice is often broken by leads, ice cracks, and patches of open water, even in the middle of winter, as our observers record:

Figure 5-5. An aerial view of Gambell, February 2001. The shore is ice-free from both sides of Cape Sivukak (Photo by Igor Krupnik).

> The first sea ice that I noticed [i.e., that left a deep impression] was when going on a single-engine airplane. [It was] on my first trip to Nome, on April 25, 1943. We flew from Gambell toward King Island. From the end of the Island to the north and northeast, right up to Nome, there was just solid ice all over. There were but a few breaks—not open leads but just lines across the ice in between the mainland and the island.
>
> This I believe, maybe at that time it was always like that, because back then we had more fair weather and cold weather. Today—as Caleb [Pungowiyi] told us about his trip to Nome from the island when he looked down—it was all broken ice everywhere, all the way to Nome. And nowadays it is most often like that (Conrad Oozeva, 2001).

When I flew over the same route in February 2001, there was open water with scattered ice floes all the way from St. Lawrence Island to Nome. In the middle of the arctic winter, Nome's waterfront was ice free and Gambell had no shore-fast ice whatsoever, from both the northern and western side of the cape (Figure 5-5). In Savoonga, large areas of open water were clearly visible across the narrow patch of the shore ice (Figure 5-6). According to the sea ice chart produced by the National Ice Center for

Figure 5-6. The shore at Savoonga, with the darker patches of open water off the island (photo by Igor Krupnik, February 2001).

February 23, 2001 (Figure 5-7), the edge of heavy winter pack ice (0.9-1.0 cover) was positioned some 60 miles north of St. Lawrence Island or at least 300 miles (500 km) further north than its usual position at this time of the year (Figures 5-8). In fact, to the south of St. Lawrence Island, the Bering Sea was essentially ice free (Figure 5-7).

Although the sea ice eventually returned to stay until late spring, this was indeed a special year to remember, as both our key observers recalled:

> Warming also made the difference, because the warm air kept coming. This is the third time in my memory that this happened, but it is also the longest time of all I remember. It is quite usual, as I said, that we have warm weather on the island for 2-3 days in midwinter; even some snow thawing may start, but the ground is still frozen. This past winter, it was really very long period of warming and one in February that made all this difference. It is quite normal to have warming waves in December, even in January, but not in February (Conrad Oozeva, October 2001).
>
> Last winter's snow (of 2000–2001) was so high I thought it was the second highest snow on my memory. At some places, we had up to fifteen feet of snow, particularly near the houses. The other time we had that much snow was about twenty-five years ago. ... Because some of the houses were buried with snow and snow drifts almost half way to the roof height.
>
> As it was snowing heavily, the water was wide open in November and most of the month of December. We had some little sea ice in late November and early December—mostly mixed ice: *siku, kagimlegh, umestaghaq, unenghelnguk.* But then we got wide-open water once again for most of December, and it lasted for couple of weeks or more. This is the second time on my life when this hap-

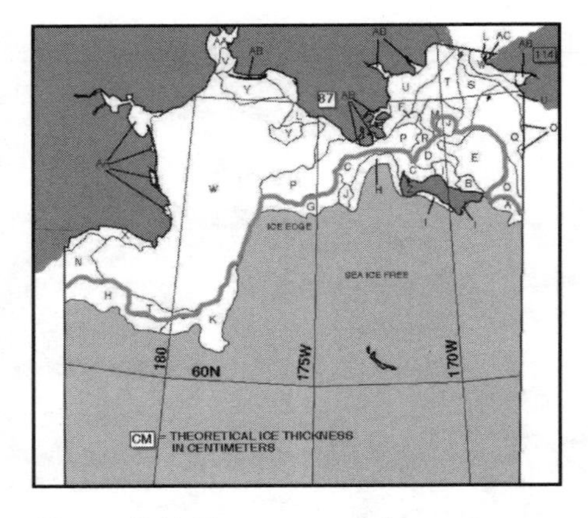

Figure 5-7. Fragment of the sea ice satellite chart, Western Bering Sea area, February 23, 2001

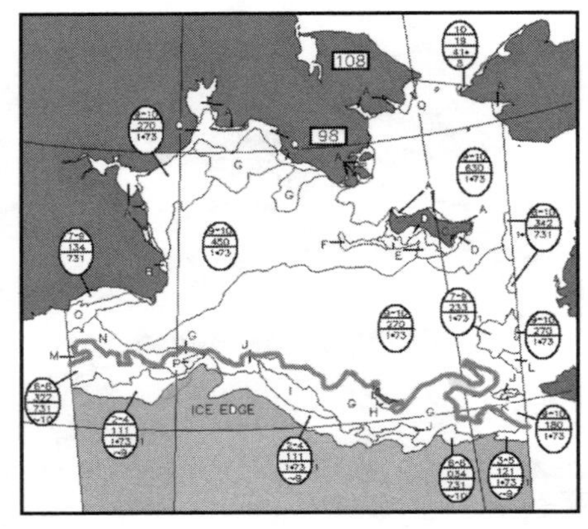

Figure 5-8. Fragment of the same chart for February 22, 2000.

> pened, because usually we have a solid, ice-covered Bering Sea at this time (mid-late December). This past December (2000), the water was wide open, and as far as I could see, there was a bluish color (on the sky) to the northwest to west, toward Gambell. It indicated open water over Gambell and even further than Gambell. Straight north and northeast from here, the entire area was whitish in the sky—this indicates some ice in the sea but very, very far away.
>
> ... Then, the second ice came in at the very late part of December. The walruses were very close to this ice, so the hunting was much better than years before. We even saw a few bowheads off the village, but we did not try to follow them, because it was already heavy ice. The ice was almost similar to the first one—good ice. People could walk on this ice.
>
> Then in the mid-part January, we got wide-open water again. The shore here was almost free—just the gravel and the sand, no ice. We did not get the fast-shore ice until about our third wintering (ice coming) in February; so, there was no ice near the shore. The ice moved up north as far as one could see. If the weather is clear, I often go up to the mountains to look for the ice, usually, once or twice a month in winter. That past January, I also went to the mountains twice but could not see anything in the sea, except for few whitish spots, glares *(qupaq)*. That indicated that the ice was very far away from shore.
>
> It was mostly southerly winds, also southeast and southwest, for a couple of months in winter, moving back and forth, back and forth. This last February, open water lasted for a couple of weeks or even longer. This is my first time I remember the sea that way. It might have been some open leads here and there, some openings, cracks. But normally we have closed ice at this time.
>
> Finally, the Bering Sea was covered with winter ice for the third time after mid-February. This was already good winter ice. Late February, March, and even early April were quite normal here, compared to usual years (Chester Noongwook, October 2001).

Although the ice returned in late February and the winter resumed until at least late April, the unusual course of events triggered a chain of impacts upon arctic wildlife, particularly on marine mammals. In his summary of winter 2000–2001, Chester Noongwook reported that:

> In the last week of March we started to move to our (spring whaling) camps at Pugughileq, at the Southwest Cape. This was our normal time. When we came there, there was no shore-fast ice and not much of other ice. Just open water on the southern side (of St. Lawrence Island). I think, it was because of these big waves from south, we call them *ughqaghtaaq*. We usually have some shore-

> fast ice in there; so, this year was different compared to years before. Otherwise, it was like a usual first half of April: normally cold, some snow, very windy, with southerly and northerly winds.
>
> The bowheads were there and there were many of them, as usual. But it was very hard to get to them because of the weather. Weather was always a problem: windy all the time, high waves, young ice forming on the water, like *sallek* [thin young ice]. I think it was because there was too much open water—we could see the packed sea-ice but it was too far away. For a good whaling weather (in spring) we need a fast-shore ice for about a mile or less. ... So, this past spring was not very good for whaling. We were at the camp for about a month, maybe a little longer. ... In that spring, we got two whales on open water: one very close to the whaling camp, about five or six miles and we got the second whale well far off, about thirty miles from the camp. Not very good hunting because of weather, but nothing very special. The whales were also like in a normal year, but they were sounding all the time. Maybe, they were disturbed by the noise of the snowmobiles over the open water.
>
> ... When we came back to Savoonga from the camp in mid-May, it was ice and open water here. The fast-shore ice was broken up in early May or mid-May—about a week or even two weeks earlier than usual. I guess it was because of so much open water around. This open water in spring is the main thing that affects the ice. Spring walrus hunting here lasted much shorter, because the ice moved up north very quickly. Hunters had to go much further to get the walruses in late May, and then they just stopped going, because the ice was so far away. We continued our ice observations here until mid-June (June 12), but the ice was already gone here, the sea was wide open, as far as one could see.

The same thing happened in Gambell, where, because of the early spring break-up, spring hunting season for walruses was much shorter than usual:

> After that, we started walrus hunting, in early and mid-May. Mostly in the southern area off Gambell, where there was open water. ... There was still snow on the ground in the village, but it already started to melt down. The ice was moving up north. It opened up about mid-May, and then it cleared up around Gambell, both on the northern and western side. This happened because of very strong winds from the south in May. Even the shore ice on the northern side was broken away and gone far away, as far as you can see. But we could still see the edge of the ice from the top of the Gambell Mountain. When the wind calmed down, several boats went hunting up north towards the ice for walruses and *maklaks* (bearded seals). They had to travel some twenty miles to get to the ice front in mid-May; then more and more as the ice moved further north. By late May there was almost no ice around. It would be possible to travel to Sibe-

ria in mid-May—we would have no problem to get there at this time (Conrad Oozeva, October 2001).

Chester Noongwook also acknowledged that during the spring and summer months of 2001, far fewer gray and minke whales were observed off St. Lawrence Island than are usually seen. On the other hand, there were numerous reports of an unusually high number of polar bears seen on the island in spring and summer time. They were obviously left behind by the fast retreat of sea ice and had to survive on local food resources until the next winter.

The unusual winter of 2000–2001 offered a stunning model of what the ice conditions off St. Lawrence Island may look like if the current warming trend continues. It also presented a good illustration of the potential impacts on the Native economy and annual subsistence cycle. Here, the consequences were far less dramatic than one might expect from the ice charts that showed heavy winter ice pushed some three hundred miles farther north than usual, with open water all around St. Lawrence Island in the middle of the arctic winter. Native spring whaling in 2001 was only moderately successful (Gambell and Savoonga got two spring bowhead whales each) and far fewer walruses were killed in May and June of 2001 than during the last several years. As a result, by fall 2001 both villages experienced a shortage of walrus meat and very little frozen *mangtak* (whale skin with blubber) was left to last until the next spring hunting season of 2002. However, this shortage was quickly ended, when, due to much lighter ice conditions, two more bowhead whales were killed off Savoonga and Gambell, each in late November and December 2001. The Gambell whale was a particular novelty, as no fall bowhead whaling ever took place in Gambell in the memory of today's elders. Thus, the changed ice and weather situation obviously puts pressure on the established round of annual subsistence activities and its overall productivity. But it also offers some new resource opportunities, which Native residents are quick to exploit.

Yupik Ice and Weather "Watch": Some General Remarks

This section reviews some individual comments by the project participants and other local elders and let the voices of the Yupik people be fully heard. By no means does it intend to be a summary of Yupik environmental expertise ("traditional ecological knowledge") related to the sea ice and weather patterns. For more extensive and eloquent representation of how the Yupik people themselves perceive and discuss their environmental knowledge the reader is directed to the published contributions of the Girdwood workshop (Noongwook, 2000; Pungowiyi, 2000; see also Krupnik, 2000b); transcripts of the proceedings of local Elders' Conferences (EHP, 1982), and to published collections of oral stories from St. Lawrence Island (Akuzilleput Igaqullghet, 2000; Sivuqam Nangaghnegha 1985-89; Silook, 1976). Several earlier studies of sea-ice and weather-related knowledge in other Inuit communities across the Arctic offer valuable comparative data (see Freeman, 1984; Lowenstein, 1981; McDonald et al., 1997; Nakashima, 1993; Nelson, 1967; 1981; Riewe, 1991).

Another purpose of this section is to offer an anthropologist's perspective on how knowledge about ice and weather is collected, transmitted, and regarded by members of the Yupik society and how Yupik monitoring for sea ice and weather is organized. This is, again, not a comprehensive list of the basic features of "indigenous ecological knowledge," but rather some more obvious observations on the nature of "other science" that come from our project.

Experience with sea ice and climate conditions is accumulated through generations of observation and daily encounters with the moving ice, storms, currents, and rapid weather changes. It is held in very high esteem in northern communities, such as Savoonga and Gambell. Elders commonly refer proudly to the "traditions of our ancestors," whereas the younger people express the same feeling in more modern terms:

> Sivuqaq, they call it St. Lawrence Island, since it is our home, we know about its ice and its movements. We have this knowledge about the ice that was taught by our ancestors. Our ancestors studied all about life here. They were always studying on things like that. So, they knew what to expect, and it always was right. Some years it can be different, so it was put in their minds like it was written (Frank Oktokiyuk, EHP, 1982: EC-GA-82-072-T#3)

> We have our Native versions of knowledge about all these wind and ice patterns, and it has been passed down through the generations. And people, like Conrad [Oozeva], they have lived with this knowledge all their life. I mean they have knowledge on everything that pertains to the ecosystem, to the iceberg change, to the resources, and marine biology. I guess we have our own versions, in terms of understanding the comprehensiveness of our environment and all the necessary terminology. For every conceivable condition of snow and ice pattern, which is either annual or through generations. And we have stories of all these changes based upon comparative experiences and observations (John Waghiyi, February 2000).

Watching ice and weather is a critical task for every arctic community and the key factor that guarantees its prosperity and survival. This is why it is the most common pursuit and still one of the most respected duties of all men, particularly of the elderly people. It is a lifelong and a twenty-four-hour passion, since there is always someone in the community checking weather, sea, and ice at any given moment. In a critical time—when men go out hunting, during the spring whaling season, or when the weather is shifting rapidly—several people spend hours scanning the horizon and discussing signals (indicators) related to the status of weather and ice. It used to be this way always, as people remember, and it is still a common practice in both communities:

> That high land, *Aatnequsiq,* at the point northwest of Gambell [village], belongs to the elderly men. When all the men are out hunting, the elderly men

used to get out early morning and go there to watch. They would watch closely for the hunting men. The ice can move away so dangerously; so, they watch for that, when the ice just gets there. They call this early ice *akitaaghhaak* (Steven Aningayou, February 1982, EHP: EC-GA-82-71-T2)

I took my grandchildren down to school (about 9 a.m.) and I looked for water and ice near the school site at the shore—whether there is any open water out there (far at sea). I could tell it because there was some dark(ness) in the sky far over there; that dark sky is over open water. I stayed at the store for some time to get information from other people, who were there earlier in the morning. We always have a few people of my age gathering at the store, the side that faces the water and the beach—they just stay there for some time, watch the weather and ice, and talk (Chester Noongwook, February 19, 2001).

However, beyond these general attitudes, several specific factors are to be considered. *First,* whereas many people are knowledgeable and almost everyone is watching ice and weather, some are reportedly better than others. Both our senior observers, Conrad Oozeva and Chester Noongwook, are deeply respected for their special knowledge and weather forecasting skills. This is, again, a well-established pattern, as such experts are well remembered, their names praised through the generations, as seen from the following story:

Clarence Irrigoo: Who was an elderly man at Pugughileq, my father and others often told about?

Frank Oktokiyuk: Gugwiingen. Gugwiingen, our "weather bureau."

Clarence Irrigoo: He knew it (weather) better than weather bureau does. They would even be asking him about the weather, when they want to know. He would just look around from his house door way up at the sky and tell the others to wait a while and not (to) go on hunting trips. He would say, "It may not turn out." And they would just obey, or did as he said. And sure enough, it soon would not be fit for hunting.

And other times, when other men thought (it was) not good, he would say, "All right, see if you can get out now; it may be all right." By knowing, see, he knew it all from watching (February 1982, EHP: 82-072-T#3).

Second, the most remarkable feature of the Yupik watch of weather and sea ice conditions is that it is primarily wind- and ocean current-oriented—unlike the scientific (that is, instrumental) observation, which is first and foremost focused on changes in temperature and atmospheric pressure. Temperature is indeed of low importance to Native observers, unless it shifts rapidly or is highly unusual for the season, and atmospheric pressure has no special meaning in Yupik terminology. Atmospheric pressure did not become a factor until weekly and daily weather reports started to be transmitted over the radio on a regular basis. Nowadays, many people, particularly the elders and active hunters, have indoor barometers, which they duly watch. Nevertheless, there was no reference to atmospheric pressure (or its change) in any record made by our Savoonga and Gambell observation teams.

Instead, Yupik hunters use an extremely sophisticated system of wind terminology that identifies some ten or twelve types of winds by specific direction and other features. Each wind is known to bring a certain type of weather, snow, or ice movement. By identifying or referring to the wind, an observer can make a quick judgment of the situation and even make a basic forecast of upcoming conditions. Chester Noongwook started his record of observations in Savoonga (and my first orientation in the Yupik patterns of weather watch) by listing the most common winds in his native village:

> We have several winds here in Savoonga: *Aywaa (Aywaapik)* is a direct north wind from the sea. *Nakaghya* is northeasterly wind, it comes from Nome. *Kenvaq* is a northwesterly wind; this is the old name, and we now call this wind *Naayghiinaq* ("that from Siberia"). There is also another northerly wind, *Quutfaq,* that can come from anywhere between northwest and northeast. *Asivaq* is a direct east wind; *Ikevaq*—south wind from the island to the sea; with another wind from inland, *Ikevaghlluk,* from southeast. We call the southwest wind *Tapgham Ketaanganeng,* because it comes from *Tapghaaq,* which is the place on the coast between Savoonga and Gambell. There is also west wind, *Pakfalla,* that comes from the sea.[3]

Conrad Oozeva produced a similar list of winds for Gambell, with the extensive characteristics of affiliated weather conditions (Yupik Language, 1989: 19–20; see Appendix 2). Very similar lists of winds, with assigned weather regimes, are known to the Yupik hunters in nearby Siberia (Krupnik, 2001: 398–402; Vakhtin, 1988: 30). Similarly organized wind classifications are used in ice and weather prediction in many other Inuit communities across the Arctic (cf. Nelson, 1967: 41–53; 1981:7–10).

A reference to the wind is commonly the first and most important feature of the daily observation, and it typically leads to many conclusions:

> In the morning look up at the clouds, if any, and observe 360 degrees—get the idea of wind direction, wind speed from observing the cloud conditions. From these observations you can determine where and how to travel, whether to go by boat or on land, or whether to stay because of weather. In windy but clear conditions, without the snow or other visibility restrictions, we tend to stay within the village boundaries or in close proximity of the village, even if we travel along the coast.

> You could often determine the forecast for the next day from observing the high elevation clouds. Generally very nice conditions but could get windy if it is easterly (Chester Noongwook, December 2000).

It comes as no surprise that most weather records by Chester Noongwook and Conrad Oozeva, as well as many other documented stories by local hunters, commonly start by referring to the prevailing wind direction:

> *February 15, 2001: Today's Weather*
>
> Today's weather is still the same, with the wind from the southeast *(ikevaghllugmeng)* generally a little east of Mount Ateq; but there were some boats that went out hunting today. It has cleared up as the day progresses. It has been weeks since the wind has blown from the southerly direction (Chester Noongwook, February 2001).

The next crucial factor to pay attention to is the ocean current(s), the high tide-low tide cycle, and whether the sea is calm enough for people to go out hunting:

> The wind direction (today) is northeast *(Nakahgya),* something between twenty and twenty-five knots. I can feel it by my face: I always measure the wind by face, because I used to be airline agent in Savoonga for many years.
>
> The sea is rough, big waves. If this wind continues for two or three days, it will bring the ice on from the north. Regular sea ice that was blown out earlier by the south wind *(Ikevaq).* ... This is a bad weather for hunting—not very bad but rather dangerous. Too rough to go in boat and the wind is too strong: it won't bring killed animals ashore. The current is always moving eastward or northward, and in a few hours there'll be low tide. (Chester Noongwook, February 19, 2001).

This combination of critical indicators creates a very solid pattern of weather monitoring, which consists of certain established steps or actions. In acknowledging Chester Noongwook's effort to document these Yupik monitoring practices for the younger generations, I call it "Chester Noongwook's Rules of Weather Observations":

> First thing: get out early in the morning, and check the wind and the sky conditions, whether the sky is cloudy, and also whether it is cold or warm in terms of your body feeling;
>
> In the old days, we always used to go down to the seashore every morning—to check the ice and weather conditions at the water (sea level), how the current is moving, and where is the tide;

- Always talk to other people about weather and ice conditions, listen to other people's mind to see whether it is good to go out hunting;

- Check for any change in wind and weather condition; we were told to watch out for weather all the time, either we are on the ice or on shore—every hour, every minute or listen to other boats what they are saying.

- Keep watching for any change in water, because of currents or clouds or waves—any sign of water change is very important.

- You can never make a good forecast for tomorrow if based upon today's weather. Better go out and check it in the evening. Make a guess and check it next day: it is better to see whether it is correct or not (Chester Noongwook, 2001).

Third, the use of wind directions, associated with certain weather conditions as key designators, allows Yupik observers to collect and pass on information by highly meaningful environmental packages. A similar practice is recorded for the Iñupiat hunters in Wainwright, northern Alaska (Nelson, 1967: 41-53). Hence, it is not the observation itself that makes an impression of Native knowledge being holistic, intuitive, and multifaceted but rather the whole cultural "package" that is associated with each specific ice and weather term it uses. This makes a critical difference from the scientific (i.e., instrumental) type of weather observation, which is based on following the temperature and pressure "curves," and on recording their current trends. Unlike scientific weather monitoring, the Yupik watch is focused upon specific signs that signal shifts from one phenomenon, condition, or weather and ice regime to a different one that can be defined by a different term. This is the primary motivation for the use of and value in multiple specific terms for every combination of environmental conditions. The more words (and combinations) one knows, the more precise one's observation and forecast can be. On the other hand, as the use of Yupik words for specific patterns of ice and weather by younger people declines and the Yupik terms are replaced by English words, with a different (and often much more simplistic) meaning, the hunter's overall awareness of his environment fades away. This is why Yupik elders are very proud of, and so keen on passing on to the younger people, their extended Native terminology of ice and weather conditions.

Such culturally rich "environmental packages" are used to the extent possible, including the most general characteristic of the overall winter regime. The Yupik people on both St. Lawrence Island and the nearby coast of Siberia (cf. Krupnik, 2001: 396) use the terms "man-winter" *(yuguluni uksughtuk)* and "woman-winter" *(aghnangyaq),* to identify cold, heavy-ice winters and milder, lighter-ice winters, with periodic ice openings (polynyas) respectively. There is an early reference to such practice in the early 1900s (Moore, 1923: 340), and the terms are still in use today:

> Yes, I heard people saying that in Gambell too: "woman-winter," aghnangyaq. When the temperature is above 10°F [in winter], I like it. Because you can tell

> it even without seeing the thermometer. The sea between the ice looks black, no smoke, nothing. (Conrad Oozeva, 2000)

It goes without saying that the winter of 2000–2001 was a model "woman-winter," according to the Yupik terminology, whereas the winter of 2001–2002 has been already named the "man-winter," because of its much colder heavy-ice conditions (George Noongwook, personal communication). As this paper is written in January 2002, the heavy winter pack ice (with 0.9–1.0 cover) extends some three hundred miles (five hundred kilometers) to the south of St. Lawrence Island, or almost four hundred miles further south in the Bering Sea than it was in February 2001. This is almost as far south as in January 1975 and 1976, during the two coldest winters on record (Burns et al., 1981: 784; Cavalieri and Parkinson, 1987: 7,148–151; Parkinson and Cavalieri, 1989: 14,507; www.natice.noaa.gov/pub/west_arctic/bering_sea/bering_sea_west/2002).

Fourth, Native observers are also very keen to document unusual ice and weather patterns, but they have their own ways and means of memorizing and documenting such events. They look for certain and, often, very specific indicators that are meaningful to them, both culturally and individually. To a scientific observer, the resulting story may seem "intuitive" and even eclectic, but it is no less solid, since it is based on the very same practical indicators followed over many years. This is clearly reflected in both Conrad Oozeva's and Chester Noongwook's record of the highly unusual winter of 2000–2001:

> Let me add what I remember in the past on the condition of the weather. First time it happened on my memory, I was a little boy. Maybe, it was in the mid-1930s, because the men like Wamquun, Mangtaaquli, and my dad were still very active hunters. That year, we did have very much the same weather as we had this last year. This is the first time I remember this happening. This unusual weather lasted until the first week of January. Boat hunting was good when the weather would subside. All kinds of *tepaq* (this is various kind of seafood we collect on the beach) were washed ashore. Even the *mamaghwaaqs* (a kind of *tepaq*) were getting soft from moldering. On the west side, the sigughneq (the buildup of the slush ice by shore waves) was formed about fifteen feet high or more. Later that year the weather became quite normal.
>
> The second time it happened in my life was in 1965. Again, in the month of January the weather became very warm. This is the second time I remember that happening. The Troutman Lake was originally frozen and the ice was thick enough. That time, we did not have an airfield, so the motorized airplanes landed on the ice on the lake. Then, somehow, it started raining. The ice around the lake melted on the edges. As it happened, there was one plane left on the lake that had been stranded for several days due to the weather. It was not suitable to fly. Finally, the weight of the airplane broke the thinning ice.

> The airplane was halfway drowned in the water. The snow on the ground melted. There was no more snow elsewhere. Somehow, there was still the *tuvaq* (shore-fast ice) around Aqeftapak (the bay on the other side of the Mountain), although it was very watery. It seemed that the weather would not get cold again. ... And then it really got cold again (Conrad Oozeva, October 2001).

Fifth, despite a very popular perception that Native knowledge is generally intuitive and holistic, Yupik ice and weather watch is not scanning for every environmental signal possible. It is very well organized around a few key factors—such as wind, current or ice movement—and is focused first and foremost upon conditions for maritime hunting and the related behavior of critical game species. Therefore, Native experts usually have a very coherent—one may say, "fully scientific"—vision of the annual sequence of weather and ice regimes, the migration patterns of major marine mammals, and how these two cycles are related. This vision can be articulated by many details and arguments based on generational observations:

> There is much difference in sea ice that I see during my lifetime. Every year the first ice we see is mostly of this iceberg type—the floating icebergs coming from the north, mainly from the month of September. The wind always starts blowing from the north, almost regularly. I think that same high wind makes the ocean flow, too.
>
> And now we have more westerly winds. What I learned from my elderly people, as the days grow longer, the northwest throw [stream] gets stronger in between the island and the mainland of Siberia. But now with those more westerly winds we have more ice on the other side, the Anadyr side of the island, the ice is packed over there (Conrad Oozeva, February 2000).
>
> The ice is coming here (to St. Lawrence Island) in the fall as long as the wind is blowing from the north for several weeks to a month. One way of wintering to start faster, the faster way for the ice to come in, is by a couple of weeks or so with the wind blowing northwest to west. So, when the wind suddenly moves to north-northeast, the ice comes faster here in the fall. I always think that our ice is coming here mostly from the Siberian shore.
>
> It hits the island first at Gambell—so, they get the slush ice before us here, in Savoonga. The small icebergs, *kulusiit,* also come there first. These small icebergs start breaking off the polar ice some time in midsummer or early fall way up north, in the Arctic Ocean. And they start flowing down to come this way, straight down south. When the wind pushes it south, it comes straight here, to our island.

> So long as the wind is from the north, the ice stabilizes for a number of days to a week to a month. If the northerly wind continues, the ice keeps going around the island along its western and eastern side—way down south. And it melts down there. We do not know how far south it goes, at least, I do not know. Our ancestors told us that this ice that goes down south starts melting up there.
>
> In spring the current goes in an opposite direction and this southern ice gets melted and scattered around St. Lawrence Island. So, the remaining ice we have at Savoonga is the only (solid) ice we have here until spring and for the spring time. By the time it is scattered up, it has nothing to support it (from the sides) and the current pushes it up quickly and it melts much faster. This northern ice lasts longer at the northeastern section of the island until it gets melted (Chester Noongwook, October 2001).

It is popular these days to praise the power of Native knowledge on ice and climate change. I would argue that, nonetheless, Native concepts of the relationship between marine mammals, ice regimes, and currents are rather underestimated, since they represent fully developed scenarios and not just "generational observations":

> The walrus hits the island in the fall, when it comes down from the north with the ice moving south. I believe the routing of walruses stay from the very old days; they just follow. These big winter walruses, *anleghaq,* they stay here for the whole winter. I do not know why: maybe, it is a good feeding area for them or maybe they used to stay here in winter in the old days, before we had this village at Savoonga. ...
>
> So long as we have leads of open water early in winter, this is the best time for fall walrus hunting. In November or December this ice also comes with the bowheads. This is early winter here. Still, this weather and ice is not very good for bowhead whale hunting, though we recently got a whale in December right in front of Savoonga. But more often the whales hang near the northeastern section of the island, between here and Northeast Cape. Because a couple of time in the past we have seen bowheads passing in front of Savoonga. When it really opens up (a lead of free water along the shore) they travel from Northeast Cape all the way around to Gambell and then turn south, feeding down south of St. Lawrence Island in winter time. ...
>
> The current we have here moves almost around St. Lawrence Island; it is a clockwise current. Most of the animals we have here follow this clockwise movement around St. Lawrence Island. The bowhead whales do this in spring, when they move from the Southeast Cape to Pugughileq at the Southwest Cape to Gambell, and then through the strait between the island and Siberia. The walruses also follow this clockwise movement; this is why we have walruses always

> after they have them in Gambell. Even the birds travel this way from west to east, but only in the morning. I have seen them so many times flying from the side of Gambell and further northeast. In the evening they come back and they travel counterclockwise, from here back to Gambell area. I do not know, why they do this (Chester Noongwook, October 2001).

Sixth, Yupik hunters—unlike climate and ice scientists—think generally in terms of alternating ice and weather regimes rather than some sort of "average" (or "normal") condition interrupted by periodic oscillations or more unusual "extremes." One of the established alternating conditions is the dichotomy between very cold "man-winters" and warmer "woman-winters." The high tide-low tide currents, north wind-south wind as well as high ice-low ice, and other similar alterations are common or daily occurrences. This prevents people from relying upon a framework of abstract "normal" conditions that feature so prominently in western mind and the way our daily weather and temperature forecasts are delivered.

Unlike physical scientists, the more elderly and experienced Yupik hunters are generally quite reluctant to talk in terms of average meanings or to compare certain conditions with "normal" or to call something "abnormal," even when the current ice and weather are clearly far from "normal" for this time of year. The observation record of both Chester Noongwook and Conrad Oozeva of the unusual winter of 2000–2001 is extremely reserved in its wording. Both men stressed repeatedly that they did not consider those conditions as "extremes," since they have seen it before. This comes as no surprise, as people living in the highly variable arctic ecosystems are quite accustomed to dramatic shifts in their familiar environment. Nowhere in the Arctic is there such thing as a normal or "typical" condition; and it is even misleading to conceptualize the Arctic in such terms (Krupnik, 1993:156–158). Here nothing escapes the process of change, as game and birds, climate and ice, tundra and sea are in ceaseless motion. Therefore, by the time any arctic person approaches old age, he or she necessarily garners memories of numerous events in one's personal environmental "history." From this, I believe, derives the above-mentioned reservation and the sense of recurrent short- and medium-term environmental shifts.

Yupigestun Kelengakellgha Sikumllu Eslamllu

In October 2001, I traveled to St. Lawrence Island to meet with local team members and to prepare a full-size project report. After preliminary talks in Gambell, the team gathered in Savoonga, where we spent two weeks processing and checking observation records, filling the gaps in our data, and making additional interviews with local experts. The aim of this last effort was to transform the 2000–2001 ice- and weather-observation journals from two communities into a more thorough summary of local Yupik knowledge about sea ice, a sort of a "Yupik Sea Ice Sourcebook."

This was probably the first time that a book about St. Lawrence Island was not only produced mainly by the islanders themselves, but was actually assembled and prepared right on the island. We ended up with an illustrated PageMaker file on a laptop computer, accessible for demonstration and desk-printing, but not with the final book, which was to be produced later. Nevertheless, this was a tremendous experience and highly emotionally rewarding to observers, translators, and elders, who generously shared their knowledge. To all the people engaged in the Yupik sea ice observation project, it offered the first chance to see how their stories and records, personal drawings, and old photographs are making their way into a bilingual book soon to come back to their communities as the main outcome of our joint effort.

The volume called *Yupigestun Kelengakelghha Sikumllu Eslamllu. Watching Ice and Weather Our Way,* will be produced by five co-authors—Akulki[4] (Conrad Oozeva), Tapghaghmii (Chester Noongwook), Mangtaaquli (George Noongwook), Sungqaanga (Christina Alowa), and Igor Krupnik, under the general editorship of Krupnik, George Noongwook, and Henry Huntington. It will be a bilingual and heavily illustrated large-size book of some 180 pages, comprising an introduction, five chapters ("parts"), and an extended conclusion. Part One, "Definitions of Marine Ice," is a bilingual dictionary of some ninety Yupik ice terms prepared by Conrad Oozeva in the 1980s and incorporated to the Yupik Language Curriculum (1989) used at both island high schools since the late 1980s. For our volume, Chester Noongwook and Christina Alowa prepared an English translation of Conrad's dictionary, while Vadim Yenan made pencil drawings of all the types of ice identified in the dictionary (Figures 5-9 and 5-10).

Part 2, "Ice and Weather Observations: Savoonga and Gambell," features the full bilingual observation record of 2000–2001 for both communities, accompanied by Oozeva's and Noongwook's general summaries of the 2000–2001 ice year. Part 3, "Yupik Stories About Ice and Weather," presents excerpts from earlier elders' conferences and some life-stories about personal experiences with ice that were recorded in 2001. Part 4, "How We Learned About Ice and Weather," is made of interviews with six of today's Yupik elders from St. Lawrence Island about their own youth training in hunting on the ice and observing ice and weather conditions by following elders and ice experts of the 1920s and 1930s. The interviews were recorded in Gambell, Savoonga, and Nome during my visits in 2000 and 2001.

Part 5, "Old Stories, Recent Memories," presents some historical records of former ice and weather conditions on St. Lawrence Island, with comments by today's elders. Its core section offers records of weather and ice conditions in Gambell exactly 100 years ago, extracted from the journals ("log-books") of local teachers and missionaries during the winters of 1898–99, 1899–1900, and 1900–01 (Doty, 1900; Lerrigo, 1901; 1902). Most of the daily entries in teachers' log-books are rather short and, in a purely "western" pattern, they record primarily the temperature, snow or rainfall, and the wind direction, with a few local details (e.g.: "March 15, 1899: School. 30°, 32°: northeast wind, light; snowed until evening. A number of schoolboys were absent in canoes hunting"—Doty, 1900: 241).

Figure 5-9. Qenghuk, refrozen crashed ice. Pencil drawing by Vadim Yenan (October 2001)

Figure 5-10. Chester Noongwook and Vadim Yenan are working on drawings of the Yupik sea-ice terms (photo by Igor Krupnik, October 2001).

However, exact dates for almost two dozen important events during three annual winter ice-cycles (1898–99, 1899–00, 1900–01) can be extracted from teachers' journals, such as the appearance of first sea ice in the fall, the freeze-up of the lake near Gambell, the formation of the solid winter ice and its breakup in spring, the beginning of spring whaling season, the arrival of first boat from Siberia in spring, the general opening of the coast around Gambell, and so on. There were obviously remarkable variations in winter conditions one hundred years ago—as much as there are today, including days with open water and fairly warm weather in the middle of winter (like on December 30, 1898; January 31 to February 3, 1899, with temperature above freezing; March 13 to 27; 1899; December 31, 1900 to January 2, 1901, with rain, above freezing, etc.). These old journals offer an intriguing comparative sample to the observation records of 2000–2001. They show clearly that even in the "good old days"—before any large human-induced changes in the Arctic and signs of global warming—the winters on St. Lawrence Island were often marked by bouts of snowstorms, warmer temperatures, and unexpected ice breakups, although these swings probably arrived with a different frequency and in a different time they do today.

We hope that when the book is completed, it will follow the pattern pioneered by an earlier volume on St. Lawrence Island Yupik heritage produced in cooperation with the local communities (*Akuzilleput Igaqullghet,* 2000). First, a few laser-printed copies will be produced in the spring 2002 for careful review by the volume contributors and other community members. Then, a more expanded print of several dozen copies can be released, so that the book can be distributed at school and within the communities as a Yupik curriculum and Native heritage material. With additional funding, we hope to have the book printed for other interested readers, to preserve the data of the *Watching Ice and Weather* project as a monument to the dedicated efforts of local people in sustaining their ecological expertise and ways of ice observations for the generations to come.

Matching Scientific and Local Knowledge—Lessons Learned

Through ages of adaptation, Native residents of the Arctic have built a comprehensive and elaborate stock of knowledge about their environment and have developed survival techniques to cope with the constantly changing arctic climate, sea ice, and weather conditions (Berkes, 1999; Krupnik, 1993; 2000b). In the "good old days," there was no "bad weather" and success in hunting and availability of animals was always interpreted as a result of proper human behavior, of following established spiritual guidelines, rites, and beliefs. Today, however, the knowledge and legacy of the ancestors is not bestowed automatically upon everyone who is a Native or who lives on the coast or even is a sea-hunter. Whereas formerly people believed they shared the land and the sea with spirits, hunters of today speak increasingly in terms of "ecosystems," animal "behavior," contamination, and stock "health." They are now used to checking the local radio for daily and weekly weather reports, and to using the sea ice charts and long-term weather forecasts. Hence, the body of indigenous expertise about ice and climate is changing, and will not remain intact for an indefinite time.

The sea-ice observation journals, stories, and personal comments of elders and hunters from two closely related communities on St. Lawrence Island, Alaska, illustrate the enormous richness of what is commonly called "traditional ecological knowledge" (or "local knowledge"—see definitions in Berkes, 1999: 5–8). To the Yupik people of today, this knowledge is their special treasure, the best of their scholarship, and a pinnacle of many generations of experience and achievement in their harsh Bering Sea environment. To scientists, these papers and records offer an invaluable new vision on how changes in weather, sea ice, and marine life can be documented by "another form" of observation and expertise.

Of course, what has been recorded during the project is just the tip of the iceberg of what is actually known about sea ice and weather by the people of St. Lawrence Island. Unlike scientists, hunters are not bound in their observations by any defined "project time" and their stories do not refer to any specific research period. They are going on the ice almost every day, year after year; and they preserve their memories, listen to elders' stories and share their observations with others. This is the body of knowledge that has been praised highly by many experienced anthropologists and natural scientists for years (e.g. Berkes, 1993; Freeman, 1984; Nakashima, 1993; Nelson, 1969; Riewe, 1991).

It is almost trivial these days—at least, on the part of scientists—to talk about "barriers" and "hurdles" to the ways Native or local knowledge can be matched with the data collected by the scientific community. Those obstacles most commonly listed arise from the presumption (which more often that not remains untested and never fully examined) that traditional knowledge is assumed to be intuitive, holistic, qualitative, and orally transmitted while the academic or scientific knowledge is primarily analytical, compartmentalized, quantitative, and literate (Eythorsson, 1993; Lalonde, 1993; Nadasdy, 1999). While there is some truth to these differences, Native elders and environmental experts can effectively operate with both types of knowledge, and they often do it more skillfully than polar scholars. This was clearly illustrated by the many statements, stories, and observations collected during this project and, particularly, by the two senior participants of our project, Conrad Oozeva and Chester Noongwook. I regard both our senior experts as true scholars and I respect them deeply. They fully demonstrate the very best qualities that are commonly associated with the scholarly mind: analytical perception, an inquisitive drive for continuous observation and recording, the eagerness to cross-check their data with other people's views and references, and openness to what Conrad Oozeva called "your scientists' usual question marks."

It is not a different nature, but rather a different focus of scientific and "local" knowledge, that commonly keeps these two types of expertise looking in different directions. Modern scientific studies of environmental change are unmistakably time-focused, in that scientists are primarily looking for well-documented series or samples of otherwise uniformly organized data, like annual or seasonal temperature and ice series, ice charts, satellite photographs, ice core samples, and others. This time focus allows scientists to operate with both the average and the extreme characteristics of the environment that are easily and thoroughly positioned in time and marked by fixed calendar dates or years. For scientists, such things are regarded as "statistically reliable." Scientific knowledge of climate change is basically aimed at expanding the timing and reliability of the various sets of data, in order to build better explanatory models with which it can operate. The actual nature of those data may be of secondary importance.

Local knowledge, on the other hand, is first and foremost detail-focused, in that it prizes specific and very detailed information about the characteristics of the environment observed, including sea-ice and weather change. There is no issue of statistical reliability, and every personal observation is considered sound and equal, as long as it relates to the environment, which is familiar to the observer. The age of the observer is probably the closest equivalent to the scientific concept of "reliability," as changes reported by elders are always considered more valid than those observed by younger people. But even among elderly experts there are people like Conrad Oozeva and Chester Noongwook—very much like some elders from the past, such as the renowned Gugwiingen, nicknamed "the weather bureau"—whose expertise of ice and weather is considered exceptional and is highly praised in the communities.

Nonetheless, as local knowledge documents in amazing detail the many possible facets of sea ice changes, as well as all exceptional weather phenomena, there is hardly an issue of precise timing of these events. Current stories and elders' memories are not focused on absolute dating or on any source of precise timing. People may say that "such and such thing happened twice or three times in my lifetime," but there is not a traditional practice to define exactly when it happened. Native observation of the environmental change does not have a dating method, unless queried specially. That is why many scientists have problems using records from local observations. To them, local knowledge contains too much data that is very hard to organize properly in a familiar standardized time series. The events or phenomena are usually reported with the excessive detail, but they are rarely dated in acceptable terms and, thus, are hard to compare with time series available through scientific records and/or instrumental observations.

There is also a problem of scale in comparing locally detailed Native observations and the more generalized scientific models based upon satellite images, multiyear observation records, and computer simulations (see also paper by David Norton, this volume). Lewis Shapiro, the polar ice specialist, who compared St. Lawrence Island observation records with contemparaneous scientifically organized sea ice data, notes that:

> Clearly Chester (Noongwook) and the NIC folks [from the National Ice Center that produces summary ice charts for Worldwide Web distribution] are looking

> at things on different scales. In fact, the thickness of the lines between different ice types on the (NIC) charts may scale to be comparable to the distance offshore that Chester could see standing on the beach. Also, I don't have any idea of the size of the smallest area of a particular ice regime that the observers at the NIC would map independently. This creates some problems in making comparisons between the observations. If you want to see an example, go ... to the chart from the time that Chester indicates the ice was far offshore (on February 9, 2001).
>
> However, in area "G" on the chart, the area off Savoonga is mapped as 0.7–0.9 ice covered. ... The remaining area of 0.1 to 0.3 is left for open water, but there is no information about how the ice and open water are distributed over the mapped area. The contradiction is obvious, but at the same time, they are probably both right. Chester certainly didn't see any ice off Savoonga during this time, and the NIC observers, who made the chart, didn't see a large enough area of open water to map it as a separate unit; so, they folded it into their unit "G." That seems like a reasonable conclusion, but it would be nice to know how far Chester really could see (from the shore), and how finely the NIC folks discriminate their ice zones. (Lewis H. Shapiro, letter to Henry Huntington and Igor Krupnik, December 10, 2001).

Therefore, in order to be compatible, both types of observation must be substantially modified to accommodate each other's specifics—in the same way that the data from social and natural sciences have to undergo certain accommodations to be used in an interdisciplinary or joint study. The same process takes place regularly in Native science, as experts from distant areas have to adjust the sets of observations from other communities, to be able to compare them with their own expertise and to benefit from other people's knowledge.

One can also see from the observation records and, particularly, from the summary statements of project participants, that local environmental experts are often far more advanced in mastering the terms, data, and approaches developed by scientists than vice versa. There are very few polar scholars, who can distinguish and name at least a few Native definitions of sea ice (although there are some who have produced papers and books about Eskimo knowledge of ice). Today's hunters, by contrast, can talk at length in purely scientific terms about the effects of global warming on sea ice and marine animals migrations well beyond their local areas of daily travelling and observation. Some elderly experts, like Conrad Oozeva and Chester Noongwook, have an extensive knowledge of major ice and current circulation cycles in the general northern Bering Sea-Bering Strait area, from the Pribilof Islands in the south and up to Barrow and the Chukchi Sea, in the north. This may be a unique blend of knowledge shared by today's Native elders, from which polar scientists and younger generations of St. Lawrence Islanders could learn in both more traditional and in fairly modern terms:

> People like Conrad [Oozeva]—he is a life-long observer, a living database. He is very articulate in his first language [Siberian Yupik]. He is very well versed in knowledge about how the [ocean] currents are flowing, and he knows how to get out when the ice is moving. He knows about the winds in the wintertime and about ice formations, and different weather patterns—he has his own equivalents for all these terms plus some sixty years of his personal knowledge. ... Now, Conrad knows—and he is a living database—maybe thirty different conditions of ice and snow around our island. And I just talked about two different conditions: one is slush ice and the other is flowing sea icebergs. Conrad can give you many, many different variations, with their specific terminology from our Native culture (John Waghiyi, February 2000).

Despite certain reduction in shared ice and weather expertise lamented by the elders, the people of St. Lawrence Island will almost certainly maintain this body of local knowledge. As long as they keep listening to their elders and continue to go hunting and travelling on ice, this expertise will be sustained by practical needs and new experience. Unlike Native experts and practitioners, scientists learn and maintain their knowledge through "projects" and "research initiatives" (or by reading the results of other people's studies). Therefore, the only way polar science can learn to be more open to local knowledge is by developing special research on just how to do this. This was the main purpose of our pilot project in documenting the sea ice and weather "the Yupik way" and of other similar projects presented in this volume. We believe that in both cases that initial goal has been achieved.

In order for the two types of science—developed by northern residents and scholars, respectively—to be compatible, a lot of further mutual adjustment is required. This approach differs from the perspective, which is popular among natural scientists and which looks for the ways to incorporate the data from indigenous observation into scientific research. To my mind, the gap is too big for an easy scanning from "other science" and it is obvious at every meeting that brings academic scholars and Native experts together.

Scientific knowledge about recent climate change in the Arctic has to become more specific, in order to interact productively with local expertise shared within the northern communities. It has to find ways to be projected from the more general (global?) models down to the regional and even to the individual village level. For local knowledge, on the other hand, a timing mechanism has to be created to make Native record of past and present events compatible with the time series used by academic scholars. This is a complicated enterprise, as each local community has to build a "beach-ridge chronology"[5] of its own, one based upon its particular history, available documentary records, and memories shared by the most elderly experts.

To accomplish these and other goals, intensive new learning and the sharing of knowledge and data is required. Thus, future long-term efforts should be focused first and foremost on charting the ways for mutual adjustment of the two types of knowledge about arctic environmental change.

Conclusions

To the polar science community, the study of Native practices in the observation of sea ice and weather conditions offers an opportunity to learn how a record of ice and climate change is accumulated and transmitted by "other" ways of knowledge, that is, by the shared expertise of northern residents. Physical scientists may be particularly interested in the potential observational and explanatory value of Native terminologies and databases on sea ice, like the one developed by the Yupik people of St. Lawrence Island. The ability to "read" through the records, both oral and written, of Native monitoring of sea ice and climate change is another obvious asset to the study of climate change. Social scientists, from their side, are to benefit from the increased awareness about the structure and functions of Native scholarship and from a better understanding of the role environmental knowledge plays in the complex social fabric of modern community life.

To the many ongoing impact assessment studies of arctic climate warming, such projects provide practical tests of collaboration between the two most involved groups of stakeholders, polar scientists and local residents. Such collaboration and mutual correction is critical at all stages of the impact assessment process, including observation, documentation, and analysis of the various signals of climate change. As scientists and northern residents alike experience rapid shifts in arctic weather and ice regimes, both constituencies express their concerns about the scope of potential change they may face in future. Climate scientists, in order to attract public attention to climate change issues, often respond by articulating some extreme scenarios under their computer-simulated models. Many such scenarios, like those featuring the 40% decrease in the volume of arctic sea ice over the next 20 years and the ice-free Arctic, with a complete disappearance of the permanent summer ice pack by the year 2050 (e.g., Vinnikov et al., 1999; Gough and Wolfe, 2001, etc.) may have dire or unprecedented consequences on Native residents. It is hard to say what practical worries about the future of the Arctic are shared by the science community; but their expression projects a message of extreme anxiety, to which a much more engaged Native political leadership responds with great alarm (Fenge, 2001; Ford, 2001; Whitehorse Declaration, 2001)

The response from the Native community, on the other hand, is a mixture of two clearly distinctive motifs. One is the message of confidence and endurance, based upon the legacy of survival in the ever-changing arctic environment and upon decades of personal experience by the elderly experts. Native elders are normally quite reserved and they rarely boast with confidence regarding the power of their expertise. Nonetheless, whenever they comment about the value of accumulated knowledge in addressing the climate change, they stress it eloquently:

> This change (of winter ice and weather) is visible. *Anleghaq* [going out hunting at sea, particularly to the east—IK] has changed. When I was younger, there were these migrating walruses coming on the eastern side of Gambell first; and then they were going around the point, and headed south. That used to happen in November, first part of November. ... That did not happen like that anymore. Now they do like that for very short time in December, which is late. ... Something is changing—we noticed this.
>
> [However] with the warmer weather, the animal behavior is still the same. They just migrate earlier. And during wintertime, despite all these changes of ice and weather, there is no effect on them. We can get more walruses in winter now, if it's a good weather for hunting. Because of that, I am not that much concerned.
>
> I have lived through at least three periods [winters—IK] of exceptional warmth, including the latest one, but in between it was always like more or less normal winter. Still, I have a big question mark, like you scientists always have. It may get back to normal but it may not. If it does not change rapidly, it is not going to change our hunting and all the knowledge we have will be still valuable (Conrad Oozeva, 2001).

> It is probably safe to say that changes are going to become a daily part of our lives. In light of more unstable weather conditions, I think that it would help if we can anticipate these changes based on observations and data collected by the scientific community and also observations by local hunters. Then we can make a collective prudent decision of where and when to concentrate our efforts in order to feed our souls, physical, and biological needs.
>
> We cannot change nature, our past, and other people for that matter, but we can control our own thoughts and actions and participate in global efforts to cope with these global climate changes. That I think is the most empowering thing we can do as individuals (George Noongwook, 2000:24).

This motif of endurance ("resilience"), however, is often accompanied by the message of grave concern, as northern residents watch rapid shifts in their environment and struggle with explanations. Some of their statements clearly follow paradigms borrowed from the scientific discourse; others reflect their own monitoring of the habitat that has been familiar for decades:

> I am concerned about this happening to our weather and these changes—the global warming, the erosion on our island. If it erodes more, the permafrost is melting, our island might get smaller. Hunting is different, too because of this changing weather and ice. And the animals are different because they are living along with the ice, like the bowheads, walruses, seals, polar bears, maklaks

[bearded seals]. They all come close to our island because of the ice. This year even minke and gray whales are fewer than before—I do not know, why.

To my knowledge, the whole globe is turning the other way. I had a strange vision at my camp last year, when I looked at the Big Dipper. It always used to be straight to the North Star. It looked like slightly off this past year. I was scared. It could have been my own sighting but I am concerned about it. And I am not trying to look that way anymore.

When we are building something, moving or mining, we are disturbing the permafrost and it is melting. The water is rising because of this, all these waves, erosion. That is why I am thinking about global warming. It could alter our way of life as well (Chester Noongwook, 2001).

Modern arctic ice and climate change studies can benefit tremendously via the contribution of properly collected local observations, done by Native people themselves and from their cultural perspective. As Native people are given a full chance (and appropriate means), to document how they observe weather and sea-ice in their own words, in the course of their daily lives, a whole set of new records will be available for a thorough and comparative examination. Then, and only after then, we may start thinking about how this knowledge can be matched with the data collected by scientists and used in scientific models, maps, and in the overall discussions of arctic climate and ice change in the course of recent global warming.

Acknowledgements

This is the first preliminary review of the project Watching Ice and Weather Our Way, to be followed by further accounts and, hopefully, by a bilingual volume to be prepared for the communities of Savoonga and Gambell, in cooperation with my research partners—Henry Huntington, the overall project pirector; and George Noongwook, Savoonga project coordinator. Henry's role in preparing the many project documents and in offering advice and criticism on the draft of this paper should be specially acknowledged. The sea ice and weather observation in 2000–2001 was accomplished thanks to its endorsement by the communities of Savoonga and Gambell and to the financial support provided by the Marine Mammal Commission (Executive Director Robert Mattlin, Chairman John Reynolds). In my trips to St. Lawrence Island in 2000 and 2001, I also received substantial assistance from the Arctic Studies Center, Smithsonian Institution; the National Park Service, Western Arctic National Parklands office in Nome, Alaska; and the Eskimo Heritage Program of Kawerak, Inc. in Nome (Director Branson Tungiyan). My colleagues in arctic research, Fikret Berkes, Dyanna Jolly (Riedlinger), Gary Kofinas, Lewis H. Shapiro, and Carleton G. Ray, offered valuable comments that were received with gratitude. Finally, it was the greatest pleasure to work with the observation teams from the two Yupik communities, particularly with our two senior experts, Conrad Oozeva and Chester Noongwook. Their dedication and sharing of expertise was the key contribution that made our joint venture in sea ice and weather observation "the Yupik way" and this paper as its first summary, possible.

Notes

1. This is the most extensive account so far of the project team activities in 2000–2001 (Igor Krupnik, principal investigator; Henry Huntington, project director; George Noongwook, local coordinator). See earlier reports in Krupnik, 2000a; Krupnik and Huntington, 2001.
2. The communities included Gambell, Savoonga, Mekoryuk, Unalakleet, Elim, Nome, Deering, Kotzebue, Kivalina, and Barrow in Alaska, plus Inuvik, in Canada's Northwest Territories.
3. In Gambell, the basic names for the winds are the same but their meanings are slightly different, which is a common pattern among Inuit (Eskimo) groups throughout the Arctic (see Fortescue, 1988).
4. Following the pattern preferred by the volume contributors, the Yupik portions of the text go under the Yupik names of the authors, whereas English portions are under their documented English names.
5. This refers to the dating method developed by geologists and archaeologists, whereby events or objects of the past are dated by callibrating the series of beach ridges that have been built on shore by the sea surf through time. Scientists call it a "beach-ridge chronology": the ridges closest to the sea are the latest to be built up and, thus, the youngest; and those farthest from the sea are the oldest ones, and so on.

References

Akuzilleput igaqullghet. Our words put to paper: Sourcebook in St. Lawrence Island Yupik heritage and history. Krupnik, I., and Krutak, L. (comp.), Krupnik, I., Walunga, W., and Metcalf, V., eds. 2000. Washington, D.C: Arctic Studies Center, Smithsonian Institution.

Berkes, F. 1993. Traditional ecological knowledge in perspective. In: J. T. Inglis, ed. *Traditional ecological knowledge: concepts and cases,* pp. 1–9. Ottawa: Canadian Museum of Nature.

Berkes, F. 1999. *Sacred ecology: TEK and resource management.* London and Philadelphia: Taylor and Francis.

Bielawski, E. 1996. Inuit indigenous knowledge and science in the Arctic. In: L. Nader, ed. *Naked science: Anthropological inquiry into boundaries, power, and knowledge,* pp. 216–227. New York: Routledge.

Burns, J. J., L. S. Shapiro, and F. H. Fay. 1981. Ice as marine mammal habitat in the Bering Sea. In: D. W. Hood and J. A. Calders (eds.). *The eastern Bering Sea shelf: Oceanography and resources.* Vol. 2, pp. 781–97. Seattle: University of Washington Press.

Cavalieri, D. J., and C. L. Parkinson. 1987. On the relationship between atmospheric circulation and the fluctuations in the sea ice extents of the Bering and Okhotsk Seas. *Journal of Geophysical Research 92*(C7): 7141–7162.

Doty, W. F. 1900. Log book, St. Lawrence Island. *Ninth annual report on introduction of domestic reindeer into Alaska 1899:* 224–256. Washington: Government Printing Office.

Eskimo Heritage Program. 1982. Eskimo Heritage Program. Proceedings of Elders Conference. Tapes EC-GA-82-71-T29, (transcripts). Eskimo Heritage Program: Nome.

Eythorsson, E. 1993. Sami fjord fishermen and the state: traditional knowledge and resource management in northern Norway. In: J. T. Inglis, ed. *Traditional ecological knowledge: Concepts and cases,* pp. 133–142. Ottawa: Canadian Museum of Nature.

Fenge, T. 2001. Inuit and climate change: Perspectives and policy opprtunities. *Silarjualiriniq 10:* 1–14.

Ford, V. 2001. From consultation to partnership: Engaging Inuit on climate change. *Silarjualiriniq 7:* 2–4.

Fortescue, M. 1988. Eskimo orientation system. *Meddelelser on Grønland. Man and Society 11.*

Fox, S. 2000. Project documents Inuit knowledge of climate change. *Witness the Arctic 8(1):* 8.

Freeman, M. M. R. 1984. Contemporary Inuit exploitation of the sea-ice environment. In: A. Cooke and E. Alstine, eds. *"Sikumiut": People who use the sea-ice,* pp. 73–96. Ottawa: Canadian Arctic Resource Committee.

Gough, W. A. and E. Wolfe. 2001. Climate change scenarios for Hudson Bay, Canada, from general circulation models. *Arctic 54*(2): 142–148.

Hughes, Charles C. 1984. St. Lawrence Island Eskimo. In: D. Damas, ed. *Handbook of North American Indians, Vol. 5, Arctic,* pp. 262–277. Washington: Smithsonian Institution.

Huntington H. P. 2000a. Native observations capture impacts of sea ice changes. *Witness the Arctic 8*(1): 1–2.

Huntington, H. P., ed. 2000b. *Impact of changes in sea ice and other environmental parameters in the Arctic.* Report of the Marine Mammal Commission workshop. Girdwood, Alaska, 15–17 February 2000. Bethesda, MD: Marine Mammal Commission.

Huntington, H. P., Brower, H. Jr., and Norton, D. W. 2001. The Barrow Symposium on Sea Ice, 2000: Evaluation of one means of exchanging information between subsistence whalers and scientists. *Arctic 54*(2): 201–204.

Krupnik, I. 1993. *Arctic adaptations: Native whalers and reindeer herders of northern Eurasia.* Hanover and London: University Press of New England.

Krupnik, I. 2000a. Scientists and Inuit exchange observations on arctic climate and sea ice change. *ASC Newsletter 8:* 15–16.

Krupnik, I. 2000b. Native perspectives on climate and sea ice changes. In: H. Huntington, ed. *Impact of changes in sea ice and other environmental parameters in the Arctic,* pp. 25–39. Bethesda, MD: Marine Mammal Commission.

Krupnik, I. 2001. *Pust' govoriat nashi stariki. Rasskazy aziatskikh eskimosov-yupik. Zapisi 1975–1987 gg. [Let our elders speak. Oral stories of the Siberian Yupik Eskimo, 1975–1987].* Moscow: Russian Heritage Institute.

Krupnik, I. and H. Huntington. 2001. Documenting arctic climate 'The Yupik way.' *ASC Newsletter 9:* 16.

Lalonde, A. 1993. African indigenous knowledge and its relevance to sustainable development. In: J. T. Inglis, ed. *Traditional ecological knowledge: concepts and cases,* pp. 55–62. Ottawa: Canadian Museum of Nature.

Lerrigo, P. H. J. 1901. Abstract of journal, Gambell, St. Lawrence Island, kept by P. H. J. Lerrigo, M.D. *Tenth annual report on introduction of domestic reindeer into Alaska 1900:* 114–132. Washington: Government Printing Office.

Lerrigo, P. H. J. 1902. Abstract of daily journal on St. Lawrence Island kept by P. H. J. Lerrigo, M.D. Log book, St. Lawrence Island. *Eleventh Annual Report on Introduction of Domestic Reindeer into Alaska 1901:* 97–123. Washington: Government Printing Office.

Lowenstein, T. 1981. *Some aspects of sea ice subsistence hunting in Point Hope, Alaska.* Barrow: North Slope Borough, Coastal Management Program.

McDonald, M., Arragutainaq, L., and Z. Novalinga, comps. 1997. *Voices from the bay: Traditional ecological knowledge of Inuit and Cree in the Hudson Bay bioregion.* Ottawa and Sanikiluaq: Canadian Arctic Research Committee.

Moore, R. D. 1923. Social life of the Eskimo of St. Lawrence Island. *American Anthropologist 25*(3): 339–375.

Nadasdy, P. 1999. The politics of TEK: power and the "integration" of knowledge. *Arctic Anthropology 36*(1–2): 1–18.

Nakashima, D. J. 1993. Astute observers on the sea ice edge: Inuit knowledge as a basis for Arctic co-management. In: J. T. Inglis, ed. *Traditional ecological knowledge: concepts and cases.* Ottawa: Canadian Museum of Nature. p. 99–110.

Nelson, R. K. 1969. *Hunters of the northern ice.* Chicago: University of Chicago Press.

Nelson, R. K. 1981. *Harvest of the sea: Coastal subsistence in modern Wainright. A report for the North Slope Borough Coastal Management Program.* Barrow: North Slope Borough.

Noongwook, G. 2000. Native observations of local climate changes around St. Lawrence Island. In: H. Huntington, ed. *Impact of changes in sea ice and other environmental parameters in the Arctic,* pp. 21–24. Bethesda, MD: Marine Mammal Commission.

Parkinson, C. L., and D. J. Cavalieri. 1989. Arctic Sea Ice 1973–1987: Seasonal, regional, and interannual variability. *Journal of Geophysical Research 94*(C10): 14,499–14,523.

Pungowiyi, C. 2000. Native observations of change in the marine environment of the Bering Strait Region. In: H. P. Huntington, ed. *Impacts of changes to sea ice and other environmental parameters in the Arctic,* pp. 25–28. Bethesda, MD: Marine Mammal Commission.

Riedlinger, D. (see Jolly, D.). 2001. Inuvialuit knowledge of climate change. In: J. Oakes, R. Riewe, M. Bennett, and B. Chisholm, eds. *Pushing the margins: Native and northern studies,* pp. 346–355. Winnipeg: Native Studies Press.

Riewe, Richard. 1991. Inuit use of the sea ice. *Arctic and Alpine Research 23*(1): 3–10.

Silook, Roger S. 1976. *Seevookuk: Stories the old people told on St. Lawrence Island.* Anchorage.

Sivuqam Nangaghnegha. Lore of St. Lawrence Island. Echoes of our Eskimo elders. 1985–89. In: A. Apassingok, W. Walunga, and E. Tennant, eds. Unalakleet: Bering Strait School District. Vol. 1, Gambell (1985); Vol. 2, Savoonga (1987); Vol. 3, Southwest Cape (1989).

Vakhtin, N. B. 1988. *Practicum po leksike eskimosskogo iazyka [Asiatic Eskimo Language Lexical Curriculum].* Leningrad: Prosveshchenie Publ.

Vinnikov, K. Y., A. Robock, R. J. Stouffer, J. E. Walsh, C. L. Parkinson, D. J. Cavalieri, J. F. B. Mitchell, D. Garett, and V. F. Zakharov. 1999. Global warming and Northern Hemisphere sea ice extent. *Science 286*(5446): 1934–37.

Whitehorse Declaration. 2001. Whitehorse declaration on northern climate change. Adopted by the Circumpolar Climate Change Summi, held in Whitehorse, Canada on March 19 to 21, 2001. *Silarjualiriniq 7:* 5.

Yupigestun kelengakellgha sikumllu esclemllu: Watching ice and weather our way. In prep. Oozeva, C., C. Noongwook, G. Noongwook, C. Alowa, and I. Krupnik, comps. I. Krupnik, G. Noonwook, and H. Huntington, eds.

Yupik Language. 1989. *The St. Lawrence Island Yupik language and culture curriculum. Grades K–12.* A. Apassingok, project director. Unalakleet: Bering Strait School District.

Appendix 1: Chester Noongwook's Daily Weather Story, December 8, 2000

I usually go out every morning to check the weather, about 8 or 8:30 in the morning. I got up and went out of my house at about this time this morning, and I looked around, to see how the weather would be today. And it was going to be a good hunting day. Sky condition was very thin—just a few clouds here and there. I could even see the moon and the stars from the breaks in the clouds. The wind is going to be light and it is from the south *(iqevagatawa)*. It is a good wind to go out hunting in boat.

Then I took my grandchildren down to school (about 9 a.m.) and I looked for water and ice near the school site at the shore—whether there is any open water out there (far at sea). I could tell it because there was some dark(ness) in the sky far over there; that dark sky is over open water. I stayed at the store for some time to get information from other people, who were there earlier in the morning. We always have a few people of my age gathering at the store, the side that faces the water and the beach—they just stay there for some time, watch the weather and ice, and talk.

It was very slushy ice—we call it *qenu*. The ice from the bottom moving up and starting to make it a very slushy young ice. Many of us can see some scattered ice cakes. We call this ice *sikupik:* it can be one foot or two feet or even three feet high. This is what we were told by our fathers, *ayemqutet* or *uughhutet*—pieces formed from the main (polar) ice being blown out to the island by the northern wind. These *ayemqutet* are making scattered ice floes from the main ice formations. It is mostly wind not the current (that brings this kind of ice) and also the icebergs from the north. We call these icebergs *kulusiit*.

This ice now is not good—neither for foot (walking) nor for a boat, because it is very hard to travel by boat (at this type of mixed and slushy ice). And it breaks very easily under the foot. My grandfather always told me that we have to wait until somebody was starting to go out hunting. It is always better this way; that what he used to say.

In the afternoon it got much warmer and the ice was moved in by the current. We have to be very, very watchful for this type of ice (because it moves quickly). As long as the weather stays warm, the hunting won't be very good for us. This is because of the new ice, *qenu*, slush ice. It is dangerous for travelling in boat—

it has to get frozen first. When colder weather comes in, it may solidify, so that we can walk over on foot.

The weather for tomorrow: from my guessing, so long as the wind is from the south *(iqevaq)* or whatever my barometer says, the condition of the sky, the weather might be better of than this one by tomorrow. But we never know for sure. My grandfather used to tell me that you would never tell the weather for tomorrow until you get up, go out, and check the weather yourself. We are just thirty-five miles from Gambell, but we do not even know how the weather is up there today.

I could see a little bit of moon through the sky now (in the evening) and so I can guess that the weather might be a little bit better tomorrow.

This time of year (early December) the walrus is usually going from west to east. We call this walrus *angleghaq.* They are mostly females mixed with few bulls and calves from this year or a year before. Early in the morning, when the weather is calm, we can hear walruses in the water making noise. Seals and maklaks (bearded seals) here are almost the same as walrus. My grandfather used to call it *katawsaq'a*—like, pouring out with the ice. Ice moves in here with all these animals. My brother killed a walrus today, so everybody in the village has some fresh meat. No seals, no *makllaks* so far.

The day before yesterday (December 6) we have seen two gray whales—*angtughaq.* We do not hunt them here until we are really pressed (by shortage of food). We still have a chance to see or hunt a bowhead (whale) nearby. Usually around Stolby Rocks, at the place called Kitnik. And later on they may come by and going westward, toward Gambell, and then all the way around the island to the Southwest Cape. So, we still hope, maybe we can get one (bowhead) this year—we already started looking for one now. From this day on and until late winter, we will be carrying (whaling) gear in our boats.

Appendix 2: "*Anuqem Nakengutanga*"—Wind Directions in Gambell, St. Lawrence Island

Compiled by Conrad Oozeva. Translated by Branson Tungiyan

Aywaapik	(North)	*Pakfalla*	(West)
Nakaghya	(Northeast)	*Naayghiina*	(Northwest)
Kiwavaq	(East Northeast)	*Naayviinaq*	(South Southeast)
Asivaq	(East)	*Saallghaghta*	(Unexpected Storm)
Ikevaghlluk	(Southeast)	*Quutfaq*	(Northeast)
Ikevaq	(South)		
Tapgham Ketanga	(Southwest)		

1. ***Aywaapik* (North).** The weather (with the wind coming) from the true north is good. Because the weather is always good from that direction, travelling anywhere during the wintertime is considered safe because it is clear. (With this northern wind) the weather is often clear in summer as well, even if there is some wind blowing.
2. ***Nakaghya* (Northeast).** The true direction for this wind is in between north *(aywaapik)* and east *(asivaq)*. This wind direction often brings a lot of fog. The fog associated with this wind does not clear up quickly, but often has light winds without big swells in the sea.
3. ***Kiwavaq* (East Northeast).** When the wind is from east-northeast, it is often coming from the direction of the tip of the Gambell Mountain. This direction has a special rule that comes with it, in regards to going out onto the sea ice for hunting. The men are told not to go out for hunting on the ice because this direction comes with stronger winds.
4. ***Asivaq* (East).** This wind comes directly from the (Gambell) mountain, and it often comes with snow, and even blowing snow, when this wind continues to increase. There are strong downdrafts with this wind around here (in Gambell). On the other hand, the ice is taken out from that direction in winter (the shore opens up in Gambell for boat hunting). Also, if the winds are light, they tend to stay light.
5. ***Ikevaghlluk* (Southeast).** At wintertime, like this, the weather usually warms up; but it gets very stormy, with blowing snow and strong winds. Also, when the wind comes from that direction, it opens up ice leads in the southeast direction. The ice that has been brought by this wind from the south *(ivgaghutkak)* can be often seen out here (right in front of Gambell, on the western side). When the wind is light, it tends to stay light; but when it is packing strong winds, it is always stormy and blowing snow. In the summertime, it is always wet and rainy, and windy.
6. ***Ikevaq* (South).** This wind comes from the south, from the southern mountains direction. With this wind in the summer time, the travel along the northern part of

the island and all around Gambell is usually good. The weather also warms up with the southern wind. But mostly in the springtime, it brings a lot of fog with it. Still, it is often (considered) a good weather because it does not have strong winds.

7. ***Tapgham Ketanga* (Southwest).** This wind also brings a lot of heavy fog because the fog is brought over from the warmer air [over the open sea], and when [it is] brought to the cooler air, it creates foggy conditions. And, it sometimes gets drizzle. It almost always gets foggy when the wind is from that direction. This weather does not have strong winds. Travel conditions (on the sea) are sometimes quite good, particularly along the northern part of the island.
8. ***Pakfalla* (West).** This wind comes from the true west direction. When the weather becomes of "the normal wind strength" (*igiighta*—is kind of hard to explain in English), it often clears up, although there might be some clouds. If the shore ice has gone out, this wind often brings it back to the shore. In this case, even though there have been a lot of ice around, the shifting of ice by the western wind causes the ice to 'change places' for a while. This wind also tends to bring strong gusts, especially in the fall if the snow begins to set in. Then, it often creates snow showers, but then it clears up. That is how it tends to be. In the summer time, the rains that come with this wind are pretty much the same as the snow showers are in the fall.
9. ***Naayghiina* (Northwest).** This wind direction comes directly from the Siberian Mountains (that is, across the strait between St. Lawrence Island and the Siberian mainland). We hardly ever get real strong winds from that direction (in Gambell), but if it gets windy, it gets windy. These west and northwest wind directions do not stay long; it usually changes to a different direction shortly. These winds are favorable (for hunting and travelling) and the weather tends to be clear. All these wind directions that I have just told, the winds can change quickly and it could get windy from another direction. So, now that we have begun to understand, if [big] storms go through close to us, and the winds shift to certain directions because of the low pressures, the winds can become very strong from that direction.*
10. ***Naayviinaq* (South Southeast).** That south-southeast wind direction comes right in between *Ikevaq* (south) and *Ikevaghlluk* (southeast). It gets very windy as well, but the wind does not stay in that direction for long; it shifts either to the southeast or to the south. (Overall) it is very similar to the southeast wind, but it shifts direction much quicker. When this wind increases, it gets foggy, but it does not have much fog condition. Even though it gets foggy, the lower level at the ground tends to be clear. It could also be quite wet, because it can warm up (the temperature) in winter. This wind is just the same in summer, with wet weather conditions, like the drizzle.

11. ***Saallghaghta* (Unexpected storm from any direction).** These unexpected strong wind conditions could occur from any wind direction, and in the winter can become very whiteout (blizzard) conditions quickly. Also in the summer, it can be very wet. Mostly, when heavy dark clouds rise from the southern direction, the gale force winds will arrive from the northeast direction and it can get very stormy real fast. Unexpected blizzard conditions (may come) from the east, southeast, west, or from any wind direction. In most cases, the wind and weather condition that is called *saallghaghta* is like that.

---

* This is the only (though a very clear) indication that traditional Yupik knowledge of wind patterns is now influenced by more 'scientific' explanations of atmospheric circulation, driven by the high pressure-low pressure alterations.

Figure 6-1. Looking westwards across Bathurst Inlet at Ikalulialuk Island from the community of Umingmaktuuk. As it is July, there are still many caribou nearby as suggested by the meat in the foreground.

6

Nowadays it is Not the Same:
Inuit Qaujimajatuqangit, Climate and Caribou in the Kitikmeot Region of Nunavut, Canada

Natasha Thorpe,[1] Sandra Eyegetok,[2] Naikak Hakongak,[3] and the Kitikmeot Elders[4]

We do not know the weather, it has a caretaker of its own (Analok, 1998).

Inuit observations of a warming climate, and how climate change has influenced caribou, were recorded as part of the *Tuktu* (caribou) and *Nogak* (calves) Project (TNP) between 1996 and 2001. This community-driven effort sought to document and communicate Inuit knowledge, known as *Inuit Qaujimajatuqangit* (IQ), of the Bathurst caribou and calving grounds by working with elders and hunters from four communities in the Kitikmeot region of western Nunavut, Canada.

This chapter begins by detailing the underlying goals, objectives and methods of the TNP. With this necessary context established, we share the major findings related to climate change and impacts on caribou. Since many Kitikmeot Inuit[1] observations on environmental change in the Kitikmeot region are already presented in our recent project volume (Thorpe et al., 2001), this chapter addresses in greater length the methodological grounds of our study. It also offers some general comments regarding the nature of IQ and its compatibility with the scientists' knowledge of caribou.

The Need for a Study

Numerous circumstances led to the TNP.[2] First, individuals who were conducting the Naonayaotit Traditional Knowledge Study (NTKS)—a regional IQ study with a focus on all wildlife and fish species—realized their study was too broad to provide a detailed account on caribou and calving areas in the Bathurst Inlet region. Given that many NTKS interviewees described the primary importance of caribou to local

1. Tuktu and Nogak Project, 231 Irving Road, Victoria, BC
2. Government of Nunavut, Department of Education, Cambridge Bay, NU
3. Government of Nunavut, Department of Sustainable Development, Cambridge Bay, NU
4. c/o Enekniget Katimayet, Cambridge Bay, NU

subsistence, identity, culture, and tradition, Gerry Atatahak from Kugluktuk (principal researcher for the NTKS), suggested initiating another research project, one that centred on caribou and calving areas. In 1996 he began asking for community support and input to the TNP.

Wildlife resources are critical to northern peoples who depend upon healthy populations of animals, especially caribou. For the Kitikmeot Inuit, the Bathurst caribou are of particular relevance since the herd migrates through and calves in areas nearby. The amount of previously documented IQ about this herd was minimal, especially relative to the importance of this caribou herd to community members.

The need to preserve elders' knowledge was a third factor contributing to the TNP. In Nunavut, elders are passing away while youth, more fluent in English than Inuinnaqtun, struggle to live on the land, speak their language, and learn from their elders. IQ held by elders represents intergenerational wisdom that spans many spiritual, spatial, and temporal boundaries. Loss of this understanding would be detrimental to Inuit culture in general, and to the sustainable management of northern lands, resources, and wildlife in particular. Before elders who have lived most of their lives on the land pass away, the TNP sought to record their expertise and experience such that it could be accessible to current and future generations.

Mineral exploration and the potential for mine development provided a fourth impetus for seeking a comprehensive understanding of caribou in the Bathurst Inlet area. The early 1990s were marked by an unprecedented surge of exploration activity in the region while at the same time there was a relative paucity of documented baseline IQ to inform sound and sustainable decisions related to mineral development, lands, resources, and wildlife. The TNP sought to ameliorate this information gap.

Through the TNP, thirty-seven elders and hunters were interviewed at least twice during semidirected and semistructured interviews conducted within communities, on the land, and during an elder-youth camp. The study area included historical and current hunting grounds of the communities of Umingmaktuuk (Bay Chimo) and Kingauk (Bathurst Inlet). Individuals from Kugluktuk (formerly Coppermine) and Ikaluktuuttiak (Cambridge Bay) were also consulted to include former residents of these primary communities who have hunted, trapped or traveled in the range of the Bathurst herd (Figure 6-2).

In the Beginning: Establishing an Advisory Board and Research Agreement

With the reasons for a study clearly established, Gerry Atatahak connected with Natasha Thorpe in 1996 to discuss the possibility of starting the TNP. Gerry and Natasha already shared social capital based on their early experiences working together on a water quality monitoring network in the north in 1994 and were prepared to work together on another project. Thus began the slow process of community consultation, both formal and informal, throughout late 1996 and early 1997.

During the summer of 1997, community members from Ikaluktuuttiak, Kingauk, Kugluktuk, and Umingmaktuuk volunteered their time to refine funding proposals, establish the project methods, and nominate representatives to sit on an advisory board called the Tuktu and Nogak Board (TNB). With equal representation of men and women and proportional membership from each of the four study communities, the TNB established their mandate: to provide direction and oversee all activities related to the TNP. With the TNB in place, Gerry's role in the TNP became one of background support.

The TNB gave the study a name, defined its goals and structure, and suggested locals who could be trained as staff. Once funding[3] was secured, the TNP began in earnest: staff members were hired and trained and then began collaborating with the TNB to write research agreements detailing the ownership, use, access and storage of research materials; how and where the research would proceed; how funds should be spent; and what would be done with the final products.

After several days of discussion in workshops, a ceremonial signing of the research agreement took place in *Umingmaktuuk* (Bay Chimo) in June 1997. This Agreement set out the goals, objectives, and principles that would guide the TNP (see Table 6-1).

High levels of community consultation, adaptability and flexibility to local input, and opportunities for youth involvement were identified as integral to the research process. For example, the TNB insisted that youth participate in all phases of the research. Youth helped to interview elders, record findings, make drawings, take photos, and shoot videos that were used in numerous communication materials such as reports, posters, and a book.

From the outset, youth, community members, and researchers reported a feeling of empowerment gained through the TNP because of their involvement in training, planning, researching, writing, public relations, and other capacity-building activities (Eyegetok, pers. comm., 1999). Further, the high value placed on IQ by researchers, funding agencies, and other supporting organizations, meant that people felt proud that their knowledge was seen as valuable. This was a welcome change from a history of many people not recognizing the importance of IQ (Wolfley, 1998; see Fox, this volume).[4]

Figure 6-2. The study area for the Tuktu and Nogak Project (1996–2001). The study area was defined as the range of the Bathurst herd commonly used by residents of Umingmaktuuk (Bay Chimo) and Kingauk (Bathurst Inlet).

Establishing ownership of the research was fundamental before any of the research began. The TNB decided that each and every interviewee should hold ownership of their information as contained in the transcripts, and that the TNB would collectively represent the interests of the interviewees. Every interviewee supported this decision. The elders of the TNB were adamant that they wanted to retain ownership rather than surrendering it to an Inuit, Nunavut, educational or government agency.

The TNB insisted that *Nunavummiut*[5] and both Inuit and Nunavut agencies have free and open access to all research materials. The local heritage society and designated Inuit organization were granted the role of gatekeepers. Part of the profits gained from sharing the information with "outsiders" was designated for elder-youth initiatives and other IQ projects. The TNB insisted that this information be shared rather than held under lock and key. This is possible since Inuit have a settled land claim and do not have to fight for ownership of their land or resources in the same way as required by other Canadian native groups. There is less of a threat of IQ being used against Inuit because of the strong political and social framework within this territory of de facto Inuit self-government (Légaré, 1998).

Building the Partnership: Stages and Approaches

An emic methodology, one that evolves with community direction, formed the foundation of the TNP and was enhanced by a combination of participatory methodologies: participant observation (Spradley, 1980; Burgess, 1984; Whyte, 1991), participatory action research (Greenwood et al., 1993; Dene Cultural Institute, 1994; Chataway, 1997), participatory rural design (Chambers, 1991; Theis and Grady, 1991; Holden and Joseph, 1991; Webber and Ison, 1994), and experiential data (Strauss, 1987; Glesne and Peshkin, 1992; Maxwell, 1996).[6]

Table 6-1. The goals, objectives, and research priorities of the Tuktu and Nogak Project (TNP) as defined by its board

Goals	Objectives	Research Priorities
• to collect and share IQ of caribou and calving grounds in order to improve caribou management in the Bathurst Inlet area for present and future generations in Nunavut.	• protect caribou during calving seasons • protect caribou from development activities • encourage youth to speak their language • encourage elders and youth to interact with each other	• behavior • changes in weather and climate • habitat • health • historical changes/trends • interaction of different herds • location and habitat of calving grounds • migration • protection and management • seasonal changes • traditional use and hunting areas

Our methodology began with participant observation then moved towards participatory action research. Insofar as we continued to conduct qualitative research in partnership with community members, we adopted elements of participatory rural appraisal. Participant observation, participatory action research, and participatory rural appraisal are all closely related and are inevitable in a social science setting. They are, at once, modes, tools, techniques, and methods.

One of the key successes of the TNP was in the way that the research process and partnerships evolved throughout the eight stages of the project, detailed in the following sections. The TNP research process was designed to be inductive, pliable, and responsive to input from community members. Such qualities are critical in community research because "any significant pre-structuring of the methods [could have led] to a lack of flexibility to respond to emergent insights and [created] methodological blinders in making sense of the data" (Maxwell, 1996: 63). The advantage of using an adaptive approach was that it allowed flexibility to focus on particular emergent phenomena. Community members guided the research from the outset such that certain methods were adapted according to local input. For example, more interviews were conducted out on the tundra rather than in the communities because people suggested this was a more appropriate and thought-provoking environment. Flexibility to local input meant that both the research process and findings were more meaningful.

Given the theme of the volume and space limitations, we limit our discussion to Kitikmeot Inuit observations of the effects of climate change in relation to caribou, rather than attempt to discuss overall findings of the TNP. Other components of the TNP, along with this topic, are discussed further and in more detail in Thorpe et al., 2001. Several sources of evidence informed the analysis presented in this chapter:

- A representative subset (n=27) of a set (n=37) of semistructured and semidirected interviews conducted between 1997 and 2000.
- Unstructured interviews and personal communication (email, telephone, and written correspondence) with key informants from Kitikmeot Inuit and Nunavut agencies (n=5).
- Memos, journal entries, and research notes from frequent and extended periods of qualitative research conducted from September 1996 to July 2001.

Stage One: Consulting and Communicating

With community consultation at the heart of the research process, we frequently visited each community and held lengthy discussions to determine ways in which people wanted to contribute and how the TNP should progress. Radio announcements, websites, posters, meetings, reports, and word-of-mouth were some of the ways in which additional communication was carried out. We held formal meetings on a regular basis, at least four times per year in each community. Informal visits with TNB members and other elders maintained the key relationships, trust, and communication necessary for carrying out meaningful and respectful research. This was particularly true for meeting with the elderly who were no longer mobile enough to easily get out of their houses.

Stage Two: Hiring and Training Local Researchers

The second phase of the TNP involved training researchers in the ways of IQ research. This occurred through a series of workshops held in Ikaluktuuttiak, Kingauk, and Umingmaktuuk during 1997 and 1998. Topics included interviewing, photographing, writing, and computer skills, as well as sharing important tips on appropriate ways of conducting community research. Sandra Eyegetok, Naikak Hakongak,[7] Margo Kadlun-Jones, and Natasha Thorpe worked as full time staff and reported directly to the TNB, as illustrated in Figure 6-3.[8]

Stage Three: Conducting Interviews with Caribou Experts

With help from the TNB and other respected elders, we identified local caribou experts. Between 1997 and 2001, we interviewed thirty-seven[9] elders and hunters and made over 100 hours of both audio and video recordings. Each of these elders and hunters were interviewed on at least two separate occasions.

We used semidirected interview techniques (Huntington, 1998) and conducted interviews mainly in Inuinnaqtun with two researchers and one student, either in elders' centres, at people's homes or on the land. An elder-youth camp held for one week during the 1998 summer turned out to be a main component of the research (Thorpe, 1998; Thorpe and Eyegetok, 2000a; 2000b). This natural environment facilitated storytelling, capacity building[10] and elder-youth interactions. On the land, it was both easy and enjoyable for elders to share their knowledge of and experience with caribou and calving areas, demonstrate caribou and people interactions (e.g., hunting, butchering, skinning), and spend quality time with youth (Figure 6-4). Elders led the camp by deciding which lessons would be taught each day and sharing what they felt was most im-

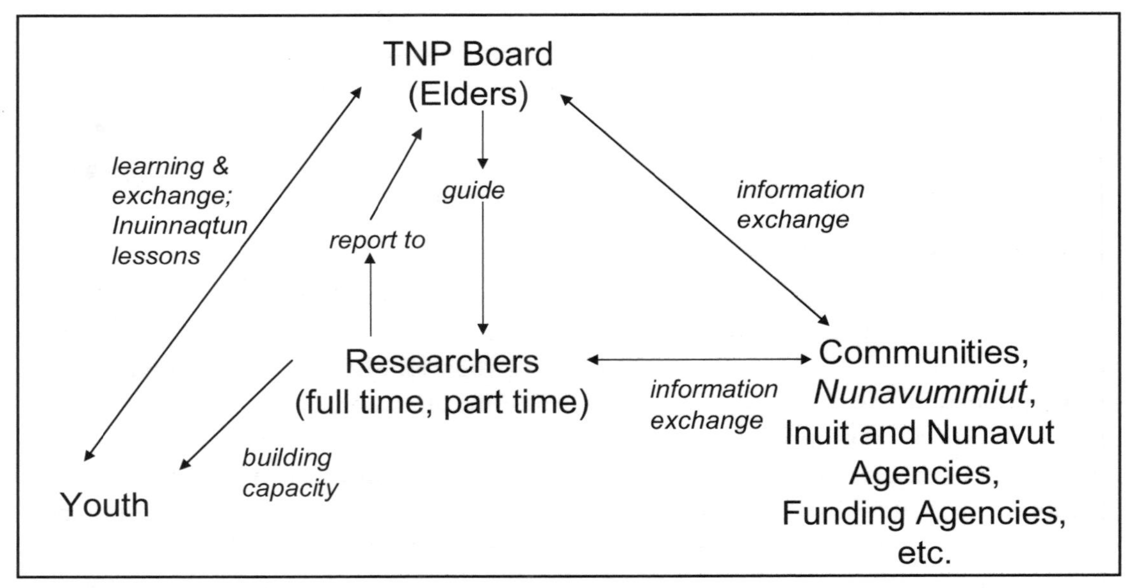

Figure 6-3. Organizational framework for the TNP.

portant for the youth to learn. They achieved this through proactive demonstration, rather than simply responding to what other people determined was valuable as can be the case when using a questionnaire or a ready-made camp agenda.

We developed the interview questionnaire with local hunters and elders who revised and tested the questions. Once the interviews began, the questionnaire was again revised when we realized it was too long and difficult to use as a guide rather than as a list of questions that needed to be systematically asked. Ironically, our lengthy questionnaire often hindered more free conversation, rather than facilitating discussion. With such a long list of questions it was sometimes easier to simply skip to the next question when the elder or hunter answered yes or no. This was especially true when we were new at interviewing and did not want to be responsible for breaking the rules of the questionnaire. When it was not comfortable or possible to carry out a more relaxed conversation, the questionnaire became a crutch. However, as the research progressed, we became more at ease and had more informal and comfortable conversations with interviewees.

Visual aids such as photographs, maps, drawings, and books helped to trigger discussions during the interviews. With clear plastic (mylar) overlays taped to 1:250,000 map sheets of the study region, elders and hunters used felt pens to mark out hunting

Figure 6-4. Elder Mary Kaniak of Umingmaktuuk joins two youth, Koaha Kakolak and Pitik Niptanatiak, to help them set up a caribou skin tent along the shore of the Hiukkittaak River. It took twelve caribou skins for Ella Panegyuk of Umingmaktuuk to make this tent.

areas, caribou migrations, calving grounds, and whatever else related to their stories (Figure 6-5). These markings were digitized and incorporated into a geographic information system (GIS) for viewing on a computer or printing out maps.

Of thirty-seven people, nineteen marked on maps, the majority of whom were men. One reason why many elders did not like using maps is that they traditionally relied on mental maps rather than paper maps. Some people had difficulties moving from images in their minds to representations on paper. Of those who did use maps, many did not feel comfortable marking areas that they had not been to before. Instead, they referred us to other people who might know more about a particular region or phenomenon.

Stage Four: Transcribing and Verifying the Interview Transcripts

Once the interviewing was finished, the tapes were transcribed.[11] Although most interviews were conducted in Inuinnaqtun, a small number were in English. Regardless of the language used in the interview, every transcript was translated so that it could be available to all community members in either language.

Interviewees verified their transcripts and accompanying maps in different ways depending on their age. Younger hunters read through the transcripts quickly, made necessary changes, and responded to additional questions that arose. However, verification for elders meant that we had to read aloud each transcript to them. This was time consuming, but necessary to ensure the accuracy of the transcripts and to respect the elders. Reading aloud gave us the opportunity to ask additional questions about either confusing or remarkable issues. Recordings were made of the verification so that new information could be added to the existing transcripts.

Stage Five: Sorting and Organizing the Transcripts Using a Database

Transcripts were then entered into a computer database that was engineered by BHP Diamonds Inc., Kitikmeot GeoSciences Ltd., the Kitikmeot Inuit Association, and the Kugluktuk *Angoniatit* Association. In use today, this database is like a giant library that classifies and stores all of the interview transcripts and maps in a manner that provides for easy retrieval of specific information. Sections of the transcripts were coded

Figure 6-5. Elder Nellie Hikok of Kugluktuk shows Margo Kadlun-Jones some of the places where she used to camp while hunting for caribou.

by keywords, such as "traditional use," "calving," or "migrations," so that like sections could be grouped together.

Continuing with the metaphor of the database as a library, each transcript section is filed under several different headings on various bookshelves. In this way, anybody using the database can learn more about a particular topic by simply entering a keyword that triggers a search through all of the interview transcripts.

Within the database, interview transcripts are linked with the map overlays created by the interviewed elders and hunters. The database allows the interviewees' knowledge to be connected with places on the land through the use of maps and text boxes that pop up when one points and clicks the mouse at a specific feature on a map. For example, a story about caribou migration is linked to a series of arrows on a map. A traditional camp, marked with an *X* on a map, is linked to an elder's story about how he spent the summer hunting caribou with his family at this camp. An example of this map (Figure 6-6) shows the overlap between caribou migration routes and Kitikmeot Inuit travel routes.

As with all research materials, access to the database is free and open to Inuit as well as Inuit and Nunavut agencies. In the future, access by outside agencies may be controlled by a password, although this has not yet been developed. Although the TNP is complete, the TNB and elders have revisited the research agreement signed in 1997 and are currently in discussion with the local heritage society and designated Inuit organization as to exactly how the research materials will be shared.

Three key advantages of this database[12] are that it is expandable, can be regularly updated, and that it is compatible with the NTKS. In this way, the two IQ projects can easily supplement one another.

Stage Six: Deriving a Data Set: SemiStructured Interviews about Climate

Using the database, we developed a list of keywords (n≈200) in order to sort and group similar sections of each interview transcript such that the interviews were systematically categorized by themes (e.g., climate change, hunting techniques, migration routes), and subthemes (e.g., warming trends and cooling trends, traditional hunting techniques and modern hunting techniques, spring migration, and fall migration), as shown in Figure 6-7. These keywords (i.e. themes and subthemes) emerged throughout the research and roughly mirrored major headings of the interview questionnaire.

With the interview transcripts now coded, we commanded the database to sort and output the transcript sections by theme. At the end of the day, we were left with twenty-one piles of interview transcripts each representing a theme. Judging by the thickness of the pile, climate emerged as a key topic. It was discussed by twenty-seven of the thirty-seven elders and hunters.

Next we systematically categorized the sections within each interview transcript according to six climate keywords: climate change, warming trends, cooling trends, environmental effects, vegetation, and caribou. The necessary condition to include a transcript in our analysis was that any given section had to contain any combination of at

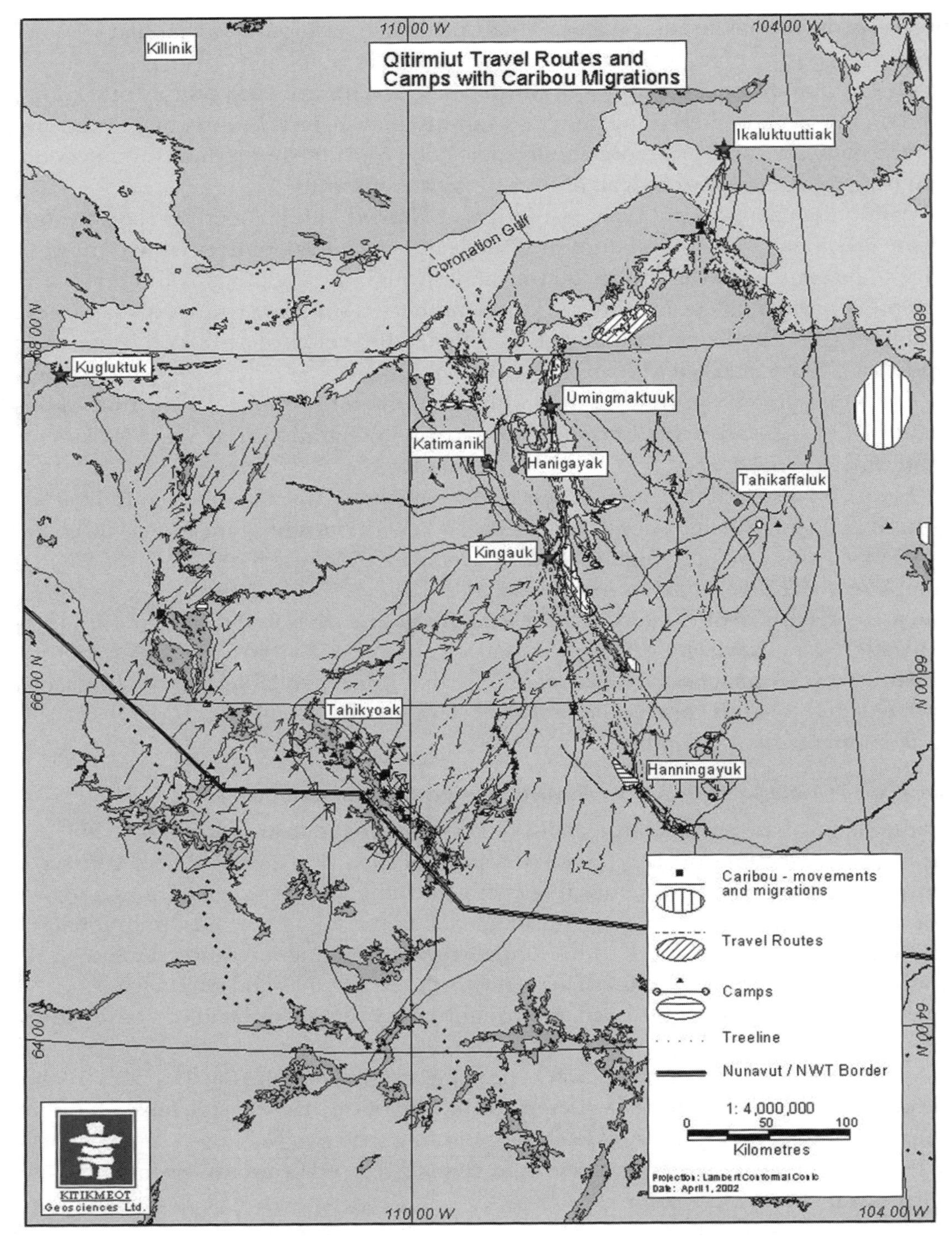

Figure 6-6. A map generated using the TNP database to show all spatial references to Kitikmeot Inuit (Qitirmiut) travel routes and caribou migrations. Note the overlap between these two that illustrates how Inuit traditionally followed caribou migrations.

least two of these six keywords: for example, "caribou" and "cooling trends" or "vegetation" and "climate change." The twenty-seven transcripts that met this criterion are of interviewees who generally represented the age and gender of the thirty-seven interviewees: they contain individuals from all four study communities and range in birth years from 1911 to 1956 with an average of 1934. They were typical of a cross section of the communities in terms of family groups and traditional hunting areas.[13]

Stage Seven: Communicating Results

After hearing community concerns about numerous earlier researchers who failed to return to communities to share their results, the TNB and TNP staff made the appropriate and effective communication of results a priority. Community members directed the research team to present results in a user-friendly format with "lots of photos, colour" and "not too much *qaplunaaq*" (in this case, meaning western or English influence). Elders and hunters emphasized the importance of sharing IQ through non-written means. Accordingly, a book, several videos and the database were created.[14] Our book, *Thunder on the Tundra: Inuit Qaujimajatuqangit of the Bathurst Herd,* is perhaps the best way in which the results from the TNP are communicated. It is a collection of IQ based on interviews and contains illustrations made by elders and youth as well as photos of every participating elder or hunter. Through the book, we attempted to celebrate IQ by demonstrating its richness and importance to our greater understanding of caribou and climate and to both honour and empower all Inuit.

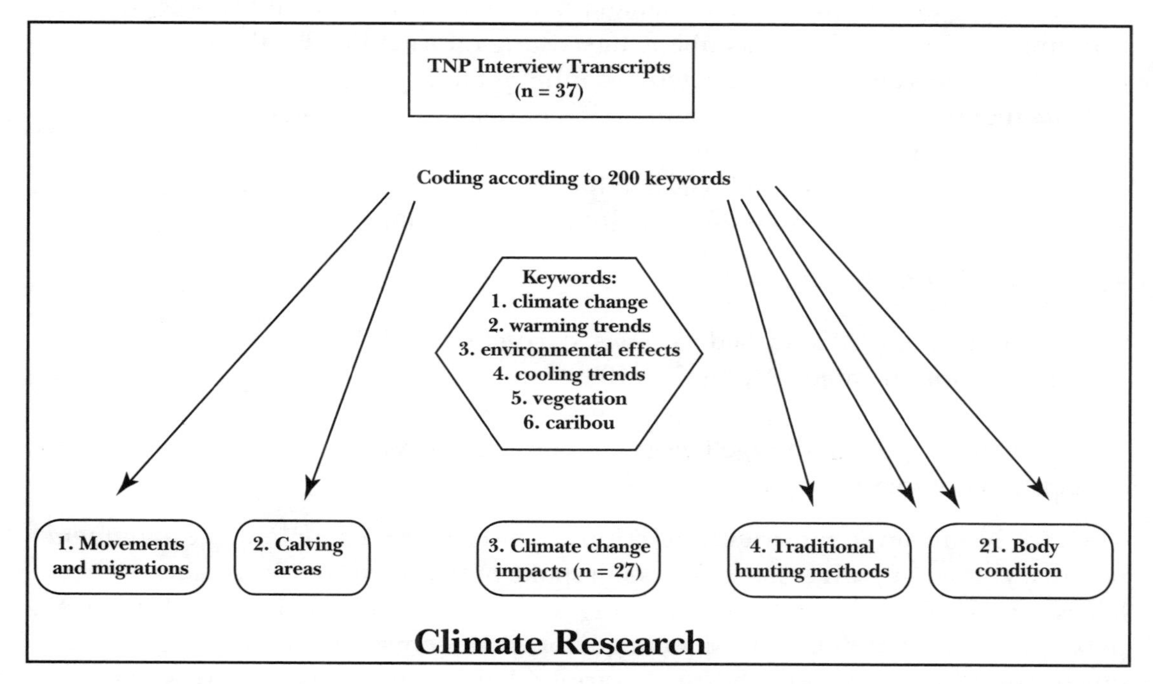

Figure 6-7. Using the database to identify climate change impacts on caribou as an emerging theme.

Stage Eight: Storing and Distributing Research Materials

Copies of the original transcripts and tapes were distributed to the Hunters and Trappers Organizations in Umingmaktuuk and Kingauk, the Kitikmeot Heritage Society in Ikaluktuuttiak, and the Kitikmeot Inuit Association in Kugluktuk. Until the new cultural centre is completed in Ikaluktuuttiak in May 2002, the original tapes are being stored at the Prince of Wales Heritage Centre in Yellowknife.[15]

The Weather Gets Too Hot

> Weather, it is hard to say, but I find the weather from long ago is getting hotter, getting hotter every year. I noticed the weather is getting warmer and hotter from long time ago (Kavanna, 1998).

For generations, Kitikmeot Inuit have observed correlations between the weather, land, and caribou to enable them to survive in a harsh and often unforgiving environment. These relationships are changing, however, due to a combination of earlier spring melt and later fall freeze-up. Higher temperatures led to a longer period of summer-like conditions, particularly in the late 1990s, that have influenced the migrations, body condition, and population levels of local caribou.

In this section, Kitikmeot Inuit observations of a warming climate and associated benefits and costs are presented to set the context for the rest of the chapter, in which we discuss the effects of earlier spring-melt and later fall freeze-up on caribou. Also, we discuss the contemporary unpredictability and variability in weather patterns, particularly as they relate to caribou. In our analysis, we demonstrate the utility of IQ in contributing new information and enhancing existing information. Furthermore, we illustrate the richness and interconnectedness inherent to IQ.

Note that this framework is a combination of relationships articulated by Kitikmeot Inuit as well as the researchers' interpretations of observed linkages. Figure 6-8 is an oversimplification of the complexity and interconnectedness of these ecological variables, but it serves to illustrate some identifiable associations.

Temperatures are Getting Warmer

> It never gets as cold as it used to...it used to get real cold in the past but nowadays it is not the same. (M. Algona, 1999)

> A long time ago, Cambridge Bay was usually cold. Nowadays ... the weather gets too hot sometimes. (Komak, 1998)

Of the twenty-seven interview transcripts reviewed, eleven people spoke about temperatures increasing while just one person spoke of temperatures decreasing. Other locals commented on recent climate changes during informal discussions (Hakongak, pers. comm., 1998, 1999, 2000; M. Omilgoetok, pers. comm., 1998). Temperatures during the 1990s were said to be much warmer than in earlier decades (B. Algona,

1999; Hakongak, pers. comm., 1998; Corey, pers. comm., 1999; Kadlun-Jones, pers. comm., 1999; Stern, pers. comm., 2000).

> It was so much warmer the last few years, in the 1990s and the late '80s. (B. Algona, 1999)

> It has been quite warm, though, [compared to] other winters ... in the last four or five years I guess (Kapolak Haniliak, 1998).

Some people (M. Angulalik, 1998; Komak, 1998) complain that these warm years have brought "too hot" (Komak, 1998; Kailik, 1999) temperatures that are "not so nice" (M. Angulalik, 1998).[16] Such hot temperatures are one of the deleterious effects of Arctic warming. In the following section, we elaborate on local reflections about how warmer temperatures are both positive and negative in how they influence daily life.

Some Pros and Cons of Warm Weather to a Subsistence Way of Life

> Nowadays, [the ice] goes earlier so waiting is not so bad anymore (Komak 1998).

People both like and dislike the impacts of higher temperatures. Most advantages are linked to the fact that summer-like conditions last longer now. Waiting for the ice and snow to melt can be trying because people cannot leave the communities by snowmobile or boat while the land is a mosaic of thin snow and bare patches or the sea

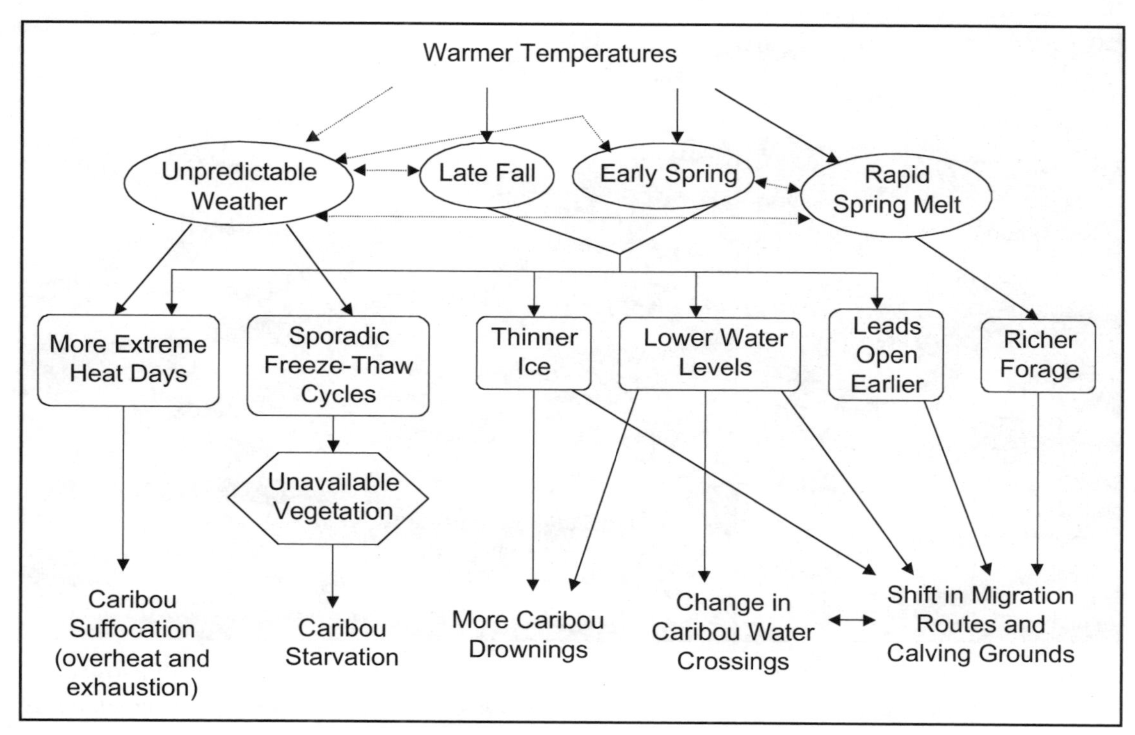

Figure 6-8. Kitikmeot Inuit observations of causal relationships resulting from warmer temperatures.

ice is a fractal maze of dangerous cracks called *uinit*. There is not enough snow and ice to travel safely, while at the same time there is too much snow and ice to go boating!

An early spring melt makes people excited to go travelling, hunting, and camping (Figure 6-9). Locals found that in the 1990s the boating season and fishing opportunities arrived earlier and lasted longer (Kakolak, pers. comm., 1998). Birds were more plentiful and arrived earlier than usual, thereby making geese hunting and egg gathering more fruitful.

Along with these positive impacts, warm weather during the summer can also bring unfavourable conditions for subsistence activities. Recently, there have been "hot hot" and "humid summers" which has made the land "drier" (B. Algona, 1999) and brought more bugs, especially mosquitoes (Analok, 1998; Kakolak, pers. comm., 1998; Kamoayok, pers. comm., 1998; Alonak et al., 1998; G. Panegyuk, pers. comm., 1998). The number of mosquitoes increases with temperature, but only until a certain threshold, at which point the mosquitoes cannot survive. This threshold was crossed during the 1998 summer when mosquitoes were a nuisance for only one instead of the usual two weeks (Kamoayok, pers. comm., 1998; Hakongak, pers. comm., 1999).

Warmer temperatures combined with the increase in mosquitoes are said to make travelling, hunting, and camping out on the land a challenge to one's stamina. During the 1998 summer, Natasha travelled with an extended family to an outpost camp sev-

Figure 6-9. Mona Keyok of Umingmaktuuk prepares arctic char at a traditional fishing and hunting camp on the eastern shores of Bathurst Inlet. This has been a popular spring camp for as long as people can remember.

eral miles outside of Umingmaktuuk. The plan was to spend the weekend camping, fishing, and hunting. Instead, the group decided that the mosquitoes and heat were so unbearable that they left after just one night. In this case, the temperatures were hot, yet not above the threshold for mosquitoes.

Seen for the First Time: Newcomers to the North

Warmer temperatures have brought profound changes to the nature and distribution of birds and animals in the region. Birds and animals are being seen for the first time in living memory. Robins and an unidentified yellow songbird have flown as far north as Kiillinik (Victoria Island) and are not reported to be common (Haniliak, 1998; Analok, 1999; Eyegetok, pers. comm., 1999). The same phenomena are described in the nearby community of Sachs Harbour and Holman (IISD, 2000; Inuktalik, pers. comm., 2000; Joss, pers. comm., 2000) as well as in the central and eastern Arctic (see Fox, this volume).

Previously unheard of, grizzly bears were seen crossing from the mainland northwards to Kiillinik in 1999 (Corey, pers. comm., 1999; M. Omilgoetok, pers. comm., 1999). Community members from Umingmaktuuk and Kingauk cited an unusually high number of grizzlies and their tracks during the spring of 2000 (Maghagak, pers. comm., 2000). This is worrisome because the bears might start visiting areas where people have their outpost camps, especially on the south coast of Kiillinik (Hakongak, pers. comm., 1999).

During the spring of 2000, hunters described seeing strange tracks around the Hope Bay area in Elu Inlet (Maghagak, pers. comm., 2000). A few weeks later, pilots from a local airline company saw three polar bears in the same area. Locals reasoned that the unusually high number of *uinit* early in the spring caused the seals to come up onto the ice sooner than usual, so that the polar bears simply followed the cracks southwards looking for seals. Although there are stories about polar bears coming this far south in the distant past, this was the first time that people could recall seeing polar bears at such low latitudes.

Kitikmeot Inuit are generally happy to see new birds because they are a source of food or curiosity, but bear movements are of concern because bears can threaten people's safety. Others are bothered simply because the rate of changes in the weather, environment, and wildlife populations seem to be happening too quickly for people and the environment to adapt (A. Kaosoni, 1998; M. Kaosoni, 1998; Hagialok, 1998).

A Kitikmeot Inuit Explanation for Warmer Temperatures

> There is a change in the distance of the sun ... from years back. The sun seems to be higher than it was a long time ago ... In the past, the sun was lower than it is now in July. It seems to shine higher than it used to. (Analok, 1999)

> It must be hotter than it used to be now ... the sun seems to be closer now. (Keyok, 1998)

Three of the eleven interviewees suggested that warmer temperatures are a result of the sun being closer to the earth and higher in the sky nowadays. This hypothesis proposed by Analok (1998), Kailik (1998) and Keyok (1998) may challenge scientists' and other peoples' conceptual frameworks. However, the elders' words show insight into some of the logic and reasoning components of IQ. For example, such hypotheses may be interpreted as coming from individuals extrapolating from their own experience to a grander phenomenon, based on the notion that the closer you stand to the fire, the hotter you will become. As this example demonstrates, Kitikmeot Inuit have extrapolated a global phenomenon from their knowledge of a local causal relationship, regardless of whether these elders were well informed of global warming by media or have observed a warming climate on their own. The ability to move from a small to large scale and to impute causal relationships is characteristic of Inuit way of thinking, as told throughout this volume. Knowing this may enable people to understand better the logic and reasoning that underlie IQ.

How Observations of a Warming Climate May be Influenced

That eleven interviewees reported warming temperatures strongly supports the conclusion that locals perceive climate to be warmer. However, three factors may have influenced people in their statements about climate change. These are important to consider because they show that, as in other disciplines, observations should not be accepted without scrutiny (Krech, 1999; Cruikshank, 1998, in press; Thorpe, in press).

The first possible influence follows from the fact that families have moved from outpost camps to communities within the last fifty years and now enjoy the warmth of a permanent home rather than a snow house or *iglu*. Consequently, people might think that the weather is getting warmer when really it may be that they are spending more time indoors and less time outdoors. This lifestyle shift may cause perceptions of temperature to change. Thus, Kitikmeot Inuit tolerance to and sensitivity regarding any change in their environment may have been altered because of new living conditions.

Second, news of global warming has trickled into the north and is another factor that likely influenced interviewees' perceptions. Recently, Inuit, national, and international broadcasting coverage has highlighted climate change issues, especially regarding threats to Canada's northern passage and sovereignty, and stalls in ratifying the Kyoto protocol. In addition to being influenced by media, interviewees may have developed an awareness of global warming through word-of-mouth that has become *a priori* knowledge.

> Five years ago now … Kugluktuk was the hottest spot in all of the North America. (B. Algona 1999)

This quote illustrates how weather and climate news have reached some interviewees and could have influenced perceptions.

A third possible influence is that people generally have better memory of more recent versus past events. Interviewees may remember climatic conditions in the 1990s better than the 1950s. Since most interviews were conducted during the record-breaking hot year of 1998 (Jeffers, 2000), people may have overemphasized the presence of a warming trend. Nobody was expecting the heat to be intense enough to melt most of the snow and ice in just a few hot days in May! Caribou meat usually kept cool outdoors in a snowbank, suddenly went bad and was unfit to eat. During the same month, two teenagers who were snowmobiling in Umingmaktuuk nearly drowned when the ice broke and they fell into frigid waters while trying to travel over ice that had been safe just the day before. Although the snowmobile sank to the bottom of the sea, the teenagers were saved because there were witnesses who acted quickly to save them. The fact that the hottest year on record was fresh in the minds of interviews may have influenced people to conclude that temperatures are increasing.

It is impossible to know how these three possible influences entered into the interviews, but the fact remains that eleven interviewees observed a warming trend and reported concomitant impacts on the local environment. Such similarities in the interviewees' observations suggest that perceptions of a warming phenomenon are not only real, but also recurrent and systematic.

Some Key Effects of Warming Weather on Caribou

Safe and successful hunting, fishing, and sealing require careful observations of snow and ice conditions, especially during times of melting and freezing. Inuit have always studied ice thickness, snow patchiness, melting rates and patterns, and changes both within and between the layers of snow and ice. These observations guide them to safe travel routes, for example, where the ice is thick enough to hold a snowmobile or thin enough to allow for ice fishing. In the past, subsistence living provided a strong impetus to monitor carefully the environment. Today, Kitikmeot Inuit do not subsist entirely off the land, yet hunting, fishing, and sealing remain significant pursuits such that careful observation of ice and snow conditions during spring-melt and freeze-up remains important.

Earlier Spring Melt

> [T]he weather is warmer now ... the snow seems to go earlier in the late spring. (Komak 1998)

> A long time ago, the ice on the ocean used to go away late (Analok 1998).

Five of twenty-seven interviewees commented that the spring-melt occurred sooner in the 1990s than in the past, and attributed this to warmer temperatures. Spring melt is said to be intensified by the heat generated by buildings, traffic and human activity, which leads to an even earlier melt in communities as compared to out on the land. Thus as communities grow, so does their capacity to generate heat.

> Here in town … the snow seems to melt earlier … because the traffic makes the snow melt … because when … people are gathered in one area, the snow seems to go quicker. (M. Angulalik, 1998)

This is an example of IQ that can contribute hypotheses, for example, to research on how a developed area may experience the effects of climate change.

Shift in Migration Routes:* Uinit *Open Earlier

Kitikmeot Inuit have observed that the warming climate in the 1990s, and the resulting changes in water, ice, and snow caused caribou to shift their migration routes. There were differences in the timing and manner that ice melted, and in the levels of water bodies that caribou had to cross.

On a large scale, warmer spring temperatures caused the ice to melt earlier, which created *uinit* (leads) that influenced caribou during their spring and fall migrations across the sea ice. If a lead is too wide to swim across, caribou will turn around and go back to shore, or they will move parallel to the lead until they find an area where the ice is thick enough to continue on their route. Sometimes caribou swim across leads if the open water stretches from the ice edge to the shore. Otherwise, Kitikmeot Inuit report that caribou rarely swim across *uinit* because it is difficult for them to get out of the water and on to the ice. When leads are particularly wide, caribou are forced to shift their migration routes.

One interviewee commented that leads along a traditional caribou-crossing route have been opening earlier in the last few years near the community of Kingauk (Kapolak Haniliak 1998). A traditional caribou-crossing route is understood as an area where people have observed caribou either swimming across a water body or walking across the ice year after year. With leads opening earlier, some caribou have not been able to migrate from the west to east and across the sea ice at the southern region of Bathurst Inlet (Kapolak Haniliak, 1998; Hakongak, pers. comm., 1999). This was the case during 1997 and 1998 when caribou remained on the west side of the Inlet to calve rather than crossing over to the east side.

> [The caribou] mainly come through there [across from Elliot Point in southern Bathurst Inlet] when they are calving on the east side. For some caribou, it is still the same. … The caribou … come around through from Portage Bay off Kuadjuk Island … they come south then go around again. They cannot hit that open water down there … That open water has always been the same … [but] it opens earlier and way bigger [nowadays]. (Kapolak Haniliak, 1998)

When rivers melt earlier than usual, they can flow with such speed and force that they can be difficult for caribou to cross. In addition, large pieces of ice floating downstream can be dangerous.

> Generally, [the caribou] have been staying to the west. Only a few coming through the last few years, mostly staying around the west. Maybe it is because the Hood and Bathurst Rivers started to go earlier and they have a hard time

getting across. All the rivers. Warmer weather these past few years. (Kapolak Haniliak, 1998)

Kitikmeot Inuit make careful note of which leads open first and how these leads shift throughout the melting period from one year to the next. Certain places along winter travel routes are known to be extremely hazardous during the spring and fall because of ocean currents that make the ice thickness unpredictable. Such expertise is valuable to travelers' safety and may also be of value to scientists who are assessing how caribou are influenced while making sea ice crossings. This may be particularly relevant to researchers looking at how traditional caribou-crossings on sea ice may be affected by icebreakers and the length of the season (i.e., shipping window) during which large ships could operate.

Migration Towards Rich and Abundant Vegetation

In addition to leads opening earlier, another possible reason for migration shifts may be the availability of richer vegetation for forage and more lush vegetation for shade. Kitikmeot Inuit noticed that tundra vegetation was more lush, plentiful, and diverse in the 1990s compared with earlier decades. Five locals reported that vegetation is larger and there are more individual plants of the same species, particularly shrubs, in certain traditional camping areas such as the Hiukkittaak River that flows into the east side of Bathurst Inlet (B. Omilgoetok, pers. comm., 1998; P. Omilgoetok, pers. comm., 1998; Kamoayok, pers. comm., 1998; Alonak et al., 1998; Koihok, 1998). The number of different types of vegetation has also increased, especially at higher latitudes such as on Kiillinik (B. Omilgoetok, pers. comm., 1999; P. Omilgoetok, pers. comm., 1999). According to Moses Koihok (1998), nowadays there are some types of lichens and flowering plants on Kiillinik that he has never seen before.

On a small scale, caribou are attracted to areas where the vegetation is rich, moist and green. Areas that first become snow-free support abundant vegetation that is attractive to cows seeking forage rich in nutrients (Figure 6-10). This vegetation is typically rich with fresh green shoots and new bark, both of which are preferred by caribou (Alonak, pers. comm., 1998; Akana, pers. comm., 1998; Alonak et al., 1998; Kamoayok, pers. comm., 1998).

> When plants sprout on the land [caribou] eat it. (B. Omilgoetok, 1998)

> They like it there. It might be much safer for them to have their young ones there and more food there for them. Older caribou must be used to it. (Kuptana, 1998)

> Must be, maybe [calving grounds are] probably a little more richer. Maybe it comes out, maybe it is exposed earlier in the spring so that it has time to grow ... they find a spot where the snow goes the soonest, the vegetation grows quicker maybe. ... The vegetation would have to be better too. I think the food would have to be a bit richer because the calves need to grow quickly because

> they start walking soon after they are born but they need to eat all the nutrients as fast as they can before the fall or winter set in. It is going to be a long winter so what they eat probably has to be a little bit richer than what they normally eat. (Hakongak, 1998)

> Maybe they know the migration pattern from the year before [or] maybe they know there is not going to be as much food as before so they move over a couple of miles so that they are sort of like lawnmowers. When you think about it, you know they go along at a certain length of an area on the way up [North] and the next year they'll move over a little bit. (Hakongak, 1998)

Vegetation influences caribou behaviour, especially their migration and selection of calving grounds.

Tall and Lush Vegetation Provides Shade

Based on Kitikmeot Inuit observations, it appears that earlier spring-melt combined with warmer temperatures promotes plant growth near water bodies, which draws caribou to forage in these areas during migration and calving. Nine interviewees who discussed vegetation mentioned the significance of moisture and water bodies to caribou. Three of these interviewees spoke of the importance of the shady areas typically found near water bodies. At the same time, Kitikmeot Inuit have observed increasingly tall and lush vegetation growing near water bodies in the 1990s. It follows that caribou may be attracted to these areas because they provide refuge from the sun.

One evening during the elder-youth camp held along the Hiukkittaak River, elders reminisced about their travels as young children. The elders commented repeatedly that the plants near the shore of the river had grown taller and lusher than they remembered seeing in the 1930s and 1940s. When probed, some elders explained that willows and alders throughout the Bathurst Inlet region grow larger nowadays in terms of the height of the plant, girth of the stem, and number of branches than they recalled during years that they travelled with their parents and as young adults

Figure 6-10. Bathurst caribou nearby the Hood River on the west side of Bathurst Inlet. It is early July and the caribou are selectively foraging for the snow-free tundra vegetation.

(Kamoayok, pers. comm., 1998; Alonak et al., 1998; B. Omilgoetok, pers. comm., 1998; P. Omilgoetok, pers. comm., 1998).

Although no interviewee articulated the relationship between warmer temperatures and taller and more lush vegetation and that these conditions have ultimately provided caribou with an escape from the heat, it is possible to extrapolate this cause-effect relationship from the elders' independent observations.

> When it is hot outside the caribou would go on the shores of the ocean where it is cooler. They stay in the shaded areas as well. (M. Angulalik, 1998)

> When the weather is warm, the caribou would stay in shaded areas during the summer. People would look for caribou in the shades during warm days. It is easier for the hunters to get close to them. (Kavanna, 1998)

> [Caribou] probably eat all sorts of vegetation. Whatever is sprouting on the land. They are constantly on the move and would often stop to eat as they are travelling ... on lake shores, whatever is sprouting. Moist vegetation. Grass. Whatever is sprouting on the shores of lakes, river sides. Wherever there is moisture. (Analok, 1998)

> Caribou eat in the shade ... by the lakes. By the lakeshores, as well as shaded areas wherever they can find what they like to eat. Moss and grass are their food. (Akoluk, 1998)

While it is clear that caribou are drawn to moist areas near water sources, it may also be true that shade provided by large and lush vegetation also attracts caribou seeking relief from the sun. During the spring and summer, these shady areas become increasingly important as temperatures rise.

Later Freeze-Up

Five interviewees reported observing a later fall freeze-up in the 1990s, and that sea ice freezes later today than in the past. Before temperatures started to warm, it was common to have freeze-up begin in late August or September, whereas in the late 1990s, freeze-up commenced in October or November (Hakongak, pers. comm., 2000).

> Nowadays freeze-up occurs in November. Sometimes there would be no ice at all and other times it would go again after freezing up. It has changed. In the past, the ocean would be completely frozen over in November. (Analok, 1998)

> Sometimes in the late summer years ago it would freeze-up. Now it seems to freeze-up late and other times it would freeze-up earlier. Sometimes it would freeze-up late. That is how it is. ... This happened for many years. A long time ago the ice on the ocean used to go away late. (Komak, 1998)

This last passage speaks to the variable timing of freeze-up, yet Kitikmeot Inuit have noted a recent pattern of a later freeze-up and earlier break-up within this variability. As described later in this chapter and in many of the papers in this volume, this variability affects how Kitikmeot Inuit predict weather and environmental change.

Thinner Ice

> Nowadays it freezes up later than usual. [The ice] does not get thick as it used to. (Kailik, 1999)

> It does not get as cold as it used to. A long time ago it would be bitter cold and the ice would be thicker ... the water gets warm now and takes longer to freeze ... the ice does not get as thick as it used to maybe because the water is warmer than it used to be ... [my] brother was sealing one day and said "Sis, come and see this. The ice is thinning. It is not even spring yet and it is thin," ... the ice does not get as thick as it used to. ... Some of the younger people have mentioned that as well. (Nalvana, 1999)

With warmer temperatures, longer summer-like conditions and shorter winters, the ice does not have as much time to become as thick as in the past. Thinner ice has implications for caribou, since they need a certain ice thickness in order to travel across frozen bodies of water, both in the fall for caribou to cross during their migration southwards towards their wintering grounds and in spring for their migrations northwards towards their summer calving grounds. While the warmth of the sun, the length of daylight, and the timing of the season together may trigger caribou to cross over a frozen lake, river, or ocean, the ice may not be thick enough to support their weight.

If the ice is thin or there is open water in leads, caribou will alter their migration routes by walking alongside the water body for several kilometres thereby wasting valuable energy. Alternatively, they may attempt to cross over the ice, fall through, and die of hypothermia.

> Last year I noticed the ice close fairly late from the years before. That is when a few caribou were trying to cross from Cape Peel, on Kiillinik [Victoria Island]. I heard from the guys that were working for the North Warning System that some caribou drowned near Cape Peel, about seventy miles west from Ikaluktuuttiak. They were trying to migrate across towards Surrey Lake and Ikaluktuuk. ... I heard not lots drowned, but ... less than a hundred I think. (Kavanna, 1998)

In the 1990s, several Kitikmeot Inuit observed an increase in the incidences of massive caribou drownings because of thinner ice. While travelling by snowmobile in 1996 from Umingmaktuuk to Kugluktuk, two community members found themselves amid hundreds of antlers that were frozen and sticking out of the ice "like an antler forest" (Kakolak, pers. comm., 1998; G. Panegyuk, pers. comm., 1998). This can happen either in the spring or fall when the temperatures are not cold enough to form a secure ice surface.

Lower Water Levels

> The water level seems to be getting lower. … In the past, the water levels were higher. Some of the rivers have gone. … During the summer, in August, people travelling by boat have noticed that the water level has dropped compared to the past. People liked the water levels of the rivers and ocean then. The islands on the ocean seem to be getting bigger than they used to be. Kanuyauyaq, an elder, used to say the rocks barely showed back then. Now the island seems longer, higher and bigger. (Akana, 1998)

> The water level seems to have gone down in the lakes and the rivers do not flow as strong as they used to. The lakes seem to dry out too. … I wonder why that is happening. (B. Algona, 1999)

In all four communities, elders noticed decreasing water levels in lakes, rivers, and the ocean around Bathurst Inlet, affecting both people and caribou. While nobody specified why this is occurring, other observations made by Kitikmeot Inuit suggest that that water levels are likely dropping for a combination of reasons related to warmer temperatures: earlier break-up, later freeze-up, increased evaporation, and decreased snow-pack.

There are two ways that lowered water levels may influence caribou according to interviewees. With less water in rivers, caribou do not have to swim as far as they used to from one side of the river to the other. On a local scale, one might conclude that not having as far to swim might benefit caribou because it conserves their energy. Dropping water levels also alter shorelines, which are key caribou habitat.

> [Caribou eat] along the shores of lakes and the ocean…close to water, where there is moisture…they would eat along the shore…by the lakes on the land. Things must be different now because the water level is dropping on the ocean (Keyok 1998).

> There would be [caribou] trails by the lakes, and along the mouth of the river. The trails look like they've been drawn in when you are flying over the lakes. (Koihok, 1998)

It is probable that such shoreline changes affect caribou, but it is unclear how. Several interviewees commented on the importance of shoreline habitat for caribou as grazing potential, and as a cool and shady refuge from both the sun and mosquitoes.

> In the summertime, in the evening, when it gets dark, they would walk along the shore and graze. Caribou like to eat on the lakeshores where the grass is

> plenty. When it is hot outside the caribou would go on the shores of the ocean where it is cooler. (M. Angulalik, 1998)

> The caribou ... stay along shore in the summer in June and July. ... To stay away from the bugs and to keep cool [caribou] stay by the ocean and the islands. (Koihok, 1998)

> The caribou usually go to the ocean along the coast. They can also be seen swimming in the lakes, staying away from mosquitoes. (Analok, 1998)

> When it is hot outside the caribou would go on the shores of the ocean where it is cooler. They stay in the shaded areas as well. They would lie on the patches of the remaining snow as well to keep cool. (M. Angulalik, 1998)

> When it starts to get dark outside [caribou] would go to the lakes. ... Night time, caribou stay by lakeshores, which makes it easier for them to escape the wolves. (M. Kaosoni, 1998)

While these examples of IQ need further investigation, they stand together in demonstrating the contributions that IQ can and does make to our understanding of climate change impacts.

Unpredictable Weather

In addition to noticing the warmer temperatures in the 1990s, Kitikmeot Inuit find it more difficult to plan for the weather because it has become so unpredictable and variable. Today, it is problematic to understand and forecast weather because weather events seem to be beyond the realm of expectation of what people consider normal.

Kitikmeot Inuit have articulated two possible cause-and-effect relationships that associate the impacts of warming temperatures on caribou. Nowadays there are more cases of freezing rain and sporadic freeze-thaw cycles that lock vegetation in frozen sleet and make it impossible for caribou to eat and so they starve to death. Days of extreme heat were also more common in the 1990s, when caribou can become overheated, exhausted, and skinny. Combined, these extreme weather conditions could lead to lower caribou population levels. In this section, we demonstrate how IQ is used to interpret and understand the effects of a changing climate on caribou and the predictability of weather patterns.

Inuit Seasons

In the spirit of their worldview, Kitikmeot Inuit believe that intangible forces may guide the weather. This concept is a tenet of animism that underlies much of Kitikmeot Inuit culture. It is owed partially to a legacy of loss and struggle that force people to not only respect social and environmental changes brought by climate, but also to provide an explanation when reasons for such changes are not obvious.

Traditionally, a respect for weather combined with the need to monitor climatic conditions led Kitikmeot Inuit to identify and group similar weather patterns into six seasons: *ukiakhaq* (early fall); *ukiaq* (late fall); *ukiuq* (winter); *upin'ngakhaq* (early spring); *upin'ngaaq* (late spring); and *auyaq* (summer). Months were identified by phases of the moon and grouped together according to environmental conditions such as weather events and wildlife activity.[17] For example, winter includes the months that are bitterly cold *(iijji)* whereas fall and early spring include the months that are just cold *(alappaa)*.

Today, Kitikmeot Inuit continue to observe weather, wildlife, and phases of the moon, but now they link these to current calendar months. Frank Analok (1998) explained how people knew the seasons by detailing an annual cycle of weather changes, wildlife migrations, births, deaths, and moon phases. A portion of his discussion is presented below:

> They used the moon only, the moon. The moon was used to tell the seasons. Like it is June right now. They used the moon as a way to tell seasons long ago. It was a way to tell the seasons. When the moon would come during the spring thaw, when there is water, the caribou are calving and the birds are nesting. That is how it was. The month of June ... the moon would go away again during the month of June. When it returns you know when the birds are moulting. You know they are moulting that time of the year which is during the month of July. The moon was the only way the Kitikmeot Inuit knew the time of the year. That is the time of year the birds would moult, when the moon returned after the last. ... We did not know it was July then. After it disappeared it would return then it would be August. That is when the caribou furs would get nice. People would mention they were good for clothing. The birds would be flying again. The young birds would have grown then. That is how they knew the seasons. That is during the month of August. (Analok, 1998)

To categorize seasons, Kitikmeot Inuit identified weather patterns through repeated observations and by following a patterned thought process (Ross, 1992). For example, Kitikmeot Inuit noticed predictable cycles in the moon phases and connected these with the timing of the geese migration or sea ice melting. Linking these patterns to wildlife activity and moon phases enabled people to combine days into months and months into seasons and seasons into years through a very specific thought process.

> When there is too much water in the spring ... when it rains too much, the number of squirrels would go down. That is how it is. (Keyok, 1998)

> I remember when Bessie [Omilgoetok] talked about weather predictions. She said, "The loons are circling high. The weather will be funny tomorrow." Later I asked her how she knew. ... She said that animals can sense weather changes

> which makes their behaviour different and that it was taught to her to use animals' unusual behaviour to predict weather. (Eyegetok, pers. comm. e-mail, 2000)

Inuit use patterns in the snow and ice, changes in the wind strength and direction, and temperature fluctuations to forecast the weather.[18] In making these linkages, they predict weather conditions associated with certain months and seasons. In the 1990s, this element of predictability changed as a function of warmer temperatures and increasing variability in weather patterns.

Predicting Weather Today

> Back in the 1960s I could almost say I could predict the weather any day. Whereas nowadays you might think it is going to rain ... or snow back then but today there is nothing. Turn back and no [snow] dump. ... That is the big difference from the 60s and today. ... Everybody I talked to said they can predict the weather in the 60s. It's more predictable and stable, whereas, nowadays it's unstable ... any kind of weather any day. You can have rain in February sometimes nowadays. Or snow in August. Hail in August. Hail or even snow in July. I can remember ... six or seven years ago, we had a snow bank outside my door in August. That one night it really snowed and froze and a big snow piled outside my tent in August. About seven years ago [1992], I think it was. ... And next day it was all gone again. (B. Algona, 1999)

Three interviewees made direct reference to weather becoming more difficult to predict, while other community members have mentioned this phenomenon during informal discussions (Hakongak, pers. comm., 2000). Even though weather seems to be more changeable, Kitikmeot Inuit are humble and cautious to point out that:

> You never know the weather. ... It is a fact that the weather is never the same ... every year is always different ... everything is always changing. Right now it seems to be getting worse. That I have noticed in my lifetime. It used to get real nice outside when I was younger, right after a storm. Right now when the weather gets bad, it seems consistent. (Analok, 1998)

Inexplicable forces control weather and all other ecological variables. It follows that Kitikmeot Inuit try not to take chances with the weather given that it can be so unforgiving. Today, with weather so unpredictable, the risks of getting lost, caught in bad weather, or stranded on bad ice can be too great.

Freeze-thaw Cycles

> It has been melting sooner than usual, [and] then freezing again. ... It's been melting and freezing. (Anonymous, 1998)

Kitikmeot Inuit have also noticed more short-term temperature fluctuations which also make the weather unpredictable. At times these fluctuations cause a repetitive and sporadic freeze-thaw cycle. This occurs when a few days of warm weather that start to melt the ice and snow are followed by a sudden cold period, causing the meltwater to freeze and form an icy layer on top of the snow. These freeze-thaw cycles can happen in both the spring and fall, particularly during the times of break-up and freeze-up. Five interviewees discussed changes in the frequency and nature of freeze-thaw cycles and their effects on caribou.

> [It was] raining heavily. That is what happened once in Bay Chimo, it must have been 1977. There was hardly any caribou to eat that time because of the ice on the snow. It was really slippery. (Keyok, 1998)

> The snow was covered in ice. It had rained after a big snowfall. That is when some of the caribou had starved to death but in another area of land, where it is not so rough, they were fine. ... Some areas were fine where it did not rain. ... The land was covered in sleet and ice and some caribou and musk-ox froze to death. When the land is covered in ice, where it is not so rough, some caribou would freeze to death. (Komak, 1998)

> They had starved to death because of sleet. They had nowhere to eat. The ice was too thick ... they could not dig through it. (Koihok, 1998)

Kitikmeot Inuit have noticed that caribou numbers decrease during and after the years of frequent freeze-thaw cycles. Thus, a decrease in caribou population levels owing to starvation or death may be an indirect effect of warmer temperatures and the concomitant unpredictability of weather conditions.

"Hot Hot" Days

As discussed earlier in this chapter, during the last few years Kitikmeot Inuit have seen that days of extreme heat are more plentiful. This was particularly the case during the 1997 and 1998 summers when temperatures were hot enough to melt the ice and snow in just a few short days both in communities and out on the land. These hot days are said to be common in this recent era of unpredictable and unstable weather.

High temperatures can cause caribou to overheat and die of exhaustion. In extreme cases, caribou can overheat and then fall unconscious or die, especially while trying to escape mosquitoes.

> We saw dead caribou; they had died of exhaustion. We saw this in the recent past. (Komak, 1998)

> When there are too many mosquitoes [the caribou] would gather and go in circles to get rid of the mosquitoes. Sometimes when they shook the flies off it would make the sound like thunder. There would be so many mosquitoes that they would look like snowflakes, you can see, even from a distance. (Koihok, 1998)

> During the summer when there are a lot of mosquitoes in the warm weather they would die of exhaustion … when the weather is too hot for them. (B. Angulalik, 1998)

Another indirect effect of these extreme temperatures is that they increase the number of forest fires in the southerly regions of the Northwest Territories and Nunavut. The summer of 1997 was the haziest that several people could remember (Akana, pers. comm., 1998; Keyok, pers. comm., 1998; Alonak et al., 1998; Kamoayok, pers. comm., 1998).

> There is always smoke. When there is a forest fires down south, it really gets smoky up here and if it is foggy you could smell the smoke. Last year [1997] it was really smoggy. You could smell forest fire, maybe for at least five days, four or five days anyway. Somewhere there was a big forest fire and it was a dry year for the Yellowknife area. (Anonymous, 1998)

> From the smog, there were a lot of dead bulls on the land. It was really hot that time and some caribou had died. (Keyok, 1998)

> Probably [caribou were sensitive to smoke]. We were. We would go out and say "So stink! Cover your nose and mouth!" (Anonymous, 1998)

Community members speculated that these same hot temperatures that raise the number of forest fires and make the skies hazy, may also contribute to caribou fatalities. This suggests that as climate generally warms and days of extreme heat become more frequent, ways to prevent dehydration and overheating become more important for caribou.

Ways That Caribou Combat Heat

Kitikmeot Inuit described how caribou drink water, eat vegetation with high water content, and eat and suck on mushrooms to avoid dehydration in extreme heat. Caribou are thought to drink water to keep cool, often tasting the water before they make a crossing at a river. Even if Kitikmeot Inuit have not observed caribou drinking water, they speculate this is important for regulating body temperature.

> Lakes are important [to caribou] for fresh water. … I do not know how much water a caribou drinks though. I have never seen a caribou drink water. (Hakongak, 1998)

Seven interviewees mentioned that mushrooms are an important part of the caribou diet as they provide nutrients as well as water. Caribou can try to prevent dehydration by sucking or eating mushrooms and other plants that are high in water content.

> [Mushrooms] are what the caribou use to keep their mouths moist when they walk. They need water and that is what they use when they are thirsty … they

> would keep these mushrooms in their mouths because they are moist inside. Wet, really wet. ... When caribou are walking around, they could smell [mushrooms] right away and they go after them. ... [They] last a long time. The caribou would keep them in their mouths at the back of their cheeks. ... Just like whale blubber ... just like snuff. (Alonak, 1998)

> When the weather is hot during the summer the caribou would have those [mushrooms] when they are thirsty. (B. Omilgoetok, 1998)

Unpredictable Weather Across Nunavut

As reported here and throughout this volume, elders in other parts of Nunavut also suggest that weather is becoming increasingly more unpredictable. According to Bessie Inuktalik (pers. comm., 2000) and Sadie Joss (pers. comm., 2000) of the Olokhatomiut Hunters and Trappers Committee in Holman, weather is difficult to predict because storms and winds are stronger and more frequent and the rate of their onset is different.

Near Resolute Bay, Inuit have complained about the weather becoming "wilder" and the sun hotter each year (Davis, 1998: 44). In Coral Harbour and Rankin Inlet, east of the Kitikmeot region, elders are saying that the weather is warmer, there are more storms, and frost does not form on top of snow like it used to (Ussak, pers. comm., 2000).

> Even if we try to predict what it is going to be like tomorrow ... the environmental indication is not what the elders said it would be. Sometimes, it is still true but sometimes it is not. In the past, when they said, "it's going to be like this tomorrow," it was. But our weather and environment are changing so our knowledge is not true all the time now. We're being told [in Hudson Strait] that maybe if we put January, February, or March one month behind, our knowledge of weather would be more accurate, because the weather in those months is not the same anymore. (Lucassie Arragutainaq, in McDonald et al., 1997: 27)

These observations point to the changeability, instability, and variability in weather patterns nowadays that make it more difficult for people to know what to expect. Weather changes more suddenly, more often, and in ways that are atypical relative to peoples' lifetime observations. For example, in mid-June 2000, there were two weeks of 15°C days which led people to expect that summer conditions had arrived. Suddenly, the winds started to gust and the temperatures dropped back to 3°C. People feared that the baby birds and caribou calves may not survive such a sudden temperature drop. Weather has always varied, but locals report that this kind of changeability is more common in the 1990s because of the overall warming trend (Maghagak, pers. comm., 2000).

Conclusion

> People listened to traditional knowledge *[Inuit Qaujimajatuqangit]*. That is how they know everything. (Maniyogina, 1998)

Although IQ cannot possibly tell us all that there is to know about the impacts of a warming climate on caribou, findings of the TNP have demonstrated that Arctic residents, as keen observers and primary resource users, have much to contribute to our local and global understanding of environmental change. The TNP was carried out between 1996 and 2001 with thirty-seven elders and hunters from four Nunavut communities. Through this community-driven project, grounded in participatory and adaptive approaches, their knowledge and perspectives of climate change and its impacts on the surrounding environment and caribou emerged as a key focus for this chapter and our entire project.

Kitikmeot Inuit presented conclusive evidence of a warming climate marked by unpredictable and variable weather and a changed environment substantiated by new growth rates in tundra vegetation, lowered water levels, increased seasonal variability and the presence of new plant, bird and wildlife species in northerly latitudes. These observations concur with earlier studies throughout the circumpolar zone (for example, Vibe, 1967; Spink, 1969; Krupnik, 1993; Nakashima, 1993; Kofinas et al., 1997; Enerk, 1994; Cohen, 1997; Bielawski, 1997; Ferguson, 1997; Fox, 1998; Ferguson et al., 1998; Fox, 2000a, 2000b; IISD, 2000; Riedlinger, 2000; Riedlinger and Berkes, 2001; and others in this volume). While a comprehensive comparative analysis of local knowledge of climate change in the circumpolar zone is certainly required, it is beyond the scope of this paper. However, a preliminary and cursory review suggests that such similar and repeated findings speak to the utility of IQ at both a local and regional scale in contributing to a national and global understanding of climate change impacts on wildlife, including the caribou/reindeer as one of its key species.

Since the beginning of time, people have always connected weather patterns with wildlife activity and moon phases. Kitikmeot Inuit are now having more difficulty making these connections due to unpredictable and variable weather. Extreme events such as temperatures above the realm of reasonable expectation are new to the region and force caribou to find ways to prevent dehydration. Temperatures over 30° C were more common during the 1990s and led caribou (not to mention people!) to overheat and mosquito harassment to increase. Inevitable forest fires and the concomitant hazy skies caused caribou to "suffocate" from the smoke, particularly in the hot year of 1998. Changeability in the weather is evidenced through freeze-thaw cycles during spring melt and freeze-up that seem to be increasing in their frequency and causing caribou to have difficulty foraging through the hard layer of ice and snow that traps tundra vegetation such as lichen. Exactly how both people and caribou will adapt to these events is yet to be seen.

A key finding of the TNP related to climate change is that Kitikmeot Inuit observed a combination of earlier break-up in the spring and later freeze-up in the fall during the 1990s which led to a longer period of summer-like conditions. This may have some benefits to caribou, for example, in providing richer forage or more abundant vegetation that provides shade. However, most locals suggested that the effects of warming temperatures and the accompanying longer period of summer-like conditions are overall deleterious because they lead to higher caribou fatalities. For example, thinner ice means that caribou often drown while crossing semifrozen water bodies during their instinctual drive northwards to their calving grounds and their return southward migration to their wintering grounds. Alternatively, they may inefficiently expend energy by being forced to take less direct routes during these migrations. Lower water levels make river crossings shorter for caribou thereby conserving energy, but at the same time, they influence shoreline habitat that is important to caribou for grazing and escaping the heat, insects, and predators.

These specific events reported by Kitikmeot Inuit suggest important cause-effect relationships that contribute to a broader understandings of indirect and direct effects of a warming climate on the environment and, in particular, on caribou. Such insight can provide hypotheses for future research that can enhance knowledge of global change.

While western-science-based research provides much information about climate and caribou, insufficient attention has been given to how these animals are observed, experienced, and explained by Inuit and other subsistence users themselves, to the extent that IQ on caribou has remained poorly documented. While sharing many common properties with science, a fact that warrants a discussion too large for this chapter, IQ is unique in the ways that it is acquired, verified, and used. IQ is "what has always been known"—Inuit knowledge, insight and wisdom that is gained through experience, shared through stories, and passed from one generation to the next. More than just knowledge, as commonly defined, IQ includes a finely tuned awareness of the ever-changing relationship between Inuit and *nuna* (the land), *hila* (the weather), wildlife, and the spiritual world. Experiential knowledge, insights, and skills that are specific to Inuit way of life have passed the test of time and remain in use today. As this chapter and volume demonstrates, IQ should be celebrated for being a holistic view that provides new insight or enhances current perspectives and theories regarding the relationship between arctic climate change and the land, animals, and residents.

In recognizing that IQ has much to contribute, future projects can now be modelled after successful research approaches such as the TNP and other joint knowledge-sharing efforts chronicled in this volume. At least five factors were critical to the success of the TNP. First, the project came from Inuit themselves and was run by and for Inuit through the TNB, in ways that encouraged empowerment, capacity building, and elder-youth interaction. Second, the research approach remained flexible to local input and direction. This meant hearing and actually *listening* and then *following through* on community direction. For example, by publishing our final report as a multimedia book or collaborating with the Inuit Broadcasting Corporation to develop videos rather than preparing a relatively dry and technical project report. Third, interviewed elders and hunters retained ownership of the TNP. Having this control meant people seemed more willing to participate. Fourth, we were *mamiahuyuittuq,* or patient, and worked at a pace most respectful of and appropriate for elders. Patience meant everything from asking a question and waiting several minutes for an elder to formulate his/her response, to being committed to a research process based on community input and feedback. Such a commitment presented a challenge, especially given the pressures of fiscal year budgets, funding agencies, and academic timelines. Fifth, we have taken IQ from the oral domain to both a written and database format, complete with spatial linkages. Through our book and database, IQ is now easily accessible to Nunavummiut communities, schools, decision-making agencies, and other organizations throughout Nunavut and Canada (Figure 6-11).

Now in a new millennium, we are poised to face continued environmental and climate change, and the associated consequences to be experienced by people and their habitats. As this paper demonstrates, to be better prepared for change, we are to be equipped with sound data from both the scientific and Inuit communities that can help us monitor and influence this change and to find ways in which we can adapt to both the positive and negative inevitabilities. We know that climate is changing at a rate that is unprecedented in living human memory. We know that change is influencing people and their relationships with both reindeer and caribou. We know that change is affecting Inuit and other polar nations in threatening peoples' health and safety, subsistence, sustenance, culture, and identity. We know that change is altering the sustainability and health of the land, wildlife, water, and the many interactions between these ecological variables. The TNP and similar projects that are currently being undertaken across the Arctic zone, present clear and unequivocal evidence that "people on the ground" see this change coming and see it coming very quickly. What is not so clear is how the global community will use what we know. And in this regard, we cannot remain *mamiahuyuittuq,* patient.

Acknowledgements

On behalf of the TNB, we wish to thank everybody who participated in the project. We are indebted to the elders and other community members who gave us the faith and support we needed to make our joint effort a success. Our project relied fully upon a unique collaboration between elders, hunters, youth, researchers, artists, and academics, who are too numerous to mention by name, from the initial stages of community consultation to the completion of our project in a book format as well as this paper. Such co-operation would not have been possible without the enormous camaraderie and hard work demonstrated by many people, especially the community interviewers: Nancy Haniliak, Eileen Kakolak, Myste Kamingoak, Eva Komak, Meyok Omilgoetok, and Karen Ongahak and our fabulous translation team of Doris Angohaitok, Martha Angulalik, Margo Kadlun-Jones, Eileen Kakolak, John Komak, Mary Kaosoni, and James Panioyak.

We are grateful to the following agencies for recognizing the value of Inuit ecological knowledge and respecting the process of community-driven research: BHP Minerals Inc.; Department of Sustainable Development, Government of Nunavut; Education, Culture and Employment, Government of the Northwest Territories; Department of Indian and Northern Affairs, Government of Canada; Kitikmeot Heritage Society; Kitikmeot Inuit Association; Nunavut Tunngavik Incorporated; Nunavut Wildlife Management Board; West Kitikmeot/Slave Study Society and local hunters and trappers organizations. Natasha would like to acknowledge the generous support from the Arctic Institute of North America; Association of Canadian Universities for Northern Studies; Northern Scientific Training Program; and Social Science and Humanities Research Council.

Drs. Gary Kofinas and Evelyn Pinkerton have also provided tremendous guidance to the TNP over the years. This paper would never find its present form without the encouragement and dedicated assistance by/from this volume's editors, Igor Krupnik and Dyanna Jolly. And finally, our deepest thanks go to our families for their unwavering support and understanding and for enduring our time away from home.

Notes

1. Inuit from this region are also known as Kitikmeot (or Qitirmiut) or Copper Inuit. We use the term Kitikmeot Inuit throughout this paper to avoid any confusion. The term Kitikmeot can be used to mean a region or people from this region. That there is no distinction in the terminology between a region and people from this region points to how closely Inuit link themselves to the land.
2. The Kugluktuk Angoniatit Association (KAA) conducts the NTKS for the Kitikmeot Hunters and Trappers Association (KHTA).
3. The TNP has been funded by over thirty funding agencies. Primary funding was received through the Government of Canada, Government of Nunavut, Kitikmeot

Figure 6-11: Kitikmeot elders with Naikak Hakongak, Natasha Thorpe, and Sandra Eyegetok (standing) during a recent book launch ceremony in Ikaluktuuttiak. Over 150 community members attended the celebration where they were able to have the elders, artists, and other authors sign their books.

Inuit Association, Nunavut Tunngavik Incorporated, Nunavut Wildlife Management Board, and the West Kitikmeot Slave Study Society. For a listing of sponsors and their websites, see www.polarnet.ca/tuktu.

4. Many historical records report on early explorers such as John Franklin who perished because they did not follow local IQ of how to survive in the Arctic.
5. Meaning all people of Nunavut.
6. Standard and accepted ethnographic techniques and ethical principles agreed to by the Association of Canadian Universities for Northern Studies (ACUNS), the *Ethical Guidelines for Research* prepared by the Royal Commission on Aboriginal Peoples, and the *Guidelines for Traditional Knowledge Research* designed by West Kitikmeot Slave Study (WKSS) further informed the research.
7. Naikak Hakongak is a wildlife officer employed full time by the Government of Nunavut. The willingness of the Government of Nunavut to support Naikak in his participation of the TNP was critical to the success of the project. He assisted in all areas of research as well as logistics, guiding, and planning.
8. We were assisted on an as-needed basis by Nancy Haniliak and Eileen Kakolak in Umingmaktuuk; Martha Angulalik, Myste Kamingoak, and Karen Ongahak in Kinaguk, and Eva Komak and Meyok Omilgoetok in Ikaluktuuttiak.
9. Thirty-nine elders and hunters were actually interviewed but records and tapes from two interviews were unusable. For this reason, the total number of interviews considered is thirty-seven.
10. Capacity building is defined as providing training and skills that can be carried on by community members.
11. Mary Kaosoni of Ikaluktuuttiak did most of the transcribing and translating. She was assisted by Martha Angulalik, Sandra Eyegetok, Mary Kaosoni, John Komak, James Panioyak, and Naikak Hakongak.
12. For more information about this database, contact Kitikmeot GeoSciences Ltd. in Vancouver, BC.
13. For a detailed account of interviewees in terms of age, gender, and community, refer to Thorpe, 2000.
14. The use of video as a key communication medium has been recently demonstrated through *Sila Alangotok* (IISD, 2000; Jolly et al., this volume) as well as a CD-ROM pilot project (Fox, 2000b; this volume).
15. "Safe" in the sense of fire proof safes and humidity-controlled rooms. Standards for safekeeping have been set by the Canadian Museum of Nature.
16. While kayaking down the Hood River for a month during the same summer, Natasha rejoiced in daily temperatures reaching the high twenties because it meant a warm swim at the end of the day. However, this was not the norm and such hot temperatures seemed to suggest that she was not in the Arctic after all!
17. See MacDonald et al., 1997.
18. In other regions, it has been reported that "when clouds obscure the sun. . . Inuit study the reflection of the ice on the underside of low clouds" (Davis, 1998: 35).

References

Akana, J. 1998. Board Member, Tuktu and Nogak Project: Hiukkittaak River, personal communication.

Akana, J., elder. 1998. Interview by N. Thorpe, E. Kakolak and D. Keyok. June 8, Umingmaktuuk. Tape recording. Tuktu and Nogak Project, Ikaluktuuttiak.

Akoluk, M., *hunter.* 1998. Interview by N. Thorpe and M. Kamingoak. May 22, Kingauk. Tape recording. Tuktu and Nogak Project, Ikaluktuuttiak.

Algona, B. *hunter.* 1999. Interview by N. Thorpe and S. Eyegetok. November 2, Kugluktuk. Tape recording. Tuktu and Nogak Project, Ikaluktuuttiak.

Algona, M., elder. 1999. Interview by N. Thorpe and S. Eyegetok, November 1, Kugluktuk. Tape recording. Tuktu and Nogak Project, Ikaluktuuttiak.

Alonak, J., elder. 1998. Personal communication, Kugluktuk.

Alonak, J., L. Kamoayok, M. Kaniak, and B. Omilgoeok, *elders.* 1998. Interview by S. Akoluk, S. Eyegetok, E. Kakolak, K. Kamoayok, N. Mala, P. Niptanatiak, N. Thorpe and J. Tikhak Jr., August 8, Hiukkkittaak Elder-Youth Camp. Tape recording. Tuktu and Nogak Project, Ikaluktuuttiak.

Analok, F., elder. 1998. Interview by S. Eyegetok and E. Komak, July 22, Ikaluktuuttiak. Tape recording. Tuktu and Nogak Project, Ikaluktuuttiak.

Analok, F. 1999. Interview by N. Thorpe and S. Eyegetok, June 12, Ikaluktuuttiak. Tape recording. Tuktu and Nogak Project, Ikaluktuuttiak.

Angulalik, B., elder. 1998. Interview by S. Eyegetok and E. Komak, July 30, Ikaluktuuttiak. Tape recording. Tuktu and Nogak Project, Ikaluktuuttiak.

Angulalik, M., elder. 1998. Interview by S. Eyegetok and E. Komak, July 24, Ikaluktuuttiak. Tape recording. Tuktu and Nogak Project, Ikaluktuuttiak.

Anonymous, *hunter.* 1998. Interview by N. Thorpe and K. Ongahak, June 5, Kingauk. Tape recording. Tuktu and Nogak Project, Ikaluktuuttiak.

Bielawski, E. 1997. Aboriginal participation in global change research in the Northwest Territories of Canada. In: W. C. Oechel, T. Callaghan, T. Gilmanov, J.I. Holten, B. Maxwell, U. Molau, and B. Sveinbhornsson, eds. *Global change and Arctic terrestrial ecosystems,* pp. 475–483. New York: Springer-Verlag.

Burgess, R. G. 1984. *In the field: An introduction to field research.* London: George Allen & Unwin.

Chambers, R. 1991. Shortcut and participatory methods for gaining social information from project. In: M. Cernea, ed. *Putting people first: Sociological variables in rural development.* Oxford: Oxford University Press: 515–537.

Chataway, C. J. 1997. An examination on the constraints on mutual inquiry in a participatory action research project. *Journal of Social Issues 53,* no. 4: 747–765.

Cohen, S. 1997. What if and so what in northwest Canada: could climate change make a difference to the future of the Mackenzie Basin? *Arctic 50,* no. 4: 293–307.

Corey, G. 1999. Wildlife Officer, Resources, Wildlife and Economic Development, Government of the Northwest Territories: Ikaluktuuttiak, personal communication.

Cruikshank, J. 1998. *The social life of stories: Narrative and knowledge in the Yukon territory.* Lincoln: University of Nebraska Press.

Cruikshank, J. In press. Uses and Abuses of "Traditional Knowledge": Perspectives from the Yukon Territory. In: D. Anderson and M. Nuttall, eds. *Cultivating northern landscapes.* Oxford, UK: Berghahn Publishing.

Davis, W. 1998. Hunters of the northern ice. In: W. Davis, ed. *The clouded leopard: Travels to landscapes of spirit and desire,* pp. 31–45. Vancouver: Douglas and McIntyre.

Dene Cultural Institute. 1994. Guidelines for the conduct of participatory community research. In: B. Sadler and P. Boothroyd, eds. *Traditional ecological knowledge and modern environmental assessment,* pp. 69–75. Vancouver: Canadian Environmental Assessment Agency, International Association for Impact Assessments, and University of British Centre for Human Settlements.

Ernerk, P. 1994. Insights of a hunter on recent climatic variations in Nunavut. In: R. Riewe and J. Oakes, eds. *Biological implications of global change: northern perspectives.* Occasional Paper 33, pp. 5–6. Edmonton: Canadian Circumpolar Institute.

Eyegetok, S. 1999. Senior Researcher, Tuktu and Nogak Project: Ikaluktuuttiak, personal communication.

Eyegetok, S. 2000. Senior Researcher, Tuktu and Nogak Project, Ikaluktuuttiak, personal communication email.

Ferguson, M. A. D. 1997. Arctic tundra caribou and climatic change: Questions of temporal and spatial scales. *Geoscience Canada 23,* no. 4: 245–252.

Ferguson, M. A. D., R. G. Williamson, and F. Messier. 1998. Inuit knowledge of long term changes in a population of Arctic tundra caribou. *Arctic 51(*3): 201–19.

Fox, Shari. 1998. *Inuit knowledge of climate and climate change.* Master of Environmental Studies in Geography, University of Waterloo.

Fox, Shari. 2000a. Project Documents Inuit Knowledge of Climate Change. *Witness the Arctic 8*(1): 8.

Fox, Shari. 2000b. Arctic climate change: observations of Inuit in the Eastern Canadian Arctic. In: F. Fetterer and V. Radionov, eds. *Arctic climatology project environmental working group arctic meteorology and climate atlas.* Boulder, CO: National snow and ice date centre. CD-ROM.

Glesne, C., and A. Peshkin. 1992. *Becoming qualitative researchers: An introduction.* White Plains, NY: Longman.

Greenwood, D.J., W.F. Whyte and I. Harkavy. 1993. Participatory action research as a process and as a goal. *Human Relations 46(*2): 175—190.

Hagialok, J., elder. 1998. Interview by N. Thorpe, M. Kamingoak, and M. Akoluk, May 26, Kingauk. Tape recording. Tuktu and Nogak Project, Ikaluktuuttiak.

Hakongak, N., hunter. 1998. Interview by N. Thorpe and M. Omilgoetok, May 11, Ikaluktuuttiak. Tape recording. Tuktu and Nogak Project, Ikaluktuuttiak.

Hakongak, N. 1998. Wildlife Officer, Nunavut Government: Vancouver, BC, personal communication.

Hakongak, N. 1999. Wildlife Officer, Nunavut Government: Vancouver, BC, personal communication.

Hakongak, N. 2000. Wildlife Officer, Nunavut Government: Vancouver, BC, personal communication.

Haniliak, N. 1998. Community Researcher, Tuktu and Nogak Project: Umingmaktuuk. Tape recording. Tuktu and Nogak Project, Ikaluktuuttiak.

Holden, S.T. and L.O. Joseph. 1991. Farmer participatory research and agroforestry development: A case study from Northern Zambia. *Agricultural Systems 36:* 173–189.
Huntington, H. P. 1998. Observations on the utility of the semidirective interview for documenting traditional ecological knowledge. *Arctic 51(*3): 237–42.
Inuktalik, B. 2000. Resource Person, Olokhaktomiut Hunters and Trappers Committee: Yellowknife, personal communication.
International Institute for Sustainable Development and Sachs Harbour. 2000. *Sila Alangotok: Inuit observations on climate change.* 14 min. Videocassette.
Jeffers, S. 2000 (unpubl.). Arctic Ocean ice cover during the last thirty years. In *Workshop on Climate Change Impacts and Adaptation Strategies for Canda's Northern Territories.* Explorer Hotel, Yellowknife, NWT, February 29.
Joss, S. 2000. Representative, Olokhatomiut Hunters and Trappers Association: Yellowknife, NWT, personal communication.
Kadlun-Jones, M. 1999. Senior Researcher, Tuktu and Nogak Project: Ikaluktuuttiak, personal communication.
Kailik, B, *elder,* 1999. Interview by N. Thorpe and S. Eyegetok, November 1, Kugluktuk. Tape recording. Tuktu and Nogak Project, Ikaluktuuttiak.
Kakolak, E. 1998. Community Researcher, Tuktu and Nogak Project, Umingmaktuuk, personal communication.
Kamoayok, L. 1998. Board Member, Tuktu and Nogak Project: Umingmaktuuk, personal communication.
Kamoayok, L., elder. 1998. Interview by N. Thorpe and E. Kakolak, August 9, Hiukkittaak. Tape recording. Tuktu and Nogak Project, Ikaluktuuttiak.
Kaniak, M., elder. 1998. Interview by N. Thorpe and E. Kakolak, August 9, Hiukkittaak. Tape recording. Tuktu and Nogak Project, Ikaluktuuttiak.
Kaosoni, A., elder. 1998. Interview by S. Eyegetok and E. Komak, July 22, Ikaluktuuttiak. Tape recording. Tuktu and Nogak Project, Ikaluktuuttiak.
Kaosoni, M., elder. 1998. Interview by S. Eyegetok and E. Komak, July 22, Ikaluktuuttiak, Tape recording. Tuktu and Nogak Project, Ikaluktuuttiak.
Kapolak Haniliak, G., *hunter.* 1998. Interview by K. Ongahak and N. Thorpe, June 5, Kingauk. Tape recroding. Tuktu and Nogak Project, Ikaluktuuttiak.
Kavanna, G., elder. 1998. Interview by S. Eyegetok, J. Komak, E. Komak, and N. Mala, July 21, Umingmaktuuk, Tape recording. Tuktu and Nogak Project, Ikaluktuuttiak.
Keyok, C., elder. 1998. Interview by N. Thorpe and E. Kakolak, July 29, Umingmaktuuk. Tape recording. Tuktu and Nogak Project, Ikaluktuuttiak.
Keyok, C., *hunter.* 1998. Ikaluktuuttiak, personal communication.
Kofinas, G. P., J. Tetlichi, C. Arey, D. Peterson, and M. Nershoo. 1997. Community-based ecological monitoring: A summary of 1996–97 observations and pilot project evaluation. North Yukon Ecological Knowledge Co-Operative.
Koihok, M., elder. 1998. Interview by N. Thorpe and J. Panioyak. May 13, Ikaluktuuttiak. Tape recording. Tuktu and Nogak Project, Ikaluktuuttiak.
Komak, A., elder. 1998. Interview by S. Eyegetok and E. Komak. July 30, Ikaluktuuttiak. Tape recording. Tuktu and Nogak Project, Ikaluktuuttiak.
Krech, S. 1999. *The ecological Indian.* New York: W. W. Norton.

Krupnik, I. 1993. *Arctic adaptations. Native whalers and reindeer herders of northern Eurasia.* Hanover and London: University Press of New England.

Kuptana, G., elder. 1998. Interview by N. Thorpe, E. Kakolak and Karen Kamoayok. June 7, Umingmaktuuk. Tape recording. Tuktu and Nogak Project, Ikaluktuuttiak.

Légaré, A. 1998. An assessment of recent political development in Nunavut: The challenges and dilemmas of Inuit self-government. *Canadian Journal of Native Studies, 18*(2): 271–99.

Maghagak, A. 2000. Executive Secretary, Nunavut Tunngavik Incorporated, Ikaluktuuttiak, personal communication.

Maniyogina, J., elder. 1998. Interview by T. Apsimik. S. Eyegetok, J. Komak, and N. Mala, July 10, Ikaluktuttiak, Tape recording. Tuktu and Nogak Project, Ikaluktuuttiak.

Maxwell, J. A. 1996. *Qualitative research design: An interactive approach.* Thousand Oaks, California: Sage.

McDonald, M., L. Arragutainaq, and Z. Novalinga. 1997. *Voices from the Bay: Traditional ecological knowledge of Inuit and Cree in the Hudson Bay bioregion.* Ottawa: Canadian Arctic Resources Committee.

Nakashima, D. J. 1993. Astute observers on the sea ice edge: Inuit knowledge as a basis for Arctic co-management, ed. Julian T. Inglis: 99–110. Ottawa: International Program on Traditional Ecological Knowledge and International Development Research Centre.

Nalvana, C., elder. 1999. Interview by N. Thorpe and S. Eyegetok. November 1, Kugluktuk. Tape recording. Tuktu and Nogak Project, Ikaluktuuttiak.

Omilgoetok, B. 1998. Board Member, Tuktu and Nogak Project, Ikaluktuuttiak, personal communication.

Omilgoetok, B. 1999. Board Member, Tuktu and Nogak Project, Ikaluktuuttiak, personal communication.

Omilgoetok, B., elder. 1998. Interview by N. Thorpe and Meyok Omilgoetok. May 14, Ikaluktuuttiak. Tape recording. Tuktu and Nogak Project, Ikaluktuuttiak.

Omilgoetok, M. 1998. Wildlife Officer, Resources, Wildlife and Economic Development, Government of the Northwest Territories: Ikaluktuuttiak, personal communication.

Omilgoetok, M. 1999. Wildlife Officer, Resources, Wildlife and Economic Development, Government of the Northwest Territories: Ikaluktuuttiak, personal communication.

Omilgoetok, P. 1998. Board Member, Tuktu and Nogak Project, Ikaluktuuttiak, personal communication.

Omilgoetok, P. 1999. Board Member, Tuktu and Nogak Project, Ikaluktuuttiak, personal communication.

Panegyuk, G. 1998. Community Member: Umingmaktuuk, personal communication.

Riedlinger, D. (see Jolly, D.). 2001. Inuvialuit knowledge of climate change. In: J. Oakes, R. Riewe, M. Bennett, and B. Chisholm, eds. *Pushing the margins: Native and northern studies,* pp. 346–355. Winnipeg: Native Studies Press.

Riedlinger, D. and F. Berkes. 2001. Contributions of traditional knowledge to understanding climate change in the Canadian Arctic. *Polar Record 37* (203): 315–328.

Ross, R. 1992. *Dancing with a ghost: Exploring Indian reality.* Toronto: Reed Books.

Spink, J. 1969. Historic Eskimo awareness of past changes in sea level. *Musk-Ox, 5:* 37–40.

Spradley, J. P. 1980. *Participant observation.* New York: Holt, Rinehart, & Winston.

Stern, D. 2000. *hunter:* Ikaluktuuttiak, personal communication.

Strauss, A. L. 1987. *Qualitative analysis for social scientists.* Cambridge: Cambridge University Press.

Theis, J., and H.M. Grady. 1991. *Participatory rapid appraisal for community development: A training manual based on experience in the Middle East and North Africa.* London: International Institute for Environment and Development.

Thorpe, N. L. 1998. The Hiukitak School of Tuktu: Collecting Inuit ecological knowledge of caribou and calving areas through an elder youth camp. *Arctic 51*(4): 403–408.

Thorpe, N. L. 2000. *Contributions of Inuit ecological knowledge to understanding the impact of climate change on the Bathurst caribou herd in the Kitikmeot region, Nunavut.* Thesis (M.R.M.), Simon Fraser University.

Thorpe, N., and Eyegetok, S. 2000a. The Tuktu and Nogak Project brings elders and youth together. *Native Journal 9* (7): 9.

Thorpe, N. and Eyegetok, S. 2000b. The Tuktu and Nogak Project elder-youth camp. *Ittuaqtuut 2*(2): 32–43.

Thorpe, N., Eyegetok, S., Hakongak, N. and Kitikmeot Elders. 2001. *Thunder on the tundra: Inuit Qaujimajatuqangit of the Bathurst caribou.* Vancouver: generation Printing.

Thorpe, N. L. (in press). Codifying Knowledge about Caribou: The History of Inuit Qaujimajatuqangit in the Kitikmeot Region of Nunavut, Canada. In: D. Anderson and M. Nuttall, eds. *Cultivating northern landscapes.* Oxford, UK: Berghahn Publishing.

Ussak, L. 2000. Member, Rankin Inlet Hunters and Trappers Organization, Yellowknife, personal communication.

Vibe, C. 1967. Arctic animals in relation to climatic fluctuations. *Meddelelser om Gronland, 170:* 5.

Webber, L. M. and R. L. Ison. 1994. Participatory rural appraisal design: Conceptual and process issues. *Agricultural Systems 47:* 107–131.

Whyte, W. F., ed. 1991. *Participatory action research.* Newbury Park: Sage.

Wolfley, J. 1998. Ecological risk assessments: Their failure to value indigenous traditional ecological knowledge and protect tribal homelands. *American Indian Culture and Research Journal 22(*2): 151–166.

Figure 7-1. In their winter travelling, hunters use many traditional navigation practices, such as stars and constellation movement, frozen grass, and landmarks. (Photo © James H. Barker).

7

Travelling With Fred George:
The Changing Ways of Yup'ik Star Navigation in Akiachak, Western Alaska

Claudette Bradley
School of Education, University of Alaska Fairbanks

This paper describes the traditional navigation skills of Fred George, a Yup'ik elder from the village of Akiachak in west central Alaska. For the last two years Fred has been my mentor as I learned about Yup'ik ways of star navigation. Traditional skills such as star navigation are highly regarded, and Fred is well respected for his knowledge, retention, and passing on of these skills. Such skills are essential to survival on the tundra and to life as a hunter.

As I learned about how Yup'ik use the environment and the stars to navigate, I began to wonder how recent changes in the weather might affect Yup'ik navigational skills. Climate change is a concern in much of the North, particularly in regions of Alaska and the western Canadian Arctic. Bethel National Weather Service Office data shows the average yearly temperature has steadily increased four degrees over the past fifty years (1949–2001) in Bethel (Figure 7-2). The accumulative snowfall decreased from 1949 to 1987. The minimum snowfall occurred in 1983, and after that year the accumulative snowfall began to increase with great instability. Beginning in 1988 to 2001 the accumulated snowfall fluctuated between fifty-eight and ninety-six inches (Figure 7-3). For Native residents like Fred George, these kinds of environmental changes make predicting the weather and using navigational skills based on snow conditions and stars more difficult. Native communities in other parts of the Arctic are already observing that the weather has become unpredictable in recent years. In the context of my work, I thought that it would be interesting to talk to Fred about the changes he had observed and how these changes might affect his ability to navigate.

At the time of writing my paper, I have only just begun to document what people in Akiachak regard as important in relation to current environmental change and their traditional navigation practices. It is still an idea in its early stages. However, what I am writing about in this chapter is a good precursor to this next stage of research. Here, I introduce my mentor, Fred George, and describe what I learned about how Yup'ik navigate using the land and the sky around them—the stars, frozen grass, windblown trees, and snowdrifts. This chapter is based on my time spent in talking to and travelling with Fred George over the last two years.

Akiachak Village, Alaska

West central Alaska is bordered to the west by the Bering Sea and stretches north to Nome and down to Bethel, the southern end of the region. There are fifty-six villages in the region, most of them located on the Yukon and Kuskokwim rivers. Between the Yukon and Kuskokwim Rivers sit acres of flatlands, dotted with thousands of lakes and marshes. The surrounding tundra consists of low bushes, tall grass, a few small trees, and occasional hills. Much of the ground below is permafrost. Decayed vegetation causes the surface to be rough and spongy.

About twenty miles northeast of Bethel is the subarctic village of Akiachak, located on the west bank of the Kuskokwim River (61° N 161' W). The village has a population of 585 people in 129 families. Ninety-five percent of its people are Yup'ik and speak their language to varying degrees. Most understand and speak English as well.

For the most part Akiachak is a strong traditional community, where people speak Yup'ik, fish for salmon, and maintain a subsistence lifestyle. It was the first village in the Yukon-Kuskokwim Delta to dissolve its city government and establish a Native village government. There is no road system connecting the villages, except for the ice highways in the winter on the frozen rivers and streams. In the winter months snowmachines are the primary mode of ground transportation.

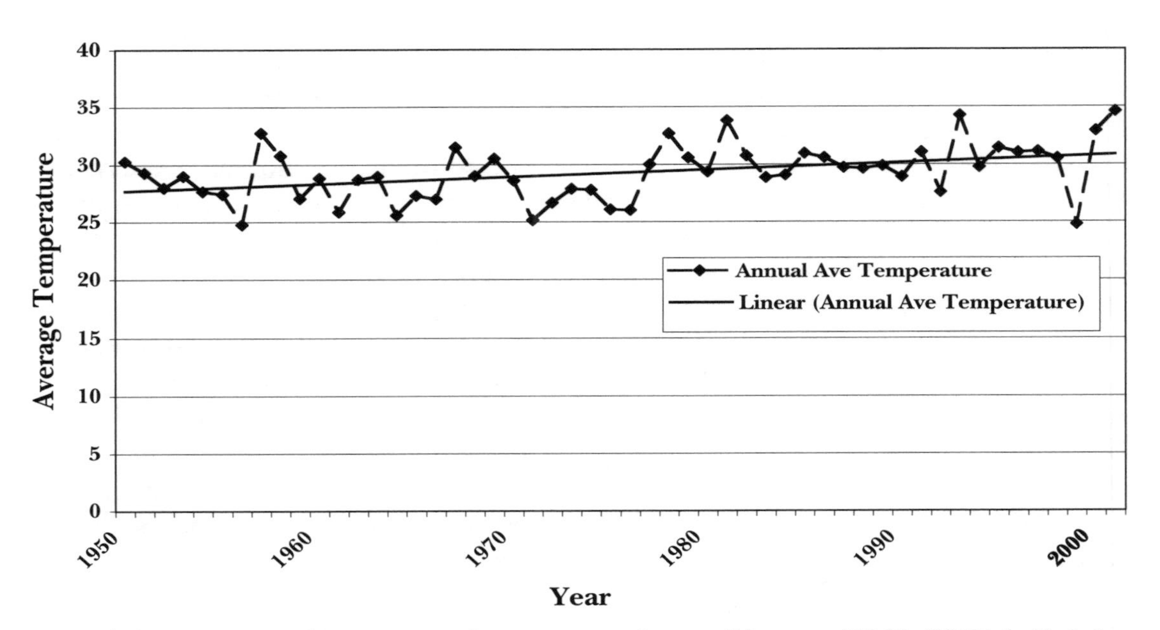

Figure 7-2: Average yearly temperature increases over the past fifty years (1949–2001) in Bethel. Source: Bethel Weather Service Office Airport Alaska, National Climate Data Center historical listing for National Weather Service Cooperative Network, 2001.

The summers are short and cool, about 50 to low 60° F. The winters are long and cold with below zero temperatures and high wind chills. The summer solstice has twenty-one hours of sunlight, whereas the winter solstice has only three hours of sunlight. Snow commonly falls early in October, but the initial fall accumulation is relatively small. Fluctuating temperatures cause the snow to alternately freeze and thaw, creating frozen snow waves on the thousands of lakes in the Yukon-Kuskokwim region. The snow waves freeze in the direction of the wind like a frozen ocean. Cold weather in December through March stays below 0° F. The snow that falls during these months remains like powder and drifts with the wind. Each year this region generally receives thirty to forty-eight inches of snow. The snowdrifts often obscure the few landmarks. Rapid changes in the weather cause low ceilings and poor visibility. Very heavy winds cause blizzards and whiteout conditions.

Navigating and Weather Forecasting Skills in Akiachak

The environment that the people of Akiachak live in demands that they understand the weather and ecology around them. At a very young age, about eight years old, Yup'ik boys are encouraged to observe the weather every day. They must learn to predict the weather days and even weeks ahead. A skilled elder can predict the nature of the coming seasons, often far in advance. There are many subtleties to learn about the environment, the snow and the stars. Just as they study the weather, they must study the stars and travel across the tundra for many years with a mentor.

In villages along the Kuskokwim River, Yup'ik hunters follow a rigorous schedule for subsistence hunting of animals, gathering of herbs and berries, and ice fishing.

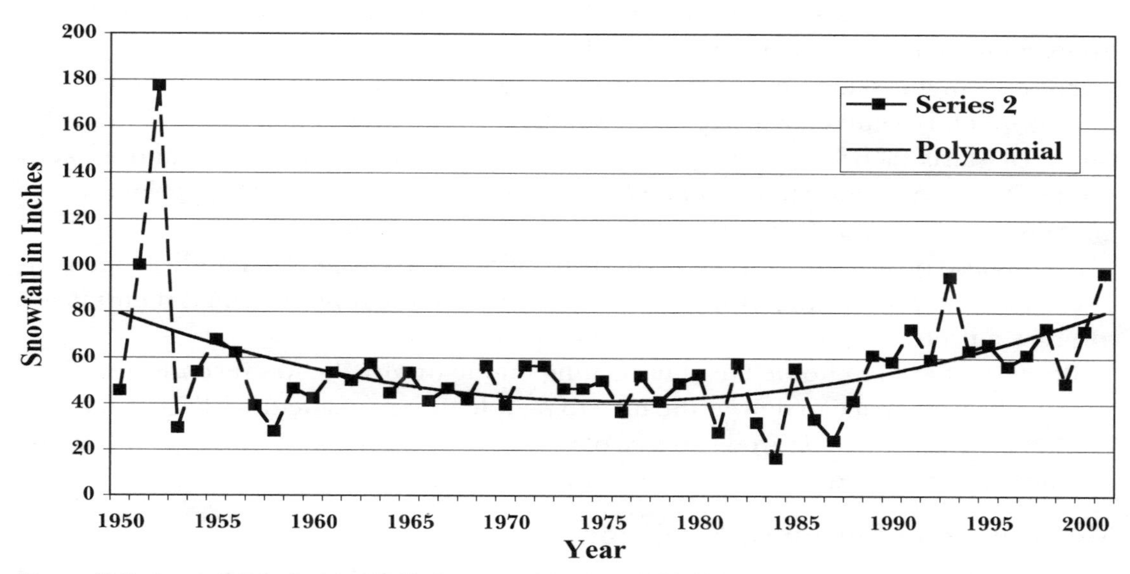

Figure 7-3. Average yearly snowfall over past fifty years (1949–2001) in Bethel. Source: Bethel Weather Service Office Airport Alaska, National Climate Data Center historical listing for National Weather Service Cooperative Network, 2001.

After October freeze-up and before April break-up they travel by snowmachine long distances across the tundra to hunt and trap. Snow covers the tundra with a white blanket, which is continually stirred up by the strong winds. During long winter nights they navigate across the tundra using stars, frozen grass, tree growth, and snow waves to guide their way.

Weather forecasting is essential for any navigator across the tundra. Survival gear must be planned and routinely cared for. High winds can occur suddenly, reaching up to sixty knots, which is a very heavy storm. Understanding and respecting the environment takes time and requires a long, trusting relationship with a skilled mentor.

The importance of navigating and forecasting skills is illustrated in the case of travelling from Akiachak to the Yukon River, a route used by many village residents. The route is over a dynamic snow-covered terrain using the stars, which are in constant motion, but which work as a clock and compass. Yup'ik men learn these skills from elders, who mentor them over many winters, as they travel together observing the sky and snow-covered terrain, sharing stories and past experiences.

Fred George

Fred George is a Yup'ik elder of sixty-five years. He lived with his late wife Mary in Akiachak for forty-eight years. Fred and Mary have eight children and many grandchildren. Most of their offspring live in Akiachak and visit their parents daily. Their three-bedroom wood frame home has a continuous flow of relatives and friends visiting, who talk, eat, and sometimes stay for an evening steam bath.

Nearly every day Fred leaves to attend to his subsistence responsibilities and returns home with lots of fish and sometimes caribou and/or ptarmigan. Fred works very hard for his family and extended family. His subsistence activity feeds not just his family, but his extended family and sometimes friends.

Fred is a highly skilled navigator of the tundra. He ventures out to the Yukon River in November, December, February, and March to travel ninety miles to his winter fish camp. He starts each year on November 10 at 10 p.m. to travel across the tundra to camp. That is the time when *Tunturyuk* ("caribou" in Yup'ik, the Big Dipper in English) is hovering over the North direction.

Fred is also one of the four search party leaders of the village. He has three men that travel with him during a missing person search on the tundra. Fred is their mentor; they must travel and listen to Fred's advice and stories.

Because of Fred's knowledge of navigating on the tundra, he was recently hired by the U.S. Postal Service to deliver the mail to people living outside the village on the tundra. Delivering mail is Fred's source of monetary income, but his true wealth is his knowledge of the tundra and his ability as a subsistence hunter.

My Training to Become a Navigator

My first encounter with navigation skills among indigenous people was in the late 1970s when I was reading a book on Micronesian canoe voyages across the Pacific Ocean, where travelers used the stars and waves to find their way to tiny islands spread over three thousand miles from east to west. When I become employed at the University of Alaska Fairbanks in 1989, I became a part of a research team working on a project entitled Adapting Elders' Knowledge and Pre-K to 6th Instructional Materials Development. The project (1997–2000) was designed to create math education modules for Elementary school children in the Kuskokwim and Bristol Bay regions. These modules would enable teachers to teach mathematics in a cultural context. One of my roles in the project was the study of navigation strategies of Yup'ik hunters and fishermen, particularly those associated with the use of stars. This began a journey in 1997 to learn about Yup'ik navigation, particularly from Fred George, who would be my mentor.

The project brought together a group of Yup'ik elders, teachers, and teacher aides, along with University of Alaska Fairbanks faculty members, to discuss and to document (to the extent possible) the stars and the Yup'ik practices in subsistence hunting and fishing in the Kuskokwim and Bristol Bay region. The project included several meetings throughout the year, both on the Fairbanks campus and in village schools. Several Yup'ik school-teachers and teacher aides from local communities also participated. The project was an opportunity to learn from their elders about traditional navigation techniques. They also provided translation during the workshops, translating between Yup'ik and English. Many elders had difficulty speaking English, and the university faculty did not speak Yup'ik. Everything that was said had to be translated into the other language.

In these meetings, Fred would describe his travel across the tundra, and his words would be translated into English. When a question arose from an English speaker it was translated into Yup'ik for Fred and the other elders. In general, elders have a working knowledge of basic, everyday English, but more detailed discussions were always in Yup'ik. With the help of a translator, I could ask additional questions of Fred to gain more clarity in my understanding or to have him explain what I perceived as gaps in my understanding of his words. For example, when he said he measured the distance of *Tunturyuk*[1] (the Big Dipper) to the horizon. I asked which star was he using in *Tunturyuk.* His answer was unclear at first. When I presented him with a *Tunturyuk* made from a pipe cleaner, he was able to point to the position of the star in the constellation that was used to measure the distance from the horizon to *Tunturyuk.*

Making concrete models of the stars and their constellations was very helpful in having Fred explain and demonstrate how he navigates. I mounted the pipe cleaner constellations onto the inside of a black umbrella. I carefully placed them in position using their latitude and longitude locations in the sky. Polaris was the very top of the umbrella, so I placed *Kaviaraat* ("Little Foxes" in Yup'ik, Little Dipper in English) at the inside top of the umbrella. *Tunturyuk* is 30 degrees south of Polaris (North Star). So is *Qengartarak* ("Nostrils" in Yup'ik, Cassiopeia in English) Since Akiachak's latitude is 61 degrees north, I placed *Tunturyuk* halfway between the inside top and the umbrella's edge. I decided the umbrella's edge was 60 degrees south of Polaris. *Qengartarak* is also 30 degrees south of Polaris; it is directly opposite *Tunturyuk.*

Fred enjoyed rotating the umbrella as the sky rotates. He would point to the position of stars relevant to the earth and turn the umbrella as the time progressed.

In March 1999, I had the opportunity to return to Akiachak and go out on the tundra with Fred George and James, his brother-in-law and search party trainee. We traveled thirteen miles out of Akiachak the first day in March. We traveled forty-five miles out the second day. In the daytime he was able to show me the frozen grass, wind-blown trees, the snow waves, and how he navigates. This began my real training with Fred as my mentor.

Fred's skills as a navigator are best understood in the context of his yearly subsistence calendar—the time of year he travels, the routes he takes, and the places he travels to. In the next sections, I describe Fred's yearly subsistence activities, and then tell the story of one of Fred's trips, a ninety-mile journey to the Yukon River guided by the stars, the snow, and the tundra grass.

Fred George's Yearly Subsistence Calendar

The following is a yearly subsistence calendar given to me by Mary George, the late wife of Fred George.

Family ancestral campsites lie along rivers and beside lakes between Kuskokwim River and ninety miles north to the Yukon River. A Yup'ik man will have a campsite on a river or lake given his father's name. When his father dies the lake or river is given his name.

After freeze-up, which normally happens in October, Yup'ik Eskimo travel to their camps out on the tundra to fish for white fish and trap otter, mink, muskrat, beaver, and fox, so the women can make parkas. Earlier in September, the men hunt moose for two weeks. Over the summer the moose newborn calves of the past spring are allowed to grow, so moose are not hunted in the summer months. During moose hunting season woman and children pick blackberries and cranberries. In winter the men hunt caribou and trap beaver. They hunt, fish, and trap, while the women cut up their catches, freeze, dry, and store them for future eating.

In January everyone stays in the village to rest. In February they travel out on the tundra to their campsites to fish for pike and hunt caribou and ptarmigan and trap beaver and rabbits. In March people hunt ptarmigan and moose, fish for pike, trap beaver, and pick frozen blackberries.

In April everyone settles down in the village to wait for break-up. In May people travel by boat up the Kuskokwim River to hunt for eggs, geese, and ducks; fish for tundra white fish; and pick frozen spring berries. Men prepare for the fishing season; they mend nets and try their boats out. By June, the smelts come into the Kuskokwim River, followed by the king salmon *(Onorhynchus tshawytscha)*. The men fish for kings for two weeks and the women cut, clean, and dry the fish. In late June the red *(O. nerka)* and dog salmon *(O. keta)* run and start to slow down in July. They diminish considerably in August, when the silver salmon (or *O. kisutch)* begin to run.

In August the silvers are in and everyone is fishing along the Kuskokwim. This is the time when everyone goes to the fall camp area. The women prepare the fish for smoking and store them in basket containers or in the freezer. End of August begins the berry-picking season for cranberries, blueberries, and blackberries.

This subsistence cycle continues each year and the Yup'ik people are able to survive the harsh environment of the tundra. In their numerous trips across the tundra terrain, they use the navigation skills described here.

Travel Ninety Miles Northwest of Akiachak to Yukon River: How Fred Navigates

Reference Position

On November 10,[2] at 10 p.m. Fred George begins his first trip across the tundra for the winter. The Big Dipper is hovering over the northern horizon, like a giant spoon sitting on a table. Astronomers identify the seven stars of the Big Dipper (from right to left), as Alpha, Beta, Gamma, Delta, Epsilon, Zeta, and Eta. On November 10 at 10 p.m. in Akiachak, Gamma is directly over north on the horizon. This initial position of the Dipper is a reference position for Fred George—the position that will be used to navigate in the coming winter months. At this time, the Big Dipper marks true north and 30 degrees west of true north. Fred knows he should travel toward the Dipper handle, but to the right of Eta. He relates this angle to five minutes on his watch. Given the 12 on his watch is facing north, he travels five minutes (30 degrees) west of north.

Before leaving Akiachak, Fred carefully checks his snowmachine for gas and working parts. He loads his machine with his hunting equipment, dresses warmly, and starts his snowmachine. He heads out of the village using snowmachine trails heading north. Away from the village, Fred stops his snowmachine to see the time on his watch and check the position of *Tunturyuk*.

Hand Measurements

Fred George checks this position of the third star, Gamma, in the Big Dipper, with a sequence of hand measurements. He uses his right "hand span" with the tip of his thumb at Gamma and his pinky pointing down to the horizon. Fred keeps the tip of the pinky in stable position while closing his fingers and rotating his hand downward, using "four-finger measurement." Then he keeps the index finger in stable position and rotates his hand downward using "three-finger measurement," which puts the index finger at the horizon. The tip of his index finger locates true north on the horizon (Figure 7-4).

Itinerary

Fred chooses one of two routes that lead to his fish camps. The shape of the Yukon River makes a wide U-shape, stretching 30 degrees west of north to 30 degrees east of north. To reach the river, Fred must stay in that range. If he travels too far west, i.e. over 30 degrees west of north, he will travel for miles and reach the Bering Sea. If he travels too far east of north, i.e., over 30 degrees east of north, he will find himself travelling parallel to the Kuskokwim River.

When the sky is clear and the winds are calm, Fred will travel towards the handle of the Big Dipper, just a little bit to the right of Eta star. He sets his bearing on the snow waves. As long as the motion of his snowmachine over the waves does not change, he knows he is keeping his bearing. As described, he also uses the frozen grass and wind blown trees to maintain his direction (see Figure 7-5). If the weather is partly cloudy and mostly overcast, Fred takes another route. He travels due north for two hours, and checks his watch and the Big Dipper. He turns east of north. This time he travels un-

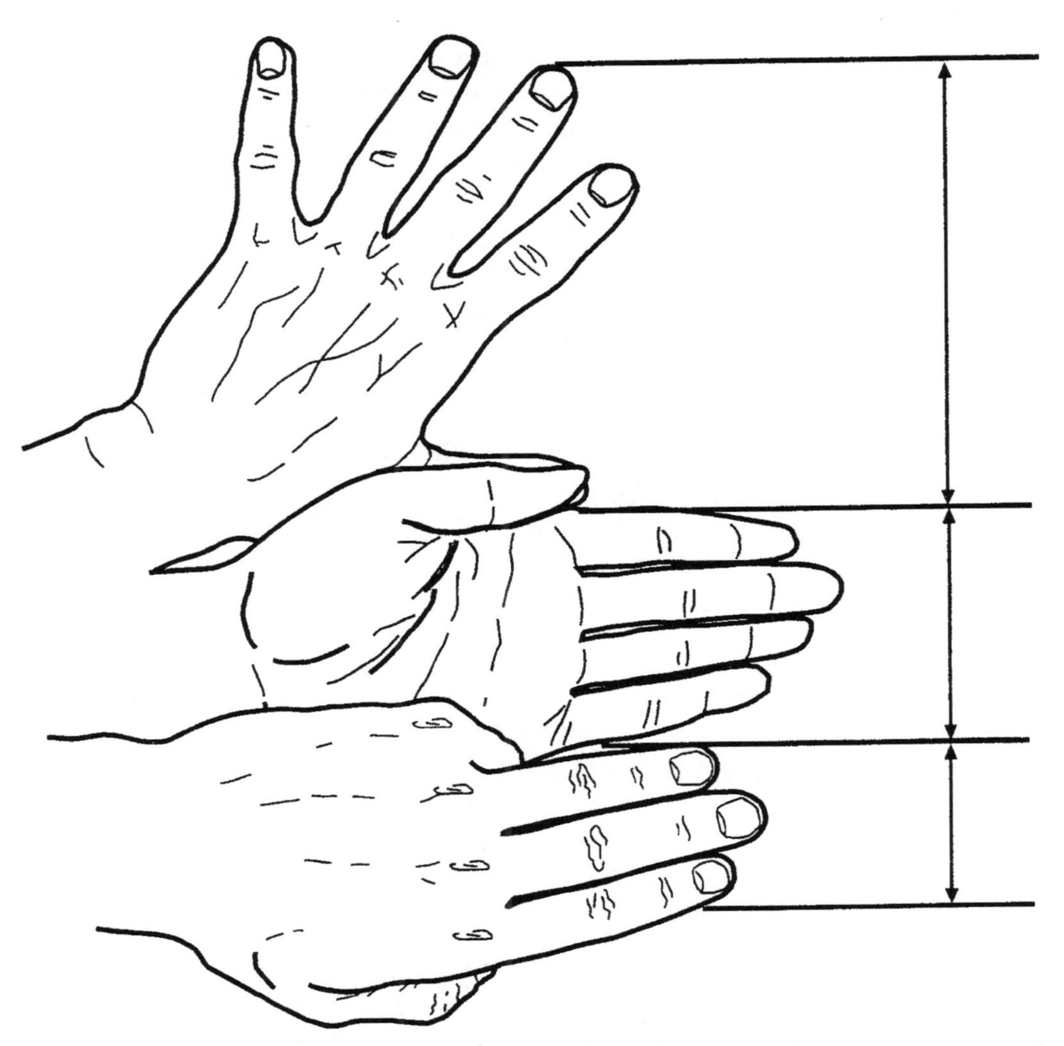

Figure 7-4. This is the hand measurement Fred George uses to measure the distance between the gamma star in the Big Dipper and true north on the horizon. The measurement is a handspan followed by four fingers, followed by three fingers. Illustration by Elizabeth (Putt) Clark, courtesy of Adapting Elder's Knowledge and Pre-Kindergarten to 6th Instructional Materials Development Project.

der the Dipper handle. The Eta star is over the north position on the horizon. His other fish camp is about 15 degrees east of north at midnight on November 10. If the weather turns bad during his journey, Fred will stop and wait until the storm is over. The elders do not travel in stormy weather. If a storm comes when they are out on the tundra, they build a snow cave to wait out the storm.

He travels by snowmachine at fifteen to twenty miles per hour. The prevailing northeast winds have created snow waves on the countless lakes between the Kuskokwim and Yukon Rivers. The lakes freeze with a wavy surface like a frozen ocean, and then the October snow piles high on the lakes. The changing temperatures cause the snow to melt and thaw frequently. This causes the snow to harden, creating snow waves rolling in a southwest direction on every lake. Fred's snowmachine rides over the snow waves in rhythmic motion, which he feels kinesthetically with his body. If the machine changes its direction, Fred can feel a change in the rhythmic motion. He learned to maintain the rhythmic motion to retain his course.

When he exits a lake and travels on land (tundra), Fred must use the frozen grass and wind-blown trees as compasses. On the first day of snow the wind blows the long blades of grass in one direction. The prevailing northeast winds mean that generally the grass is blown over in a southwest direction. The weight of the snow holds the grass in the direction of the wind and the cold temperatures freeze the grass. Frozen grass becomes a natural compass on the tundra. Fred can use the grass to determine his direction. He also uses wind-blown trees. Few trees exist on the frozen tundra, and those few trees generally grow in isolated spots. Heavy winds cause the trees to lean in the southwest direction and to grow most of their branches and leaves on the southwest side of the trunk, leaving the northeast side of the tree barren. Such trees help Fred determine his way across the tundra.

Figure 7-5. The grass is frozen in the same direction as the snow waves. Fred says the grass is always available to find his direction. Photo by Claudette Bradley.

Fred also recognizes several landmarks. His key landmarks are rivulets, streams, and small structures, or tiny log houses. Fred identified a partially fallen ten-by-ten-foot log cabin used in the old days by Yup'ik hunters. Such landmarks appear occasionally along his pathway to fish camp. Mary George said that locating landmarks is essential to navigating across the tundra. They reinforce the navigator's understanding of his position on his journey.

After riding the tundra for two hours, Fred stops his snowmachine to check his watch and the Big Dipper. It's midnight and the Big Dipper have moved to the left, i.e., counterclockwise 30 degrees. The end of Dipper handle is directly over true north. The Dipper appears to be turning upward. Fred places his watch in front of his chest with the 12 in the north direction and the 6 in the south direction. Fred checks his snowmachine is heading five minutes to the left of north, i.e., 30 degrees west of north. With the 12 heading toward the Dipper handle hovering over true north, his snowmachine must head in the direction of 11 on his watch, or five minutes to the left of north, which is 30 degrees northwest of north.

By 2 a.m., the Dipper handle has turned upward. Another two hours has past and its time for Fred to check his watch and the Big Dipper. The handle has moved to the right, now appearing parallel to a north direction. Fred will hold his two hands up so that one is in the north direction and the other is in the direction of the Dipper handle.

Fred learned the skill of navigation across the tundra from his dad. He understands it in the Yup'ik way. He can explain and describe details of his journey more clearly speaking Yup'ik. He has no knowledge of degrees. However, to explain how he travels using the stars in English it is most appropriate to use degrees. What keeps emerging is the importance of 30 degrees.

Having traveled over the tundra for sixty years, the landscape is very familiar to Fred George. He is an expert in his own right. He is a master star navigator across the Yukon-Kuskokwim Delta.

As illustrated by this story, Fred George uses the natural compass tools of the tundra and the sky to navigate and make decisions during his trip. These are described in Table 7-1.

Lesson One: A Story About Shadows

Out on the snow-covered tundra on a sunny day, Fred George showed me how he can use a shadow to find direction. With a jack knife angled at 90 degrees, he drew a perfect circle in the snow, straightened the knife, and placed it vertically in the center. The shadow of the knife was clear and overlapped the circle. We looked at our watches and saw 3 o'clock. A 3 o'clock shadow should be halfway between the north and the east directions. So, Fred marked the north, east, south, and west positions on the circle with his finger. He checked the direction with the tree (compass) and returned to the knife compass and adjusted the position of the north, east, west, and south, marking the adjustment with his finger on the circle in the snow.

Table 7-1. Yup'ik navigation practices and techniques

Frozen grass	Tundra is land covered with low bushes, and few trees. The very high grass (eighteen inches) blows in the direction of the wind. The first snow covers the tall grass, pressing it close to the ground. The cold temperatures freeze the grass into position. Fred George uses the frozen grass to find direction.
Windblown trees	The continuous strong winds from the northeast blow the trees toward the southwest, causing the foliage to grow most densely on the southwest side of the tree. The northeast side of the tree will not have any branches or leaves. Fred George uses these isolated trees to find his way.
Snow waves	The tundra has a countless number of lakes between the Kuskokwim and the Yukon Rivers. Every lake has been given a Yup'ik name. The lake names are family names given to the lake for the family's historic use of the lake for fishing and camping. The northeast winds make waves in every lake. The waves freeze in position. The October snow falls over the frozen waves and becomes hard snow, making snow waves rolling in the southwest direction over the lakes. Fred George uses the snow waves to find his direction. He feels the waves kinesthetically with the movement of his snowmachine travelling over the waves. Fred relies heavily on the snow waves when the sky becomes overcast or the weather is somewhat stormy.
Stars	February to April Fred uses Venus, which was an evening star (in 1999). He looks at his watch. Venus is at 9 and north is at 12. He then determines that his fish camp is 5 minutes west of north. He travels on snowmachine in the direction he says is five minutes before the 12 on his watch. He continues to use the snow waves to maintain his direction. The frozen grass, trees, and landmarks reinforce his journey. He can return to Akiachak in the next evening or two using Venus, the evening star. He places his watch with 12 at the south direction and 3 at the west direction, where Venus is shining brightly. He then determines that Akiachak is five minutes east of south. On clear or even partly cloudy nights, Fred uses the Big Dipper and Venus to find his way.
Landmarks	Fred uses landmarks such as rivulets, streams, and distant mountains to reinforce his understanding of his position on the tundra. Landmarks may also be the remains of old cabins used by hunters in the old days.
Lakes	The southeast wind blows over the lakes. This warms air melts the ice along the northwest bank of the lake. The water appears in the northwest bank of the lake, which makes the lake a compass for a Yup'ik navigator.

When Fred was asked where was the sun at 12 noon, he looked at the sun and the time on his watch. He pointed to the sun and moved his hand to the left, stopping at what may be interpreted as the position at 1 p.m. and moving his hand to the left again and stopping at what may be interpreted as 12 noon. I asked, "how did you know how far is 15 degrees?" His response was, "we do not know about degrees."

He also pointed out that the sun travels low in the sky about the south direction in December. Each month from December to June the arch of sun's pathway moves higher and becomes wider in the sky.

In the morning the shadow will decrease in length every hour. At the moment when the shadow points to true north the shadow will begin to increase. The end points of the shadow form an elliptical path, with the shortest distance from the center point indicating the direction of true north.

Lesson Two: A Story About Using the Morning Star to Travel Home

In fall 1999, Fred used the morning star to return to Akiachak in the following manner. He checks his watch. The morning star is in the east and should be on his left side when he faces south. Facing south, he places his watch in front of his chest so that 9 marks east, 12 marks south, and 3 marks west. Fred knows that Akiachak would be five minutes east of south, i.e., in the direction of 11 on his watch, which is 30 degrees east of south.

Observing the Sky and Learning About Constellations

Fred George talks about *Tunturyuk*, Venus, and the Milky Way as the only celestial bodies to be concerned with. However, I wanted to understand the sky's movement and the positions of the stars. I began this quest with reading and using a sky chart,[3] together with sky observations from local residents in Akiachak.

As part of our project study groups, we did some star observations with Akiachak residents in the evenings of early March. We would view the sky at dusk and see the moon and a few stars. *Tunturyuk* would not appear until an hour after sunset. Venus would appear in the western sky soon after sunset, as the brightest celestial body. If the moon is waxing, it will be visible along with *Agyarpak* (Venus). Venus has several Yup'ik names. The Yup'ik name used in Akiachak is *Unuakum Agyartaa.* People from the village of Manakotak *(Manuquutaq)* near the Bristol Bay coast call Venus *Agyarpak.* Other Yup'ik villages use the name *Agesqurpak.* Venus rotates around the sun inside the Earth's orbit. When it is to the right of the sun, it rises in the morning before the sun. It is brighter than any star, so as it rises it marks the east direction.

The first magnitude[4] stars appear about half hour after sunset. Sirius is the brightest star, one of the six brightest stars forming a hexagon. These stars are located around the constellations Orion and the Gemini Twins. Later when the sky becomes a very dark blue and then blackens to beckon in the night sky, the second magnitude stars appear. Since six of the seven stars of *Tunturyuk* are second magnitude stars, on a clear night *Tunturyuk* is most visible in those first moments of its appearance in the night sky. As the third, fourth, etc. magnitude stars appear, they compete with the second magnitude stars for the attention of the observer. On a crystal clear night around midnight there may be so many stars that the observer has a difficult time locating the major constellations. Thus, the optimum time for observing *Tunturyuk* is one or two hours after sunset.

Tunturyuk moves counterclockwise about Polaris in the sky every half hour. This movement carves out a circular path in the northern sector of the sky. Its diameter stretches from the zenith to 30 degrees over the due north in the earth's horizon. To simulate this movement with my sky chart, I would turn the chart upside down with North towards the bottom and then rotate it counter clockwise. On the sky chart *Tunturyuk* is over north at 9 p.m. Alaska Time. Alaska has four time zones, but through state legislation three time zones are on Juneau time. Akiachak is 15 degrees west of Juneau, so *Tunturyuk* in Akiachak is over due north at 10 p.m.

When the Adapting Elders' Knowledge research team went to Togiak, Alaska, I planned to observe the stars moving during the night and the sun in the day. To view the stars, I would select a spot marked by stones and a pathway for my observation sight. I would choose a spot that was near the house where I was spending the night, but not so near as to have the view of the stars obstructed by the house. As the sun began to set (6 p.m.) I would observe the sky at my observation spot and then return to the house to review my sky map and write my observations in a journal. I continued to observe the sky, consult the sky chart, and record my observations every half hour during the night.

By 8 p.m., *Tunturyuk* was in full view over the rooftop of the house at my observation site. I could use the house as a marker for observing the constellation's movement. Every half hour it would shift to the right 7.5 degrees. This shift is difficult to see with the untrained eye, but by using a marker like the house, the shift was more apparent. Each half hour *Tunturyuk* would shift to the right beyond the edge of the rooftop. After each observation I returned to the kitchen to check the sky chart for the constellations I did recognize and look for constellations I should watch for during the next observation. By midnight I could recognize and find sixteen constellations.

As the sky blackened more stars came into view. The two far right stars of *Tunturyuk* lead the observer to Polaris (these stars are called pointer stars by astronomers). If a sky observer points to the lower star (Beta star), and continues in the direction of the upper star (Alpha star) at five times their (Alpha and Beta stars') distance, he or she will find Polaris (North Star). The remaining five stars in the Big Dipper form a bow (as in bow and arrow). When an observer finds the center of the bow (Epsilon star), follows the path to Polaris and continues in the same direction, they will see the big "W" called *Qengartarak* (or "Nostrils") by the Yup'ik people. *Qengartarak* points to *Kaviaraat* and to *Tunturyuk*. Native hunters have been known to use their nose as a pointer.

As the observer's eye moves from *Qengartarak* to *Kaviaraat* and on to *Tunturyuk, Pitegcaun* ("Fish Hook" or "Arrow" in Yup'ik) becomes visible, below *Tunturyuk*.

Astronomers call this constellation Leo. To find *Pitegcaun,* the observer starts with the Delta star in the Big Dipper and continues in the direction of the Gamma star. In passing the Gamma star the observer continues in the same straight line path to *Pitegcaun.* This constellation is below the horizon when *Tunturyuk* is hovering over due north. It rises later when *Tunturyuk* is moving away from the North toward the East and rising higher above the horizon.

The Milky Way passes through *Qengartarak* and forms an arch over *Tunturyuk.* The ends of the arch extend to *Tulukaruum ayarua,* called "Raven's Cane" in Yup'ik, which is the three belt stars of Orion. The belt stars lie on the ecliptic in the sky. These stars rise in the east when *Tunturyuk* is hovering over due north and appear over the south when *Tunturyuk* is vertical and sitting over the northeast direction. They set in the west when *Tunturyuk* is in the zenith.

The handle of *Tunturyuk* points to the southern most star in *Taluyat* ("Fish Trap" to Yup'ik elders). Astronomers know this constellation as Bootes. The two stars at the far left of *Tunturyuk* (Eta and Zeta) point to *Taluyat.* To the left of *Taluyat* is *Kinguqa Akugat,* "Kuspuk Hem" to the Yup'ik and Corona Borealis to astronomers.

In a four-hour period the Dipper moves 60 degrees. However, I wanted to see a greater sweep in the sky. I arose the next morning at 5:30 a.m. to observe the constellation, which was on the opposite side of Polaris. Relative to the house it had made nearly a full swing around Polaris. I was able to see the circle path of *Tunturyuk* about Polaris.

My Own Reflections on Fred's Understanding of the Sky

Based on my school-learned knowledge of the movement of stars, I reflected back on what I learned from the people in Akiachak about their understandings of the sky. I was most interested in Fred's words and actions while teaching me how he uses the stars to navigate. Especially how, in November when *Tunturyuk* hovers over north on the horizon, Fred would move his hand in the direction of the far left two stars downward toward the horizon. He would say "west."

At that time we believed he was indicating that the Dipper handle was pointing due west, but we had no way to substantiate this. This thought bothered us for several months. We had difficulty pinning Fred down to the difference between his "west" and due west. Since that time I have realized by mathematical contradiction that the Dipper handle does not point due west (see section on understanding star movement in Akiachak). However, it does point to the west-most direction that Fred should travel in order to reach the Yukon River.

It can be explained this way. If we view the map and see the direction of the Yukon River, it bends northwest. If a traveler heads due west on snowmachine, he would never reach the Yukon River. Instead he would reach the Bering Sea. It is my conclusion that Fred's "west" is, in fact, the west-most direction that allows him to reach the Yukon.

To examine this further, I could measure the angles of the Big Dipper to see if the angle of elevation of the two left-most stars in the handle is at 30 or 60 degrees. *Tunturyuk* is setting horizontal at 10 p.m. and Dipper ladle tilts upward toward the Northeast direction, while the handle is tilted downward toward the northwest direction. The question is, what is the elevation of the Dipper handle when its far left star is over due north? The astronomy books claim the last two stars in the Dipper handle point to Arcturus, which is a first magnitude star, and southernmost, in *Taluyat*. If we know its latitude and longitude then perhaps we can determine the elevation of the far left stars (Eta and Zeta) on the Dipper handle.

Fred does not seem to refer to the stars very much when navigating in February and March. By then he uses the snow waves. I believe in November and December he uses primarily *Tunturyuk* for orientation. During those months he learns the directions of the snow waves, which are generally waving from northeast to southwest. MacDonald (1998) writes about the snow drifts in the Canadian Arctic, which are similarly created by the prevailing northeast winds. The snowdrifts rise above the flat surface shaped like chevrons with tails, with their centers pointed to the southwest. When a traveler on a dog sled learns the direction of the drifts, he can use them to determine the direction of his journey.

MacDonald (1998) also argues—based upon his knowledge of Canadian Inuit navigation expertise—that the stars used by travelers must not be too high in the sky. The traveler must be able to watch both the stars and his dogs and the landscape. Nor can the traveler be made to look too far to the right or left. Considering these factors, *Tunturyuk* is 30 degrees above the horizon when it hovers over due north and is thus in an ideal location for a traveler on a dogsled to use as a compass. When *Tunturyuk* has moved northeast and becomes vertical six hours later; it is no longer in a useful position for the traveler to use effectively. He would have to turn his head to the right and up, away from his lead dog. As the Dipper moves into the zenith, it becomes less useful as a way finder but still useful as a timekeeper. Thus, as the night progresses Fred increases his reliance on the snow waves, frozen grass, and isolated trees to find his way.

Lesson Three: A Story of Fred's Way of Finding Missing People

Fred has a search party team consisting of three young men and himself. He has been mentoring these men for several years. They travel across the tundra with Fred, carefully observing the weather and the environment, following Fred's instructions and listening to Fred's stories. When searching for a missing person, Fred gathers his team and plans the search over the tundra using a Federal Aviation Administration map, which has a grid imposed over the Yukon-Kuskokwim Delta. Each cell (square section) on the map is numbered. Fred identifies the approximate area the missing person may be found, based on the reports given to him about the missing person.

Having determined the approximate position of the lost person using the FAA map, Fred's search team ventures out on the tundra using the stars, snow waves, trees, and frozen grass to find the approximate position of the lost person. He knows the Dipper identifies north. If he is to go west he turns west so the Dipper is at his right side. Likewise, if he is to go east he turns so the Dipper is on his left side. Akiachak is on the north bank of the Kuskokwim River, so he would not have to go south unless the lost person had crossed over on the other side of the Kuskokwim.

Understanding the Movement of the Stars in Akiachak

The Big Dipper moves counterclockwise about the North Star, and the sun moves clockwise about the south direction. This may seem like a contradiction but can be easily understood if you extend your arms outward and move them forward in a circular motion, completing the full 360-degree rotation. Continue moving your arms in this circular motion. Now look to the left arm and see your hand moving clockwise. Continue moving both arms in the circular pattern. Now look to the right arm and see your right hand is moving counterclockwise.

Now go outside and face west. Extend your arms and move them forward in a circular pattern. The left arm is moving clockwise, as is the sun. Your right arm is moving counterclockwise, as is the Big Dipper.

The Big Dipper also circles the North Star along the 30-degree angle of declination. Since Akiachak is 61 degrees latitude, the North Star is 61 degrees above the north direction along the horizon. The celestial equator divides the sky into the Northern Hemisphere and the Southern Hemisphere. Stars on the celestial equator are at zero-degree declination. Angles of declination in the sky correspond to latitude on earth. The North Star, Polaris, is at 90-degree declination. All the stars that are above the 29-degree angle of declination are always in the Akiachak region sky. They never rise and set, for they are always circling the North Star, like a twirling umbrella over the north horizon, in the Yukon-Kuskokwim Delta region. The stars between 29 degrees above the celestial equator and 29 degrees below the same equator rise and set every night. These stars rotate clockwise, like a tilted barrel over the east-south-and-west horizons.

The Big Dipper has seven stars, which range in magnitude from 1.70 to 3.40. They are all second and third magnitude stars. On a clear night these stars appear brightly after the first hour and a half past sunset.

Each star has been assigned a Bayer (Greek) letter and a proper name. Alpha *(Dubhe)*, Beta *(Merak)*, Gamma *(Pheoda)*, and Delta *(Megrez)* form the quadrilateral (referred to as a rectangle) portion of the Big Dipper. Gamma and Delta are close to the twelve-hour right ascension.[5] Gamma is six minutes to the left of twelve-hour right ascension and Delta is fifteen minutes to the right of the twelve-hour right ascension. Of the four stars, Gamma sits closest to the horizon when the Big Dipper is hovering over the north direction. The declination of these stars is as follows:

Alpha	62 degrees
Beta	56 degrees
Gamma	54 degrees
Delta	57 degrees

Since Gamma is 54 degrees declination when it is closest to the north horizon it is 25 degrees (54 – 29 = 25) above the horizon. From Fred George's perspective, the quadrilateral of the dipper handle is 25 to 33 degrees above the north horizon on November 10 at 10 p.m. Gamma *(Pheoda)* is the star that is directly above true north on the horizon. He uses his right hand to measure the distance from Gamma to the horizon, which identifies the position of true north.

The Dipper handle extends from Delta to Epsilon *(Alloth)*, Zeta *(Mizar)*, and Eta *(Alkaid)*. Epsilon and Eta are very bright second magnitude stars. Zeta is a dim second magnitude star, nearly a third magnitude star, but it's also a double star. Alcor is the second star, so close to Zeta that it appears as one star. The magnitudes of these stars are as follows:

Epsilon 1.70
Zeta 2.40
Eta 1.90

Delta, Epsilon, and Zeta appear to be in a straight line. They are one degree lower than the star on the right. Eta is 6 degrees lower than Zeta. The declination of the stars is:

Delta 57 degrees
Epsilon 56 degrees
Zeta 55 degrees
Eta 49 degrees

Fred says that Zeta and Eta slant in the direction of west. He traces the line segment between Zeta and Eta and continues extending the line segment beyond the left of Eta. As described previously, for a long time we wondered if, in doing so, Fred meant due west. By process of elimination I have since deducted that Fred did not mean due west. What he meant was that the stars point to the western-most point that he can go in order to reach the Yukon River.

This can be explained mathematically. The map of the Yukon-Kuskokwim Delta region shows that Fred must travel within 30 degrees west of north to reach the Yukon River. The river bends in a U-shape, so if he travels due west (or more than 30 degrees west of north) from Akiachak he will not reach the Yukon, but eventually reach the Bering Sea. Sky charts indicate the Dipper handle points to Arcturus a first magnitude star in Bootes *(Tayulat)*. Its declination is 19 degrees and its right ascension is fourteen hours fifteen minutes. This would not be due west. Due west has a right ascension of eighteen hours.

First of all, Arcturus would be below the horizon when the Big Dipper is hovering over north on November 10 at 10 p.m. Gamma star in the Big Dipper has right ascension close to twelfth hour. (Actually, RA is 11h 53m 49.81s.)

Arcturus RA 14h 15 m
Gamma RA 11h 53m 49.81s
Difference 2h 21m 10.19s

If the Dipper handle pointed due west it would have to point close to a star in the eighteen hour. Two hours twenty-one minutes 10.19 seconds is about 36 degrees. We now have our answer that the Dipper's handle points to the west most point that Fred can use to reach the Yukon River.

Like the sun, the stars in the Big Dipper move 15 degrees in two hours and 90 degrees in six hours. The Big Dipper is hovering over true north at 10 p.m. on November 10. At midnight it will have moved 30 degrees, which Fred associates with five minutes on his watch. By 4 a.m., the Big Dipper has moved 90 degrees and is hanging vertical in the sky with the Dipper ladle tilting forward. Fred says the Dipper is emptying out. When the Dipper is emptying, he knows it's early morning. Since his trip takes five hours, he should have reached his destination by the time the Big Dipper starts to empty out.

Stars are a Compass, a Clock, and a Calendar

The meridian in the sky is the arc that passes through the North Star, Polaris, and has end points at due north and due south along the horizon. The Big Dipper carves a circle whose center is Polaris. The Gamma star crosses the meridian twice a day. On November 10, Gamma crosses the meridian at 10 p.m. and 10 a.m. However, the rising sun obscures the Big Dipper at 10 a.m.

If one can see the Big Dipper in the sky and envision its great circle path with the meridian slicing the circle in half and marking the lowest and highest position, then true north can be identified along the horizon. With the location of north the other three directions can be identified as well. This makes this constellation a compass on a clear night.

Since the Big Dipper moves 15 degrees every hour, it is a clock. If it hovers over true north at 10 p.m.; six hours later it will have moved 90 degrees or appear vertical east of Polaris with the handle pointing toward the horizon. Twelve hours later it will be upside down in the zenith; and eighteen hours later it will be vertical again, west of Polaris with the handle point away from the horizon. Since the difference in right ascension between the Gamma star and the Eta star is nearly two hours, whatever the celestial position of the Gamma star, two hours later Eta will be there.

Because the Earth is rotating around the Sun 360 degrees in 365 days, the movement of the stars shifts about 1 degree each day, which makes a 30-degree shift each month. This means on November 10 at 10 p.m. the Big Dipper is hovering over north with the Gamma star over true north, but on December 10 at 10 p.m. it has shifted 30 degrees and therefore Eta star is directly over true north on the horizon. It shifts another 30 degrees by January 10 at 10 p.m., and by February 10 at 10 p.m. it has shifted a total of 90 degrees from the November 10 position, making it Dipper vertical (and no longer as useful to Fred to set his bearings with to find his camp 30 degrees west of north). In another three months (May 10) the Big Dipper will be in the zenith, and then three months later vertical on the west side of Polaris. This pattern makes the Big Dipper a calendar.

How Environmental Changes Might Affect Yup'ik Navigation Skills

Reviewing the average temperature and snowfall data during the winter months in the Yukon Kuskokim Delta region over the past fifty years reveals some significant changes. The average winter temperature has been steadily increasing, with a change of 4° F over fifty years. Such warmer winter temperatures will undoubtedly have significant impacts on the overall ecology of the region. Annual snowfall has steadily decreased from 1949 to 1985. In the winter of 1985, there was an all-time low of 16.5 inches of snow in the Bethel region. After 1985, one can see a general increase in snowfall over the long range, but characterized by a high inter-annual variation, with low snowfall some years and high snowfall in the next year.

When I was in Akiachak, the local coordinator for the school culture-based curriculum, Sophia Kasayulie, told me that elders were talking about changes in the weather. When I asked her about such changes, her response was that the elders say the weather doesn't get as cold as it used to, and that in some years there is no snow. She told me a story about how in the early days, some years the snow would be so deep that Fred George would have to shovel under the snow to find spruce trees for firewood. In recent years the snow is not as deep, so the trees are exposed and there is no need to dig for firewood. Both Fred and Mary George have also told me about how they are seeing less caribou in the herds, and that hunters are now fishing more and bringing home more fish than caribou.

Environmental and climate-related change may have encouraged hunters to abandon dog sleds and replace them with snowmachines. Fred and his dad used dog sled for many years, up until 1985. Fred said he used his dog team to cross the tundra from Akiachak to the Yukon River for the last time in 1985. That was the year of the lowest snowfall (16.5 inches) in the Bethel region. In subsequent years he has replaced the dog team with a snowmachine to cross the tundra.

In several meetings with our research team elders have commented that "the stars are moving faster." They do not know why this is, and are very puzzled by this phenomena. I am puzzled as well. If the stars are moving faster, then the Earth should be turning faster, which would also likely have significant impacts on weather and climate. I would assume that the winds would increase. The data that I found on wind velocity in Bethel region showed that wind velocity increased from 1950 to 1975, and then decreased between 1975 and 2000. This needs more investigation. I am very interested to look at this issue further.

The day I traveled out on the tundra with Fred George there was little wind. It was a calm sunny day in March. However, the trees were all bent over, leaning toward the southeast, indicating that most of the time during the growing season the winds are very strong blowing in the southeast direction. The tundra has very strong winds. A Yup'ik navigator is skilled at predicting the weather and does not travel in stormy conditions. If caught on the tundra in a storm he builds a snow hut and uses a thistle passing through a hole in the snow hut to breath air. When the storm subsides, the wise navigator continues his journey. Strong winds on the tundra would effect the navigator's ability to hunt and fish enough for a good winter harvest.

Conclusion: Keeping Fred's Knowledge Alive

Preserving Fred's knowledge of navigating across the tundra is important to the self-esteem and cultural identity of Yup'ik people. A study of Fred's way of navigating across the tundra uncovers the wisdom, courage, and ingenuity of his Yup'ik ancestors—values that are so appealing to the modern generation of the Yup'ik youth.

Elders say that the young people of today are not spending time in listening to elders as they once did in the past, to learn their cultural ways and stories. Yet the young people drive snowmachines out on the tundra. Many get lost, run out of gas, and cannot find their way home. Fred's knowledge of the stars and the tundra environment would give them a chance to survive and to travel around safely. This is by far the most important practical aspect of preserving and disseminating Fred's knowledge within his own community that has changed so much since the time he was a young boy himself.

In order to preserve Fred's knowledge in Yup'ik navigation practices, four major strategies can be used : (a) documentation, (b) classroom instruction, (c) practical apprentice navigating with Fred George, and (d) listening to Fred's story telling. This paper and the entire project it describes is, in fact, our first effort in documentation. Documenting these practices is especially important if we want to understand how changes in the weather and environment might affect them.

In addition to this and similar mostly descriptive texts, a teacher's manual for fifth-grade and sixth-grade classroom instruction will be produced under the current project. This teacher's manual will be used at local schools and it will include Fred's way of navigating, together with the classroom activities that will help students understand how the stars move and how Fred uses the stars and the environment to find his way across the tundra. Students will also learn to study the movement of the sun using a sundial. They will study the movement of the stars through observations of the night sky and via classroom board games. This is the way we hope to keep Fred's knowledge used by Yup'ik children and young people for years to come (Figure 7-6).

With some prior understanding of the movement of the stars and the natural compasses in the environment, Yup'ik students can participate in practical navigation trips across the tundra with Fred George or other knowledgeable elders. With some previous class training and follow-up supervision, students can experience navigating with the stars on the tundra, with their teachers or experienced adults. Travel in places outside the village in the open tundra can be explored, using the stars to find the way.

Yup'ik students can also have story sessions provided by Fred George or other elderly Yup'ik experts (Figure 7-7). Such sessions can be videotaped or audio-taped and transcribed for publication in journals or put on the school web-site. These stories can be preserved and passed on to young people in their classrooms, which is the only practical equivalent of the old communal houses where the Yup'ik kids of past generations once learned the knowledge of their ancestors.

Figure 7-6. Marcus Alexie works with a sky map as Robyn Kasayulie looks on. Photo by Claudette Bradley.

Figure 7-7. Fred George holds a pipe cleaner model of Tunturyuk *and explains how he navigates in Yup'ik to a class of fifth and sixth-grade students in Akiachak Elementary School. Photo by Claudette Bradley.*

Survival of Yup'ik people as a cultural entity and as a community will depend on their ability to maintain a subsistence lifestyle and the continuing availability of animals such as the caribou, fish, and ptarmigan. If trends of increasingly variable snow cover, warming winter temperatures, and other changes in the weather continue, Yup'ik modes of navigation may be at risk and many game animals will be affected.

> Tom Kasayulie is an eighty-four-year-old Yup'ik elder in Akiachak. He went to school in Akiak and moved to Akiachak in 1933. He married a local girl and they have five children and lost two. Tom said grass grew in Akiachak in the 1930s. No grass has grown in the village for the last twenty-seven years. In the 1920s, Akiachak would have flooding at break-up in spring and in the fall when it rains.
>
> Tom said that on the tundra now there is less water from rain and snow that has caused many lakes to dry up. In the 1920s there were 100,000 reindeer in Akiachak and nearby communities. The herds have been greatly reduced by starvation due to lack of food and by the wolf population killing the reindeer. The reindeer have been taken to Nunivak Island and no longer exist on the Yukon-Kuskokwim Delta. But the caribou and moose do not seem effected by climate changes.
>
> In the past the geese would darken the sky when they returned from the south in the spring. Now just geese return and the sky never darkens. Among the various kinds of geese that return the snow geese have disappeared.
>
> The warmer weather and higher temperatures are ruining the fish drying on fish racks. We catch less salmon in the river, Tom says. Now the Fish and Game have closures and restrict fishing to three days out of seven.

As the arctic climate changes, the natural compass tools of the tundra—the wind-blown trees, snow waves, and frozen grass—may prove less reliable under scenarios of increasingly unpredictable weather. It may become more difficult for Yup'ik hunters such as Fred George to trust their forecasting and navigating skills—making it more difficult to travel safely and hunt successfully. For these reasons, I am interested in the ways that recent changes in the weather may affect such traditional practices. This is truly a frontier to be explored.

Acknowledgements

With deep appreciation and respect I thank Fred George of Akiachak for his continued patience and sharing of knowledge about the tundra and his way of navigating in the Yukon-Kuskokwim Delta. Fred is a true leader and elder in his community, who passionately shares his skills with the young people of Yukon-Kuskokwim. I also want to make a special recognition of the contributions to this project by the late Mary George, Fred's wife for over forty years. She passed away in November 2001 after her courageous five-year battle with cancer. As a translator for the navigation project, Mary shared her vast cultural knowledge and provided many useful suggestions. Mary was always with Fred and a valuable supporter of this project.

I will be forever grateful for the funding and support provided by National Science Foundation (NSF), for the project Adapting Elders' Knowledge and Pre-Kindergarten to 6th Instructional Materials Development (1997–2000), at the University of Alaska Fairbanks. Although this work was supported by NSF grant #9618099, the work reported here is that of the author and does not necessarily reflect the views of NSF or Fred George. Any mistakes or misunderstandings are my own.

This grant provided funding for airfare and lodging for our gatherings and thus countless opportunities to meet with elders and discuss their knowledge of stars and navigation. Annie Blue of Togiak deserves a special thank-you for providing the Yup'ik names of constellations, the Milky Way, and Venus. Annie shared many stories about the sky and her life as a child. Thanks also to Steve Jacobson of the Alaska Native Language Center, University of Alaska Fairbanks, for his help with the spelling and translations of these and other Yup'ik terms.

I also want to acknowledge the contribution of four elders: Henry Alayuk of Manokotak, Sam Ivan and Wassiley Even of Akiak, and Joshua Phillip of Akiachak, who shared their knowledge about navigation and weather prediction. Joshua Phillip was especially helpful with his wisdom and skill in predicting the weather. Last but not least, I want to express my gratitude to these translators and cultural experts who were invaluable to the success of the project: Sophia Kasayulie of Akiachak, Natasha Wolberg of Fairbanks, Evelyn Yanez of Togiak, and Ferdinand Sharp of Monakotak. Evelyn and Ferdinand are certified teachers, who have retired since the start of this project. Sophia is a teacher aide and local coordinator for her school district curriculum project. Natasha was a translator and staff assistant for the NSF grant.

As a final note, my deep thanks to Igor Krupnik and Dyanna Jolly for their kind assistance in getting this manuscript to print.

Notes

1. A list of Yup'ik star and constellation names used in this paper is enclosed as Appendix 1.
2. This story covers a description of an actual trip undertaken in November 2000.
3. The sky chart was from the Lawrence Hall of Science, an education research museum at the University of California Berkley in Berkley, California.
4. Magnitude is defined as the brightness of a star or other celestial body, as viewed by the unaided eye from the Earth and expressed by a mathematical ratio of 2.512.
5. Right ascension is defined as the arc of the celestial equator measured eastward from the vernal equinox to the foot of the great circle passing through the celestial poles and a given point on the celestial sphere expressed in degrees or hours.

References and Source Material

Bethel Weather Service Office Airport Alaska National Climate Data Center Station Historical Listing for National Weather Service Cooperative Network, Bethel, Alaska: 2001.

Bogoras, Waldemar. 1904–1909. *The Chukchee.* Memoirs of the American Museum of Natural History 11; Jesup North Pacific Expedition 7. New York and Leiden: E. J. Brill Publishers.

Brunner, B. ed. 1998. *Time almanac 1999: The ultimate fact and information source,* information Please, LLC, Boston, Mass.

Dickinson, T. 1998. *Nightwatch: A practical guide to viewing the universe.* Third Edition. Toronto, Ontario: Firefly Books Bookmakers Press.

MacDonald, J. 1998. *The arctic sky: Inuit astronomy star lore and legend.* Royal Ontario Museum/Nunavut Research Institute, Iqualuit, Canada.

Mechler, G., Chartrand, M., and Tirion, W. 1995. *Constellations of the northern sky.* New York: National Audubon Society Pocket Guide.

Nelson, Richard K. 1969. *Hunters of the northern ice.* Chicago IL: University of Chicago Press.

Rey, H. A. 1976. *The stars: A new way to see them.* Enlarge World Wide Edition. Boston: Houghton Mifflin Company.

Appendix 1: Glossary of Yup'ik Names for Stars and Constellations

Yup'ik Term	Meaning
Agesqurpak	The morning star Venus
Agyarpak	The morning star (in Manokotak) Venus
Erenret Quliit	Sunrise Top of the day lights
Ingricuak	Little mountain or hill
Kaviaraat	Little Dipper A group of small foxes
Kinguqa Akugat	Corona Borealis Kuspuk hem
Nayipar	The Little Dipper The small seal with one eye and two front legs and one back; usually facing south
Pitegcaun	Leo Arrow
Qengartarak	Cassiopeia The nose and two nostrils When it gets cold, they flicker
Qupnguaq	Milky Way Pretend break
Tunturyuk	Big Dipper Caribou
Tulukaruum Tanglurallri	Milky Way Raven's Snowshoes
Tulukaruum Ayarua	Orion's Belt Raven's Cane
Unuakum Agyartaa	Venus The morning star (in Akiachak)
Taluyat	Bootes Fish Traps

Figure 8-1. The snowhouse, an icon of the North, may be a thing of the past in some regions because of changes in snow composition (photo by C. Furgal).

8

Climate Change and Health in Nunavik and Labrador

Lessons from Inuit Knowledge

Christopher M. Furgal,[1] Daniel Martin,[1] and Pierre Gosselin[1, 2]

> Just a comment about global warming, you hear it all the time on radio and TV, and I don't think it makes much difference in some places if the temperature rises two degrees, but if our temperature up here rises two degrees or something, the fact that we live on the ice and snow, I don't know what would happen to us. If we ever saw a real change, a real quick change, I don't know how we'd deal with the impacts of something like that, I don't know how we'd react to it, we'd have a hard hard time. (Nain, Labrador, man age 49)

Our understanding of climate change processes and potential impacts on northern ecosystems and people has increased significantly in the last decade (Furgal et al., 2002; Ashford and Castleden, 2001; Intergovernmental Panel on Climate Change, 2001; Hughen, 1998; Cohen, 1997a; Environment Canada, 1997, 1998a, 1998b; Bergeron et al., 1997; Bielawski and Masazumi, 1994). It is in this region of the world where changes and impacts to ecosystem and human health are potentially the greatest (Intergovernmental Panel on Climate Change, 2001). Changes across the North reported to date are variable, with warming trends occurring in the western Canadian Arctic and Alaska and some moderate cooling taking place in some locations of the eastern Canadian Arctic, while other locations in the east report warming as well (Environment Canada, 1997). This variability stresses the need to take a regional approach in gaining a better understanding of the changes and direct as well as indirect impacts generated by these changes in the North.

Temperature, precipitation regime, and other changes projected for the Canadian Arctic are already being observed and reported by scientists and aboriginal residents

1. Public Health Research Unit, Centre Hospitalier Université Laval (CHUL) Research Centre, Centre Hospitalier Universitaire de Québec (CHUQ), Beauport, Québec
2. World Health Organization (WHO)/Pan American Health Organization (PAHO) Collaborating Centre on Environmental and Occupational Health Impact Assessment and Surveillance at the CHUQ, Beauport, Québec

alike (e.g., Furgal et al., 2002; Riedlinger and Berkes, 2001; International Institute for Sustainable Development, 1999; Fox, 1998; 2000; McDonald et al., 1997; Hughen, 1998; Cohen, 1997a; Environment Canada, 1997; Laborador Inuit Association, 1997; Laborador Inuit Health Commission, 1996; McMichael et al., 1996; Bielawski and Masazumi, 1994; Ernerk, 1994; Freeman, 1994; Meldgaard, 1987). Much of this information comes from the observations and traditional ecological knowledge held by local individuals living in the North and in close relationship with the land.

For nearly half a century, many anthropologists and plant taxonomists (e.g., Mayr, 1953) recognized traditional knowledge systems in the form of extensively accurate and comprehensive classification systems (Freeman, 1992; Anonymous, 1993). It is understood that these systems go far beyond simply describing—that they are ecological in nature as well. However, with regard to climate change in the Canadian Arctic, it is only recently that a number of projects have begun to engage northern Aboriginal people and focus on indigenous knowledge of climate processes, changes, and impacts. It is argued that in order to better understand the complex nature of northern ecosystems, all available knowledge must be considered and valued.

Focus of the Chapter

This chapter is a summary of a much larger project in documentation of Inuit knowledge and perspectives about environmental change and health, "Climate Change and Health in Nunavik and Labrador: What We Know from Science and Inuit Knowledge" (Furgal et al. 2002). The project was conducted in Kuujjuaq, Nunavik, and Nain, Labrador, in 2000–2001. It was initiated through discussions between a public health researcher (C. Furgal) and members of several Inuit agencies in charge of local health, environmental, and social services: the Nunavik Regional Board of Health and Social Services/Nunavik Nutrition and Health Committee and the Labrador Inuit Association. Previous and ongoing collaborative work with these regional organizations, coupled with an interest in another project documenting Inuit knowledge and climate change in the community of Sachs Harbour (see Jolly et al., this volume) provided an impetus for undertaking a similar project in Nunavik and Labrador. Further, considering the amount of attention this issue had started to gain both within and outside the Canadian North, it was surprising to the research team and northern partners that little attention had been given to the potential effects of climate on public health in northern communities. As a result of these interests and concerns, a project investigating climate change and potential health impacts was initiated in the two regions.

The project that we named "Climate Change and Health in Nunavik and Labrador: What We Know from Science and Inuit Knowledge" was aimed to develop a better understanding of climate change processes and their potential health impacts among the publics of Nunavik and Labrador. The project collected and synthesized available scientific data and documented Inuit knowledge of climate-related changes and links to human health. It was recognized that this project was something of a scoping process to begin documenting the various perspectives and potential areas of information requiring further detailed work in the future. In this sense, it provided a starting point for the regional agencies to review their current understanding about potential health impacts, to orient future resources and efforts on issues related to climate change. Also, it used various sources of information (interviews, focus groups, existing literature) in order to begin to develop a common source of information (including both western science and Inuit knowledge) on the subject. For both regions and both knowledge bases, it was recognized that other documentation existed, in part, discussing this issue. Therefore, whenever possible, these sources were sought out and included in the review conducted for this project. The information presented in this chapter focuses specifically on the Inuit knowledge documented through interviews and focus groups and reviewed in existing documentation during this project.

Organization of the Chapter

The chapter is divided into three main sections. First, the methods employed in this project to document Inuit knowledge are described, and a brief introduction to the regions of Nunavik and Labrador is presented. Then, the Inuit observations of climate and weather changes and the impacts people feel that these changes are having on them and their communities are reviewed. All observations are organized within a particular local context so as not to generalize or simplify observations as being representative of the entire Nunavik-Labrador area. We also tried to capture the variability within and between regions reported by individuals in this study. Finally, a discussion of the implications of this information for adding to the general understanding and approach to addressing the issue of climate change in the North is provided.

Methods

The collection of Inuit observations and perspectives of environmental and climate change involved both a review of existing documentation and conducting focus group discussions and key-informant interviews among hunters and elders (including both men and women) in the two regions. These methods are described briefly below.

Document Review

All available documentation including Inuit knowledge reports, interview transcripts, and workshop reports related to the issues of environment and change in the two regions were collected. These materials were reviewed for their content related to environmental and climate-related changes and individuals' reports of effects or impacts these changes have upon them. The inclusion of existing Inuit knowledge literature allowed the development of a common source of information in the two regions. Since some excellent projects and sources already exist on Inuit knowledge and environmental change in the two regions (i.e., Nunavik: *Voices From the Bay,* McDonald et al. 1997; Labrador: *Our Footprints are Everywhere,* Brice-Bennett 1977; *From Sina to Sikujaluk: Our Footprint,* Williamson, 1987) reviewing these sources allowed a more comprehensive picture to be drawn in this initial project on climate and health in Nunavik and Labrador.

Focus Group Discussions

Focus group discussions were held with a total of sixteen hunters and elders in two selected northern communities, Nain (in Labrador) and Kuujjuaq (in Nunavik). Groups were comprised of hunters and elders (both male and female). Individuals for interviews and group discussions were identified by the community research assistants and regional organizations involved in the study as people with extensive knowledge about the environment, primarily based on the time they spend on the land or their expertise related to a specific land-based activity (e.g. seal hunting). In this sense, these individuals were suggested as local "experts." They were contacted and asked to participate in a discussion group about their observations on the environment in the region. When required, an Inuktitut-English interpreter/translator assisted with the focus groups or interviews. In these cases, interpreters had been briefed beforehand to clarify the content of general discussion topics and process. Interviews were audio recorded only when permission was given by the participant(s). Notes and tapes were reviewed for clarification and participants were contacted again, if necessary, to clarify answers or provide additional information. Group discussions lasted anywhere from thirty minutes to two hours.

Key-Informant Interviews

Semidirected key-informant interviews (Patton, 1980; Bordens and Abbott, 1991; Creswell, 1994) were conducted with experienced harvesters and elders (men and women) when they were not comfortable or were unable to participate in a focus group discussion yet were still interested in talking to us. As with the focus groups, the outline, purpose and intent of the study was fully explained before the interviews began, and the participants signed a consent form. The interview process was guided as a discussion format, posing open-ended questions followed by probes or suggestions of specification (Creswell, 1994). We chose this format since it encouraged a broader survey of individuals' knowledge of topics being reviewed and allowed the discussion of other topics not anticipated by the interviewer. In this way, the interaction took the form of flowing discussions rather than question-and-answer sessions.

Data Organization and Analysis

In qualitative content analysis, as used here, the value of data groups is not reduced to simple frequencies of their occurrence. Rather, an attempt is made to consider the context, content, and value of all responses given. In this study, a process of iteratively reviewing data and developing groups or categories of related information, as described by Tesch (1990), was used for the purposes of presenting the data in the project report. General groups were then discussed and reviewed with the northern partners and revised or adapted when necessary. The format of groupings presented here is recognized as one of several potential ways of relating the information to the reader.

Kuujjuaq, Nunavik. Nunavik is the region north of the fifty-fifth parallel in the province of Québec, Canada. The region is home to approximately nine thousand Nunavimmiut

(Inuit of Nunavik) living in fourteen villages along the Ungava Bay, Hudson's Strait, and Hudson's Bay coasts in this northern region of Québec, Canada. The communities are between 1,000 and 1,900 kilometres north of Montreal and all but three of these communities have less than 1,000 inhabitants. Approximately 88% of the regional population is Inuit and more than 50% are under the age of twenty-five (Makivik, 1999). The level of language retention in Nunavik is over 95% among Inuit, and Inuktitut remains the dominant language spoken (Figure 8-2).

The regional centre, Kuujjuaq, located at the southern tip of Ungava Bay is comprised of 1,720 individuals of which 74% are Inuit [Makivik Corporation, 1999; Statistics Canada 1996 in Nunavik Regional Board of Health and Social Services (NRBHSS), 2001] (see Figure 8-3). Kuujjuaq residents have access to country foods through harvesting, sharing, sale of foods at one country food store and through some regional government subsidy programs designed to help supply country food to the community. A total of seven individuals from Kuujjuaq participated in focus groups in Nunavik for this study. One focus group with four male hunters and elders (men) and one focus group with three elders (women) were held in Kuujjuaq, between May 22 and 31, 2001.

In spite of many economic, political, and social changes in communities in this region in the last decades, Nunavimmiut have stayed very close to their traditions. They still regularly hunt caribou, small game birds, seals, walrus and beluga whales, fishing the many freshwater and marine species of the region, and sharing these items. Traditional food is still culturally and socially very important in everyday life in the region.

Nain, Labrador. Approximately 4,800 Inuit and Kablunagajuit live in Labrador, the far northeastern region of mainland Canada. Fifty percent of these individuals recognize themselves as Inuit while the others recognize themselves as *Kablunagajuit*[1] [Tanner et al., 1994; Voisey's Bay Nickel Company (VBNC), 1997]. Approximately 55% of the membership live in the communities along the north coast (Nain, Hopedale, Makkovik, Rigolet, and Postville) while the majority of the remainder reside in the communities of the Upper Lake Melville area (Happy Valley-Goose Bay, Northwest River, and Mud Lake) (Tanner et al., 1994; Williamson, 1994; VBNC, 1997). A significant proportion of the total economy of households and communities is still accounted for by the traditional economy as in other regions of

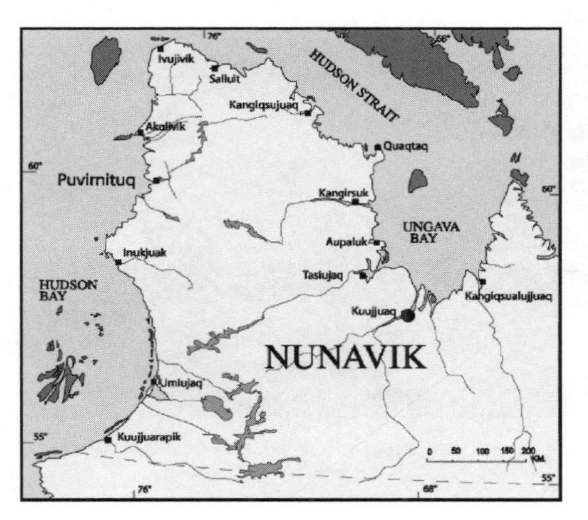

Figure 8-2. Regional map of Nunavik, with study community (Kuujjuaq) located at the southern limit of Ungava Bay.

the North (Wenzel, 1981; 1986; Rees, 1988). Forty percent of households in Nain report getting "most" of their food from the land, while 40% reported getting "some" (Community Resource Services Ltd., in VBNC 1997). In a recent dietary survey conducted by the Centre for Indigenous Peoples, Nutrition, and the Environment (CINE, 2001) Labrador Inuit consumed the least amount of country foods per capita but were the most active in harvesting activities and consumed the largest diversity of country food species as compared to all Canadian Inuit regions.

Established as a Moravian mission in 1771, Nain is the farthest north permanently inhabited community along the Labrador coast and is the regional centre for the north coast communities today. With a population of approximately 970, Nain is headquarters to the offices of the Labrador Inuit Association, Torngasôk Cultural Centre, Nain Town Council, and regional offices of the Labrador Inuit Health Commission and Labrador Inuit Development Commission. Williamson (1994) describes Nain as "the community in which Inuit lifestyle, language and values are most intact in Labrador." A total of twelve individuals from Nain participated to focus groups and interviews for this study. Three focus groups (five elders (men), three hunters (men), three elders (women)) and one individual interview with an elder (woman) were conducted in Nain from March 12 to March 22, 2001.

Inuit Observations of Change in Nunavik and Labrador

From years spent travelling, hunting, and living in close relationship with the land, residents of Kuujjuaq, Nunavik, and Nain, Labrador, have noticed many changes in their environment. Some of these changes are reported to be cyclical in nature (e.g., trends in numbers in the caribou population and some small mammals). Other changes, however, are being reported that have never been seen before, heard

Figure 8-3. Kuujjuaq, the regional centre of Nunavik, where focus groups and interviews were conducted for this study.

through stories passed down by elders, or predicted by elders and other experienced individuals. Further, many residents of Labrador and Nunavik are now eager to speak of the effects these changes have on various aspects of their lives, including their individual health and well-being. Local observations from Nain and Kuujjuaq and perspectives documented in this project are summarized as Appendix 1 and 2. For presentation here, they were grouped into the general changes associated with:

- Weather
  - Temperature
  - Weather predictability and storms
- Intensity of the sun
- Ice and snow
- Freshwater
- The land
- The sea
- Availability and access to country foods

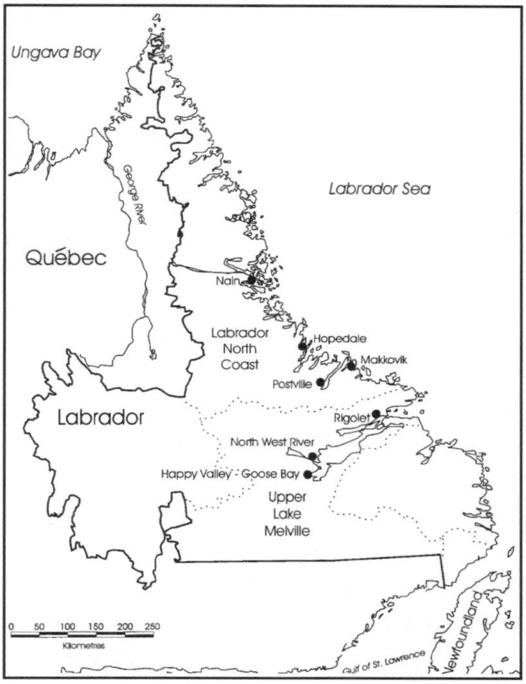

Figure 8-4. Map of Labrador, including Inuit communities of Nain, Hopedale, Makkovik, Rigolet, Postville, and the areas of Happy Valley-Goose Bay and North West River.

Figure 8-5. Nain, the centre of the north coast Inuit communities in Labrador, where focus groups and interviews were held for this study (photo by C. Furgal).

This grouping was done by the researcher to organize responses and represents one of many potential ways of grouping this information. The observations and perspectives presented below are organized on a local/community basis to emphasize the similarities and differences in changes observed both within and between regions. Further, the detailed and local nature of much of the information reported is critical to understanding the specificity of changes and thus potential impacts in each of the studied regions. What is reported in this chapter merely scratches the surface of the knowledge held by many experienced individuals and shared within the two communities, Nain and Kuujjuaq, surveyed during the project. Since individuals involved in this study did not give direct permission during the consent process to identify them in the report, we present people's place of residence, sex, and age to identify specific responses. We recognize that many of the individuals involved in this project have experience with large portions of their region through hunting and travelling on the land. Also, in some cases, individuals grew up in other communities and did not originate from Nain or Kuujjuaq. Thus they often speak of areas far from their community of residence. For these reasons, we indicate specific place names, when possible, where people have observed changes in the two regions.

Changes in the Weather: Temperature

Nain. The residents of Nain surveyed in this project report a general warming along the Labrador north coast over the past two to three decades (Appendix 1). Individuals report that there are a greater number of very hot days in the summer now than in the past. When speaking about the impacts of these very hot days, some individuals include shortness of breath and a limitation of physical activity (e.g., fishing). Alternatively, the heat is reported to be beneficial in some ways, through, for example, decreasing the number of flies during the summer. Fewer extremely cold days are also reported, allowing people to get out more in the winter.

> This past summer while we were up at OKak [location of old Labrador community along the north coast] we had one full week, if not more, of hot hot weather. I don't ever remember it being that hot. Usually, before when it was hot like that it would only last a day and then it would rain.... We were fishing and we had to go out on the boat to get away from the heat; it was shocking. (Nain, man age 71)

> I tended to my nets and my wife said she was not able to stay on land because she found it too hot. . . . I decided to stay on land and take care of our char. I should have went because I almost suffocated because of the heat; I had to keep splashing myself with water and in the afternoon I couldn't stay on land because it was so hot. I had to stay in the water to stay cool.... My breathing was even shallow. (Nain, man age 61)

> It was so hot, it was hard to breathe around our camping area where we fish. (Nain, woman age 70)

Kuujjuaq. In Kuujjuaq, residents are also reporting a shifting in temperature regimes. The weather is becoming warmer sooner in the year and colder sooner in the fall and, on average, warmer in the summer. Some people mentioned that it now gets extremely cold when it is not expected to. The unpredictability of temperature changes and the degree of change are new to the climate in this region (Appendix 2). Impacts of such changes include some disruption of individual's plans to participate in harvest activities.

> This year and last year, we have been stopped when we were going to go fishing because the weather changed at the last minute. We would have gone fishing more in the past. (Kuujjuaq, woman age 69)

It is interesting to note that some individuals attributed the impacts of these kinds of changes to their health and lifestyles to factors other than temperature change.

> I don't think the weather has changed me, but getting older only changes me. (Kuujjuaq, man age 70)

> Our bodies have changed too. We used to exercise like dogs, having to run with our dogs, we used to sweat a lot. Nowadays, you hop on a Ski-doo and never get off and can go everywhere. That makes our traditions change too. We left our traditional way of living with having dog teams. We used to run with our dogs, in all areas, travelling with them. Ever since our dogs were killed, we have been just sitting on our vehicles. That's why we feel the sharp cold now. (Kuujjuaq, man age 62).

Further, some individuals feel that the changes are simply part of natural variation in the region.

> The older people who were our parents used to talk about the weather and it looks like the weather to me is still the same. Even though some years are different, there is no special abnormal weather that I have come across. (Kuujjuaq, man age 70)

In *Voices From the Bay*, elders in communities along the Hudson Coast (Kuujjuarapik-Whapmagootsui, Umiujaq, Inukjuak, Akulivik, and Sanikiluaq) and Hudson Strait (Ivujuvik, Salluit, and Kangiqsujuaq) (McDonald et al., 1997: 46) indicated a general cooling trend for the eastern Hudson and James Bay coast and Hudson Strait regions. Colder winters with more very cold days are reported for these three areas.

Changes in the Weather: Weather Predictability and Storms

Nain. As well as changes in temperature, Nain residents report that the weather has become increasingly unpredictable compared to previous decades. There is inconsistency in observations of winter storms: some report observing more while others report experiencing fewer storms today than in the past. The number of thunderstorms is reported to have decreased and there is consensus that autumn storms have increased in frequency and intensity and that the weather changes much faster than people ever remember. This unpredictability is said to limit the ability of people going off on the land. An increasing number of people also report an increase in the likelihood of getting stranded out in storms. "Quick" changes in weather and storm intensity are reported to have significant potential impacts on animal health. For example, Williamson (1997: 32) documented north coast residents' reports of finding large numbers of eider ducks frozen in the sea ice following a sudden late autumn freeze-up.

> Now the storms are more frequent and severe ... quickly changing from mild to cold and vice-versa ... the autumns are especially windy and dangerous for speed boats. (Labrador, man; as in Williamson, 1997: 39)

> There are more fierce fall storms now. Before, it was nothing like the storms and strong winds we get now. (Nain, man age 49)

> What I don't like much is these sudden real strong winds. We rarely used to get these much, but now we're getting them, suddenly, really strong winds. (Nain, woman age 61)

> Yes, it changes so quick now you find. Much faster than it used to be.....Last winter when the teacher was caught out it was perfect in the morning, then it went down flat and they couldn't see anything. It was like you were travelling and floating in the air, you couldn't see the ground. Eighteen people were caught out then, and they almost froze, it was bitterly cold too. (Nain, man age 43).

Kuujjuaq. Similarly, Kuujjuaq residents report an increase in the unpredictability of the weather at certain times of the year. It was specifically stated that it was getting very cold unexpectedly, at varying times throughout the year which sometimes influenced people's ability to get out on the land. In interviews conducted with elders for the Voices From the Bay Project (McDonald et al., 1997), people also reported fewer thunderstorms throughout the region.

Changes in the Sun

Both Labradorimiut and Nunavimmiut in Nain and Kuujjuaq, respectively, speak of a change in the intensity of the sun's rays, observing that the sun burns their skin more now than they ever remember. This increase in intensity is said to be drying and burning the vegetation of berries and other small plants and thus impacting their availability for consumption. However, there is some question as to whether or not this is a matter of perception because of the increased awareness as a result of health messages on this issue today, as stated by one individual:

> The sun burns more now too. You can even feel it on your skin. You can tell the difference ... but I don't know, maybe you just think about it more because of all the warnings you hear, it could be that. (Nain, man age 64)

Similarly, people in Kuujjuaq report the need to wear sunglasses more often now, use protective creams and lotions, and sometimes stay indoors when it is too hot because of the change in intensity of the sun.

> The sun burns us easily, it was not very hot in the past. (Kuujjuaq, man age 62)

> The sun was not that hot in the past. Nowadays, it's really hot. My skin burns when I'm out for a while. Sometimes, we stay indoors in a shack. (Kuujjuaq, man age 70)

> The sun is very hot, it's dangerous for me. It was never like that. We did not have to put cream on and these creams were never in the store on the coast, today we have to use these sunscreens. (Kuujjuaq, man age 70)

Changes in Ice and Snow

Nain. Changes in the snow and ice are affecting wildlife and the established cycle of annual subsistence activities in both communities surveyed. In Nain, the timing of annual sea ice formation, consolidation, timing of spring melt, as well as the nature of the ice itself are reported to be changing. Ice in the bay around the town is reported to form these days approximately one month later (in January) than about fifteen to twenty years ago.

> We don't get any good ice before January now either, it's about a month later than long ago. (Nain, man age 49)

People in Nain report that it takes longer for the ice to solidify and that it appears to be *saltier* and thinner in nature and that there is much less snow on top of the ice throughout the winter each year. It was reported that the ice goes out much earlier each spring and break-up takes longer. Also, there is a decrease in the amount of pack ice coming south into the bay around the community of Nain each year.

These changes in ice are reported to have impacts on ringed seals *(Phoca hispida)* that use the ice for pupping and breeding habitat. The lack of snow on the ice is associated with the growing amount of scraped and wounded seals caught by hunters and with a decrease in the number of breathing holes and dens in the area, the latter increasing the pups' exposure to predation on the ice. Changing ice conditions were also said to make travel very dangerous during spring. A lack of certainty regarding sea ice conditions in the spring is keeping hunters and their families from heading out onto the ice. The inability to continue regular travel and harvesting routines causes mental stress at times of the year when people are used to being able to travel on the land.

> We have to wait for ice, like last fall there was no snow, the ice was bad, so you couldn't go far, just walk up on the hills … and go crazy. (Nain, man age 43)

Some participants felt that the early break-up was a positive change, since it allows ships to come into Nain earlier in the spring, and community members to go out to their fishing camps earlier in the season.

The amount and composition of snowfall and the composition of the snow has been changing over the last two to three decades as well. Many participants report that the first snow now arrives later in the year (often not until December); that there is less snow now than fifteen to twenty years ago; and that the snow today is more "granular" and lighter. They also say that the snow in the mountains north of Nain, once visible year round, now melts each year, leaving the hills bare in the summer months (Figure 8-6). During the winter months, "glitter" (frozen crust on top of snow) is observed as being more common. The presence of glitter on top of the snow during the winter is described as impacting the health of both caribou and ptarmigan, since it blocks their access to valuable food resources during cold months.

These snow-related changes restrict travel at certain times of the year (for example, not enough snow to get into the bush on snowmobiles) and thus the ability to access firewood from the forests in spring time. However, it was also seen to make walking easier, which is especially important for elders, because the snow is not as deep. Due to the changes in snow, all respondents from Nain acknowledge doubt in the ability to build snow houses these days (Figure 8-7).

> The snow is lighter [now]. As soon as it blows, the snow is gone. I don't think we can make snow houses anymore because the texture of the snow is not the same. There is a big change in the past fifteen to twenty years. (Nain, man age 71)

> ... and all of a sudden we had a storm and everyone was lost, but with the snow now, years ago you'd get the good snow for snow houses but now, you wouldn't be able to make one [snow house] if you had to. (Nain, man age 49)

Respondents from Nain also mentioned that the lack of snow on the hills in the North requires them to look for alternate sources for keeping their fish cool while fishing in this part of the region.

> We used to always stick char in the snow when we were fishing up the coast. Go up the valley to get pockets of snow or ice in the shadows, but it's hard to find now for the fish when you're up there. (Nain, man age 61)

Kuujjuaq. In Ungava Bay, Hudson Strait, Hudson Bay, and James Bay, Nunavimmiut similarly report changes in the times of ice consolidation and break-up (McDonald et al., 1997). As in Nain, the ice around Kuujjuaq is breaking up earlier and faster in the spring, and is thinner, on average, than when present-day elders were young. As a result, ringed seals are reported to leave the region earlier with the receding ice. Consequently, hunters' travel times are increasing and there is a growing concern regarding travel safety on the ice at certain times of the year.

> I felt that it was spring earlier and that the ice was thinner this winter.... Now the ice just tears open today as I see it. (Kuujjuaq, man age 36)

> I know that today that seals, it might be because of early spring break-up or that they are out on the ice floes, that the seals are nowhere. (Kuujjuaq, man age 62)

> Sometimes we just go home now because we can't go off on the land anywhere. The ice melts right away. It's a fast melt down even in the lakes too. (Kuujjuaq, man age 62)

Detailed changes in sea ice are reported in other specific subregions within Nunavik (McDonald et al. 1997: 46, 47). Fewer polynyas and poorer quality of ice were reported by elders in the Hudson Strait communities (Ivujivik, Salluit and Kaniqsujuaq), and people are seeing a greater extent of land fast-ice in Eastern Hudson Bay and Hudson Strait (McDonald et al., 1997: 46). Impacts of these changes include increased hunters' access to eider ducks, but at the same time a decrease in food sources available for these birds.

Changes in snow conditions reported by Kuujjuaq residents included those related to both the nature of snow cover and composition, similar to observations from Nain. As documented in McDonald et al. (1997: 47), elders in eastern Hudson Bay and Hudson Strait communities of Nunavik report snow melting later in the spring than in previous decades and yet melting and thawing throughout the winter, when this did not happen in the past. This was described as resulting in more "ice-like" snow covering vegetation, making it difficult for animals such as caribou to forage throughout the

Figure 8-6. Nain residents report that snowfields that used to be present year round in the Torngat Mountains in northern Labrador are now melting during the summer (photo by D. Martin).

winter. It also means that the snow is less suitable for making snow houses, as observed in Labrador.

Changes in Freshwater

Nain. The changes in temperature, snow, and rain patterns observed along the coast of Labrador have resulted in changes in freshwater systems in the region. Many individuals have observed that summers are much drier today than during past decades and that there is less snow. Further, because of the decrease in rain during the summer months, the lack of snow and the drier weather, individuals report that many brooks and small ponds are drying up or have dried up and are no longer there. These water bodies were used as sources of drinking water while people were out on the land hunting and fishing, and depending on the location of a camp, it is now more challenging to locate good natural sources of drinking water.

> ... but even where people had summer camps there is no drinking water. There used to always be little brooks around when there was snow on the hills, there are no more little brooks, just a few ponds left. (Nain, man age 64)

> Even some ponds have dried up ... we spent some time ... before the char went out to the salt water, and for the past two years we have had problems finding drinking water. The ponds and brooks are drying up. It was the worst this summer, we had to move to another camping area. We did that the summer before too ... we found a mossy area and I had to dig to find water. (Nain, man age 62)

> It's a lot drier in the summer. Up in the bays, some of the lakes and ponds, you can almost walk across where they were now. All the swamps and bogs ... they're all drying up now. Some of the ponds, about half of them in some areas are drying up. (Nain, man age 46)

Further, these changes are said to have impacts on individual's travel routes out on the land making some routes less suitable, or not passable at different times during the year because of lack of water for boat access, or very soft conditions (where ponds have drained) not passable by all terrain vehicles.

Kuujjuaq. Similar changes are observed in Kuujjuaq where many small lakes that were used as drinking water sources are now drying up. This is attributed to a lack of rain and damming in the area. These changes are reported to have damaged equipment such as boats and also limited supplies of fresh drinking water while out hunting and travelling.

> Even the small drinkable lakes have dried up, where there is no rain and because the other lake has a dam. (Kuujjuaq, man age 62)

As reported by McDonald et al. (1997: 46, 47), the water levels in many brooks and rivers have decreased considerably throughout the Hudson Bay region in such communities as Kujjuarapik, Inukjuaq, and Umiujak. Much of this is reported to be attributable to damming across the interior section of Nunavik for hydroelectric development, with decreased summer rainfall further exacerbating the situation in some years. Where summer water flow is low, residents report that the quality of drinking water has decreased (e.g., in the communities of Salluit, Kangiqsujjuaq, and Ivujivik; McDonald et al., 1997: 46).

Changes in the Land

Inuit are closely linked through their culture and way of life to the land and sea. The land is the foundation for human travel and support for vegetation and terrestrial wildlife upon which Inuit rely, so the health and stability of the land occupies a central role in the relationship Inuit have with the environment. Individuals in both Nain and Kuujjuaq report that aspects of the land are changing as a result of climate, weather, and other forces. Changes in weather, precipitation, temperature, and the intensity of the sun as described above are all impacting the local permafrost and tundra ecosystems. As a result of heat and less rainfall, the land is reported to be much drier than before. One location in Voisey's Bay (approximately thirty-five kilometers from Nain) where permafrost had been found no longer exists.

> There's a couple of places in Voisey's Bay where they used to get ice underground, the marsh I suppose used to freeze and there was ice all summer long and they used to chop it out, but that's not there anymore. (Nain, man age 54)

Further, this drying of the land is associated with some extreme events that were previously very rare in this region.

> There was even a forest fire where we never thought there would be one, the land is so dry. (Nain, man age 61)

Nunavimmiut in eastern James Bay interviewed for *Voices From the Bay* report that the land is rising, and this is reported to be decreasing water levels in some estuaries and having impacts on the numbers and species of marine animals and fish seen in these areas (McDonald et al., 1997: 46).

Changes in the Sea

Tides in some areas along the Labrador shore are said to be stronger now and the water in many bays shallower at low tide. No explanation or greater detail was provided on this issue in our discussions and interviews with the Nain residents during this study. In contrast, some Nunavik-based projects have documented decreasing water levels and weaker currents in rivers and estuaries to be a result of a combination of both changing rainfall patterns and hydroelectric development (McDonald et al., 1997: 47).

Changes in Availability and Access to Country Foods

Another group of changes observed by residents in Nain and Kuujjuaq are those related to the wildlife and plants. The impacts of environmental change on plants and animals concern Inuit communities in terms of both availability of and access to country foods, and thus may have nutritional, social, cultural, and economic effects (Fast and Berkes, 1998). Several changes are observed in the animals and plants by residents of the two surveyed communities that are related to these environmental changes. This section presents a summary of some of these accounts.

Nain. In Nain, environmental changes are discussed as having negative impacts on the health and availability of certain important country food plants and animals. Changes in the health of caribou are reported by Nain residents in terms of more "sick" caribou and caribou with bugs in their livers. The taste and texture of the meat is said to have changed in the last twenty years. A greater number of red foxes (which are potential carriers of rabies), beyond normal cycling numbers for coastal areas, are encountered today and more sick or diseased foxes are seen as well. Similarly, more "sick and skinny" seals are observed by hunters, potentially attributed to a lack of prey resources.

> It's true that animals, especially caribou, like to linger where there is snow, there weren't many places where they could cool off even this past summer, I think that might affect them. (Nain, man age 49)

> There used to be a lot of snow and when it snowed a lot during the winter, the berries and bake apples were much better and when it's too hot during the summer they dry up from the sun. (Nain, woman age 66)

> I went to OKak Bay this year and there were no berries this year, it was just too hot. There weren't even any bake apples, they were all dried up by the heat. (Nain, man age 61)

Further, changes in vegetation are said to impact caribou migration and behaviour in the region. Many of these changes are influencing the diet of Inuit through affecting the suitability of the food to eat or the access to these items.

> Years ago we wouldn't see any bugs in caribou, and we always ate the liver. But now, I don't know if it's because of the amount of caribou there are now or that we're looking for them, but there are more bugs. Every once in a while you come across one with bugs, and now you don't even look at the liver, we don't touch it ever since we started to see the bugs in them. (Nain, man age 43)

> We see more sick seals now too. Some that you kill are just skin and bones and hair. There are more "crawlers" now too [seals crawling on top of the ice looking for a hole through which to enter the water]. (Nain, man age 46)

Further, there is concern that the lack of rain and other precipitation may have negative impacts on Arctic char migrations up streams and into inland lakes during the fall.

> Where it is getting drier, some of the lakes, there are less and less lakes, and some are smaller. Where the char are going in … those brooks are just about dried up and they can't get in, they have to wait for a heavy rainfall, so they just wait … I don't know if they go to find a larger river or what, so I don't know what will happen to the char if they just wait there, or if they can change their route. (Nain, man age 49)

Changes in the number of some species and the sighting of new species in the region are discussed by hunters as well. Fewer and skinnier fish are found today and some changes in their behaviour are reported. For example, Arctic char are reported to be going directly out to sea from inland lakes in the spring, and thus hunters have to wait until later in the season than usual for them to return to the bays.

> We see moose now, and never did before. We see them more and more each year. We've seen them up as far as the bay north of here. (Nain, man age 43)

> It was maybe twenty years ago that we saw the first beaver track ever up in the bay (Webb's Bay, north of Nain) … we didn't know what it was. (Nain, man age 46)

An increase in some species (seals, caribou, fox, black bears, wolves) and decrease in others (geese and other migratory birds, fish) are reported to have impacts on individual access to country food items. Some individuals have reported not going to their traditional hunting areas and camps anymore or travelling less frequently, because of a lack of wildlife that were once there.

Changes in the extremity and unpredictability of the weather and changes in the land and sea ice stability have influenced the ability of Inuit in Nain to safely go fishing, hunting and collecting country food and firewood at certain times of the year. These changes are reported to have taken place during the last two or three decades.

> Just this past fall people were tired of waiting for the snow to arrive, people who needed firewood were tired of waiting for the snow to come so they could go out wooding. (Nain, man age 54)

> And because there's less snow and by May most of the snow is gone, it makes things drier on the land too. It's hard to get around on the land. (Nain, man age 49)

> In the spring time, you're limited in distance you can go now. Fall too, you have to wait for ice to get out hunting. (Nain, man age 43)

Positive aspects of these changes are reported as well in that they allow access to summer fishing activities or store food supplies earlier in the season because of an early ice-free season.

> I remember years ago, we used to have to wait for the pack ice to go before we could go fishing and now today we can leave in June to go to our fishing camps. I think the atmosphere is really heating up. (Nain, man age 49)

Kuujjuaq. Similarly, wildlife, plants, their health, and Inuit access to them has changed in Kuujjuaq. Changes in temperature, the general increase in unpredictability in the weather and changes in ice and snow distribution and stability have been reported to influence individuals' ability to get out on the land (Figure 8-7).

> This year and last year, we have been stopped when we were going to go fishing. The ice broke up so fast. We went fishing more in the past. (Kuujjuaq, woman age 68)

Observations of changes in animal and plant health and distribution are described by many Kuujjuaq residents and other Nunavimmiut as related to the climate, either directly or indirectly. Many of these changes have had an impact on residents' perspectives on the general health of wildlife and fauna and their suitability for consumption.

> I know today that seals, it might be because of early spring break-up or that they are out on the ice floes, but the seals are nowhere. (Kuujjuaq, man age 36)

> I have never heard of caribou being sick in the past. Caribou have toxic livers and their joints are bad. As for lake trout, before they [people] never said they caught a bad fish. But in the last twenty years, I have heard from time to time that they have caught such fish. I think that the weather is causing the animals to be like that. (Kuujjuaq, man age 70)

> The berries don't grow sometimes now. The sun burns them. (Kuujjuaq, woman age 69)

> Our plants burn before they go into full bloom now. (Kuujjuaq, man age 62)

Further, individuals reported seeing fewer whales and some bird species. This is attributed to a variety of possible causes, including climate change, regional development, contaminants, and natural variability. Changes in sea ice distribution and cover are reported to decrease access to food for eider ducks but increase human access to them for hunting in the eastern Hudson Bay region (as reported in McDonald et al., 1997: 31). Frozen and icier snow during the winter is said to have the same effect in Nunavik as in Labrador, where it is said to block access to vegetation by wildlife (caribou and ptarmigan) during the winter. Changes in fish and seal migrations due to lowering water levels in some estuaries are reported to directly influence people's diets in terms of suitability for consumption, availability, and accessibility to these country food items. Further, some people in Kuujjuaq are observing an increase in the number of cases of rabid foxes, which they consider a potential link to climate change in the region.

Environmental Change and Inuit Health: Perspectives on Direct and Indirect Impacts from Nain and Kuujjuaq

As seen from the many direct statements of residents of Nain, Labrador, and Kuujjuaq, Nunavik, Inuit in these regions are observing various changes in the climate, weather, environment and flora and fauna, and various effects of such changes on their health, way of life, and relationship with the land. According to the World Health Organization (1967), health includes aspects of physical, mental, and social well-being and is not simply the absence of disease. In the holistic Inuit vision of health, the well-being of individuals and communities is tied to the land and sea. Thus, changes in the land, weather, and sea reported here affect individual and community health and well-being in a variety of ways.

A number of *direct* impacts of climate-related changes on health were reported by residents of the two communities surveyed in this project. In this sense we refer to "direct impacts" as *those health consequences resulting from direct interactions with aspects of the environment that have/are changing with changes in local climate* (i.e., resulting from direct interactions with physical characteristics of the environment: air, water, ice, land; e.g., exposure to thermal extremes). They include such things as difficulties in dealing with heat and cold stress, alleviation of cold stress due to warmer winters, the dangers associated with travel and time on the land considering unpredictable weather patterns and ice conditions, and reports of increased incidences of sunburns and rashes as a result of increased sun intensity. Accounts of direct impacts varied within and between communities in the two regions involved in this study. The larger study we conducted used these reports, in concert with projected potential impacts gleaned from the scientific literature and discussion with environmental health experts, to develop a list of potential direct health impacts from climate change in the two regions (Table 8-1).

Similarly, a number of *indirect* climate-related health effects were identified by individuals in the two communities surveyed, as well as in the key pieces of literature for the two regions. We include under "indirect impacts" *those effects originating through indirect interactions with human health and behavior under the influence of local climate and*

Figure 8-7. A lack of snow and abundance of water on top of the ice, both of which are more common occurrences today, make travelling difficult during the spring in Labrador.

aspects of the environment changing with local climate (i.e., mediated through one of a variety of social or physical means). These indirect effects include such things as effects on diet (decreased country food species abundance and/or availability) as a result of climate related impacts to wildlife species or environmental factors influencing Inuit access to these resources (e.g., ice distribution, land stability, weather predictability); effects on health as a result of changed access to good-quality natural drinking water sources; effects on rates of disease as a result of heat and cold stress experienced by residents; and various forms of social and mental stress related to changes in the environment, lifestyle, etc. related to changes in climatic regimes. Reports from Nunavik and Labrador were used in conjunction with projected potential impacts identified in the scientific literature and through discussion with environmental health experts to develop a list of potential indirect health impacts from climate change in the two regions (Table 8-2).

Understanding the importance of climate-related factors in contributing to Inuit health is critical, as many health issues potentially influenced by climate are prevalent in these two regions today. For example, unintentional traumas (including those resulting from accidents while on the land) account for a significant number of deaths each year in Nunavik (NRBHSS, 2001), and thus recognising the relationship between environmental change and accidents is important for public health professionals. Similarly, the social, cultural, and physical importance of country foods in these two regions makes the identification of potential threats to food security paramount. Country food intake accounts for greater than 35% of the total energy in the diet of Nunavimmiut men over fifty years of age (Santé Québec, 1995). It also makes significant contributions to daily intakes of iron, protein, selenium, and vitamins A and D among some segments of the population (Santé Québec, 1995). Fish and marine mammals, especially beluga whale skin (a component of *mattak,* an Inuit delicacy), are significant sources of selenium in the Inuit diet (Blanchet et al., 2000). It is currently thought that selenium acts as an antioxidant in the prevention of artherosclerotic diseases and may reduce the risk of mercury toxicity among Inuit (Hansen et al., 1994; Salonen, 1986).[2] The consumption of other country food species (Arctic char and marine mammal fat) high in n-3 fatty acids, provides individuals in these populations with some protection against cardiovascular diseases (Dewailly et al., 2001, Middaugh, 1990) and some benefits in relation to the occurrence of some cancers, diabetes, and hyperinsulemia (Dewailly et al., 1998; Friedberg et al., 1998). Despite the known benefits associated with country food species, there is a growing shift in many Inuit regions, more prevalent among some segments of the population, towards a more "western diet" comprised of more store-bought foods and less country food items (Dewailly et al., 2000). Considering the known physical, social, and cultural risks associated with such diet shifts, any and all threats to country food security must be seriously considered. It is therefore important in terms of individual and community health to be able to properly identify and understand the impacts of any environmental factors, such as climate related changes, influencing the health, availability, or accessibility of country food species.

Understanding Climate Effects

It is important to note that not all effects related to the current climate change in the North are negative. Several positive associations to warming or cooling were reported by people from both Nain and Kuujjuaq in this study. For example, warmer winters in some areas can lessen stress related to cold exposure. Longer hunting seasons and an increased number of days available to get out on the land may improve productivity of subsistence activities and mental and social health among individuals. Climate change may also cause redistributions of animal population ranges, bringing some species into new areas and making them more accessible for human consumption. Thus, in terms of human health effects, both positive and negative aspects must be considered.

Further, it is important to note that the changes observed and the impacts experienced to date in these regions were not reported by local residents as solely due to changes in climate. They are also related to changes in such nonclimate-driven processes as economic development, pollution, change in individual lifestyle habits (such as the use of snowmobiles instead of dog sleds) and other social, cultural, economic, and natural factors (such as the change in food habits, shift from a more nomadic to more sedentary to increasingly sedentary community-based life, etc.). The vulnerability of populations to any health risk varies considerably depending on a number of moderating factors such as population density; level of social, economic, and technological development; local environmental conditions; preexisting health status (including nutritional status and lifestyle habits); the quality and availability of health care services; and public health infrastructure. Further, in the context of this study, the knowledge of and relationship to the surrounding environment is a

Table 8-1. Summary of possible direct climate-related health impacts in Nunavik and Labrador (from Furgal et al., 2002)

Mediating process	Possible Direct Health Impacts
Exposure to thermal extremes	Heat and cold related stress (especially among elders) resulting in illness etc.
Changes in frequency or intensity of other extreme weather events (storms, etc.)	Accidents resulting in death, injuries, stress related disorders, psychosocial disruption (e.g. being stranded in a storm)
Changes in ice distribution and stability, and snow composition and amount	Accidents while travelling or hunting resulting in death or injuries (e.g., accident while travelling on poor ice)
Increased UV exposure	Increased risks of skin cancers, burns, infectious diseases, eye damages (cataracts), immunosuppression

significant mediating factor of great significance in northern communities. Much more work is therefore needed to better understand the extent to which climate-related changes are influencing specific aspects of human health in the North.

The Need for Local Perspectives

It is critical to recognize and understand regional and local/community variability in order to support the enhancement of adaptive strategies and tools for remote indigenous communities. For example, a general cooling trend along the Labrador coast and in the Labrador Sea is reported by several studies of climate change (e.g., Mayer and Avis, 1997; IPCC, 2001), while people on the land—such as residents in Nain—report a general warming trend to date. For Nunavik, significant variability in temperature trends is also reported throughout the region. Nunavimmiut are experiencing warming in southern Nunavik (Kuujjuaq: this study) and a general cooling is observed in eastern Hudson Bay and Hudson Strait communities, as documented in other projects (McDonald et al., 1997; Read and Gould, 1992). These inter- and intraregional variations and the difference between scientific models and locally based knowledge highlights the value of local, indigenous perspectives in climate change research and action. As members of the modelling community argue for more regionally specific climate models (e.g., Gough and Wolfe, 2001), we need to look to local observation-based data as the scale required to accurately understand relationships and current impacts.

An understanding of environmental and climatic changes and their impacts on individuals is required if northern communities—such as Nain and Kuujjuaq—are to be able to adapt and evolve with the changing nature of their surroundings. Stakeholders, such as those in Aboriginal communities in the Mackenzie Basin, have reported feeling as though they could adapt to climate changes, if such changes occurred gradually and were predictable (Cohen, 1997b). Without the local understanding of changes and potential impacts, such as those discussed in this volume, it is difficult to know the rates of changes experienced in communities to date, and it limits the predictive ability of knowing what to expect for communities tomorrow.

Working Together to Address Climate Change and Human Health

The cooperative planning, development, and conduct of this project, involving southern-based scientists and northern environment and health professionals, researchers and local communities, was essential to its success. The project was a learning process through which knowledge was gathered and is continuing to be shared as a number of initiatives have developed as a result. Indigenous residents of the Arctic require a mechanism to voice their concerns, share, and use their knowledge, and address these issues as is being argued at the national and international levels (Fenge, 2001). This partnership is essential in order to address the broad impacts of global environmental change. Aboriginal people and their knowledge are critical in this process as they have a unique and valuable understanding of these issues developed through their relationship with their local environment.

This chapter illustrates that changes in the environment, such as those related to climate, are having an effect on Inuit health. Very little research to date has explicitly made these links between human health and climate change in northern communities. Yet, for people still maintaining a close relationship with the land and sea through their cultural, social, and economic practices and way of life, their health and well-being and that of their communities is, and will continue to be, inextricably linked to environmental change. A holistic concept of health must therefore remain central to future research and action on climate change, in partnership with indigenous peoples in the Arctic.

Table 8-2. Summary of possible indirect climate related health impacts in Nunavik and Labrador (from Furgal et al., 2002)

Mediating process	Possible Indirect Health Impacts
Exposure to thermal extremes	Infectious diseases, stress-related disorders, other health-related disorders (such as psychosocial disruption)
Change in ice distribution and stability, snow composition and amount	Dietary problems associated with availability of food sources and effect on ability to fish and hunt
Effects on range and activity of vectors and infective parasites	Changes in geographical range and incidence of vector-borne diseases and their transmission to humans
Changes in local ecology of water-borne and food-borne infective agents	Changed incidence of diarrheal and other infectious diseases—emergence of new diseases
Changes in food and drinking water availability and productivity (food and drinking water security)	Dietary problems: nutritional deficiencies and/or hunger and consequent impairment of child growth and development, cultural and social implications due to diet shift, diabetes, changed contaminant exposure, etc.
Changes in distribution and composition of permafrost	Psychosocial disruption related to damages to infrastructures and population displacement—dietary problems associated with impacts via access to country food species
Sea level rise	Increased risks of infectious diseases—psychosocial disruption associated with infrastructures damages and population displacement
Changes in air pollution (contaminants, pollens, and spores)	Increased incidence of respiratory and cardiovascular diseases—cancers etc.

Acknowledgements

This chapter is part of a larger project entitled Climate change and Health in Nunavik and Labrador: What We Know from Science and Inuit Knowledge, conducted in Nain, Labrador and Kuujjuaq, Nunavik, by C. Furgal, D. Martin, P. Gosselin, A. Viau, Nunavik Regional Board of Health and Social Services, Nunavik Nutrition and Health Committee, and Labrador Inuit Association, funded by the Climate Change Action Fund.

We would like to thank our Nunavik and Labrador partners, both individuals and organizations participating in this project, for their cooperation, guidance, and review. The knowledge, expertise, and wisdom gained through many years of living on the resources of the land and sea is the focus of this report and we therefore gratefully acknowledge this wisdom. We especially thank all residents of the communities of Nain, Labrador and Kuujjuaq, Nunavik, who participated in this study and shared their observations of the environment with us. This study would not have been possible without their involvement and openness to share their expertise. We would like to extend our appreciation to all those individuals, especially Mary Denniston, Frances Murphy, Alice Pilgrim, and Eva Papigatuk, who worked with and guided us in the two communities. We greatly appreciate the helpful comments provided by the editors of this volume and their friendly patience in bringing this paper to its final version in a timely and appropriate fashion. Finally, we would like to acknowledge the support of the Climate Change Action Fund for this work, and a grant awarded to the lead author (Chris Furgal) from the Canadian Institutes for Health Research.

Notes

1. *Kablunagajuit* refers to individuals of mixed European and Inuit descent. Because of the long history of contact between Labrador Inuit and Europeans along the Labrador coast, and the close relationship and similarities in lifestyle that developed among these populations, the decendants of European and Inuit parents *(Kablunagajuit)* were recognized as full members of the Labrador Inuit Association since its incorporation in 1972 (Tanner et al., 1994; Williamson, 1994).
2. For a detailed discussion on country foods, contaminants, and human health benefits and risks in the Canadian Arctic, see Van Oostdam et al., 1999

References

Anonymous. 1993. *Traditional ecological knowledge: The variety of knowledge systems and their study.* A report submitted to the Great Whale Public Review Support Office. June, 1993.

Ashford, G., and Castelden, J. 2001. *Inuit observations on climate change, final report.* Winnipeg, Manitoba: International Institute for Sustainable Development (IISD).

Bergeron, L., Vigeant, G., and Lacroix, J. 1997. *Impacts and Adaptation to Climate Variability and Change in Québec: Synthesis Summary.* Volume V of the Canada Country Study: Climate Impacts and Adaptation. Ottawa, Ontario: Environment Canada.

Bielawski, E., and Masazumi, B. 1994. Dene knowledge on climate: Cooperative research in Lutsel k'e, Northwest Territories. In: Riewe, R., and Oakes, J. (eds.). *Biological implications of global change: Northern perspectives,* pp. 99–103. Canadian Circumpolar Institute, Royal Society of Canada, Canadian Global Change Program, and the Association of Canadian Universities for Northern Studies. Edmonton, Alberta: Canadian Circumpolar Institute.

Blanchet, C., Dewailly, E., Ayotte, P., Bruneau, S., Receveur, O., and Holub, B. J. 2000. Contribution of selected traditional and market food to Nunavik Inuit women diet. *Canadian Journal of Dietetic Practice and Research, 61:* 50–59.

Bordens, K. S., and Abbott, B. B. 1991. *Research design and methods: A process approach.* Mountain View, California: Mayfield.

Brice-Bennett, C. (Ed). 1977. *Our footprints are everywhere: Inuit land use and occupancy in Labrador.* Labrador Inuit Association, Nain, Labrador.

Centre for Indigenous Peoples, Nutrition and the Environment (CINE). 2001. Report of dietary surveys in eighteen Inuit communities. Montreal, Quebec: CINE, McGill University.

Cohen, S. J. 1997a. *Mackenzie Basin Impact Study. Final Report.* Ottawa, Ontario: Environment Canada.

Cohen, S. J. 1997b. What if and so what in Northwest Canada: Could climate change make a difference to the future of the MacKenzie Basin? *Arctic, 50*(4): 293–307.

Creswell, J.W. 1994. *Research design: Qualitative and quantitative approaches.* Thousand Oaks, California: Sage.

Dewailly, E., Ayotte, P., Blanchet, C., Muckle, G., Bruneau, S., Proulx, J-F., and Gilman, A. 1998. Integration of twelve years of data in a risk and benefit assessment of traditional food in Nunavik. Ottawa, Ontario: Department of Indian and Northern Affairs.

Dewailly, É., Blanchet, C., Lawn, J., Nobmann, E. D., Pars, T., Bjerregaard, P., and Proulx, J.-F. 2000. *Diet profile of circumpolar Inuit.* Québec: Groups d'Etudes Inuit et Circumpolare, Université Laval.

Dewailly, E., Blanchet, C., Lemieux, S., Sauv, L., Gingras, S., Ayotte, P., and Holub, B. J. 2001. N-3 fatty acids and cardiovascular disease risk factors among the Inuit of Nunavik. *American Journal of Clinical Nutrition, 74*(4): 603–11.

Environment Canada. 1997. *Canada Country Study: Climate Impacts and Adaptation.* Canadian Arctic Summary. Ottawa, Ontario: Environment Canada.

Environment Canada. 1998a. *Arctic ozone: The sensitivity of the ozone layer to chemical depletion and climate change.* Ottawa, Ontario: Environment Canada.

Environment Canada. 1998b. *Climate change.* National Environmental Indicator Series. State of the Environment Reporting Program. Bulletin No. 98-3.

Ernerk, P. 1994. Insights of a hunter on recent climatic variations in Nunavut. In: Riewe, R., and Oakes, J. (eds.). *Biological implications of global change: Northern perspectives,* pp. 5–6. Canadian Circumpolar Institute, Royal Society of Canada, Canadian Global Change Program, and the Association of Canadian Universities for Northern Studies. Canadian Circumpolar Institute, Edmonton, Alberta.

Fast, H., and F. Berkes. 1998. Climate change, northern subsistence and land-based economies. In: *Canada country study,* pp. 205–226. National Cross-cutting Issues. Ottawa, Ontario: Environment Canada.

Fenge, T. 2001. The Inuit and Climate Change. *Isuma: Canadian Journal of Policy Research 2*(4): 79–85.

Fox, S. 1998. *Inuit knowledge of climate and climate change.* Unpublished thesis, master of environmental studies, University of Waterloo, Canada.

Fox, S. 2000. Project documents inuit knowledge of climate change. *Witness the Arctic 8*(1): 8.

Freeman, M. A. 1994. Angry spirits in the landscape. In: Riewe, R., and Oakes, J (eds.). *Biological implications of global change: Northern perspectives,* p. 3. Canadian Circumpolar Institute, Royal Society of Canada, Canadian Global Change Program, and the Association of Canadian Universities for Northern Studies. Edmonton, Alberta: Canadian Circumpolar Institute.

Freeman, M. M. R. 1992. The nature and utility of traditional ecological knowledge. *Northern Perspectives, 20*(1): 9-12.

Friedberg, C. E., Janssen, M. J. F. M., Heine, R. J., Grobbee, D. 1998. Fish oil and glycemic control in diabetes. A meta-analysis. *Diabetes Care. 21:* 494–500.

Furgal, C., Martin, D., Gosselin, P., Viau, A., Labrador Inuit Association (LIA), and Nunavik Regional Board of Health and Social Services (NRBHSS). 2002. *Climate change in nunavik and labrador: What we know from science and Inuit ecological knowledge.* Final Project Report prepared for Climate Change Action Fund. CHUQ-Pavillon CHUL, Beauport, Québec.

Gough, W. A., and E. Wolfe. 2001. Climate change scenarios for Hudson Bay, Canada, from general circulation models. *Arctic, 54*(2): 142–148.

Hansen J., Pedersen H., and Mulvad G. 1994. Fatty acids and antioxidants in the Inuit diet. Their role in ischemic heart disease (IHD) and possible interactions with other dietary factors. A review. *Arctic Medical Research 53:* 4–17.

Hughen, C. 1998. Past Environmental Change in the Arctic: A PALE contribution of Arctic system science. *Witness the Arctic, 6*(1): 1–3.

Intergovernmental Panel on Climate Change (IPCC). 2001. *Climate change 2001: Impacts, adaptation, and vulnerability.* Summary for policymakers and technical summary. Geneva: IPCC.

Intergovernmental Panel on Climate Change (IPCC). 2001. Climate change 2001: Impacts, Adaptation, and Vulnerability. Contribution of Working Group II to the Third Assessment Report of the Intergovernmental Panel on Climate Change (IPCC). James J. McCarthy, Osvaldo F. Canziani, Neil A. Leary, David J. Dokken and Kasey S. White (Eds.) Cambridge, UK: Cambridge University Press.

Labrador Inuit Association (LIA). 1997. *Environmental health study. Final report.* Labrador Inuit Association, Nain, Labrador. A0P 1L0

Labrador Inuit Health Commission (LIHC). 1996. Workshop on Social Health and Environmental Change. March 26–28, 1996, Nain, Labrador. Labrador Inuit Health Commission, Nain, Labrador.

Makivik Corporation. 1999. *Nunavik at a glance.* Kuujjuaq, Nunavik: Makivik Corporation.

Mayer, N., and Avis, W. (Eds.). 1997. *Canada country study: Climate impacts and adaptations, national cross cutting issues,* vol. VIII, National cross-cutting issues. Ottawa, Ontario: Environment Canada.

McDonald, M., Arraqutainaq, L., and Novalinga, Z. 1997. *Voices from the bay: traditional ecological knowledge of Inuit and Cree in the Hudson Bay bioregion.* Ottawa, Ontario: Canadian Arctic Resources Committee and the Environmental Committee of the Municipality of Sanikiluaq: CARC.

McMichael, A. J., Haines, A., Slooff, R., and Kovats, S. (Eds.). 1996. *Climate Change and Human Health: An assessment prepared by a Task Group on behalf of the World Health Organization, the World Meteorological Organization and the United Nations Environment Programme.* Geneva: World Health Organization.

Meldgaard, M. 1987. Human implications of Arctic animal population fluctuations: Caribou in Greenland. In: J. G. Nelson, R. Needham, and L. Norton (Eds.). *Arctic heritage: Proceedings of a symposium,* pp. 35–41. Ottawa, Ontario: Association of Canadian Universities for Northern Studies.

Middaugh, J.P. 1990. Cardiovascular disease deaths among Alaskan Natives. *American Journal of Public Health, 80:* 282–285.

Nunavik Regional Board of Health and Social Services (NRBHSS). 2001. Health statistics prepared by P. Lejeune. Unpublished. NRBHSS, Kuujjuaq, Nunavik.

Patton, M.Q. 1980. *Qualitative research methods.* Beverly Hills, California: Sage.

Read, J. F., and Gould, W. J. 1992, Cooling and freshening of the subpolar North Atlantic Ocean since the 1960s. *Nature, 360:* 55–57.

Rees, W. E. 1988. Stable Community Development in the North: Properties and Requirements. In: G. Dacks and K. Coates (eds.) *Northern communities: The prospects for empowerment.* Occasional Publication Number 25, pp. 59–76. Edmonton, Alberta: Boreal Institute for Northern Studies. University of Alberta.

Riedlinger, D. and Berkes, F. 2001. Contributions of traditional knowledge to understanding climate change in the Canadian Arctic. *Polar Record, 37* (203): 315–328.

Salonen, J. T. 1986. Selenium and cancer. *Annals of Clinical Research, 18:* 18–21.

Santé Québec. 1995. *A health profile of the Inuit: Report of the Santé Québec Health Survey among the Inuit of Nunavik, 1992.* M. Jett, ed. Montréal, Ministère de la Santé et des Services Sociaux, Gouvernement du Québec, pp. 47–124.

Tanner, A., J. C. Kennedy, G. Inglis, and S. McCorquodale. 1994. Aboriginal Peoples and governance in Newfoundland and Labrador. Report to the Royal Commission on Aboriginal Peoples. Ottawa, Ontario: Libraxus.

Tesch, R. 1990. *Qualitative research: Analysis types and software tools.* New York: Falmer.

Van Oostdam, J., A. Gilman, E. Dewailly, P. Usher, B. Wheatly, H. Kuhnlein, S. Neve, J. Walker, B. Tracy, M. Feeley, V. Jerome, and B. Kwavnick. 1999. Human health implications of environmental contaminants in Arctic Canada: a review. *The Science of the Total Environment, 230:* 1–82.

Voisey's Bay Nickel Company (VBNC). 1997. *Environmental impact statement of the proposed mine and mill project at Voisey's Bay, Labrador.* St. John's, Newfoundland: VBNC.

Wenzel, G. W. 1981. Clyde Inuit adaptation and ecology: The organization of subsistence. Ottawa, Ontario: National Museum of Man, Mercury Series, Canadian Ethnology Service Paper no. 77.

Wenzel, G. W. 1986. Canadian Inuit in a mixed economy: Thoughts on seals, snowmobiles, and animal rights. *Native Studies Review, 2*(1): 69–82.

World Health Organization (WHO). 1967. The constitution of the World Health Organization, WHO Chronicle 1967; 1: 29

Williamson, T. 1994. Labrador Inuit politics from household to community to nation. Report for the Royal Commission on Aboriginal Peoples. Ottawa, Ontario: Libraxus.

Williamson, T. 1997. *From Sina to Sikujaluk: Our Footprint. Mapping Inuit environmental knowledge in the Nain district of Northern Labrador.* Report prepared by T. Williamson for the Labrador Inuit Association. Nain, Labrador: Labrador Inuit Association.

Appendix 1: Environmental changes and impacts reported in Nain, Mar. 2001

	Focus Group—Elders (men)		Focus Group—Elders (women) + 1 individual woman		Focus Group—Hunters (men)	
	Reported Changes	Reported Impacts	Reported Changes	Reported Impacts	Reported Changes	Reported Impacts
Weather	• warmer now • hotter summers • more freezing rain than snow • much drier than before	• gets uncomfortable when fishing • no flies some summers • had a forest fire where we never thought possible	• much hotter now • more rain now and rain in winter sometimes • winter temperature more variable • get mild spells in winter now • more fierce fall storms now • stronger winds now • weather more unpredictable • get sudden strong winds (north) • get unpredicted mild weather more often	• burns berries and bake apples some years • fewer berries some years • hard to breath sometimes out on the land in summer • warm spells in winter brings sickness, flu, bad colds • limits going off on land sometimes • many people will not go off if there is a north wind	• stronger winds • stays warmer longer in the fall • fewer very cold winds • fewer very cold days • much more spring wind • fewer offshore storms • weather changes faster (and more in winter) • fluctuations in April, gets warm, then cold again • summers are much drier • not as much fog now • fewer thunderstorms • weather more unpredictable	• able to travel in wind more now • delays break up • can't get out early • dangerous to go on ice at this time • can travel through light storms now • people are getting caught out on the land • young people are getting caught out more now
Snow	• snow on hills always melts in Torngat Mountains now • snow is lighter, texture has changed • all blows away b/c it is lighter • snow comes later • less snow	• no snow to cool char in summer when fishing • caribou don't have place to cool off • can't make snow houses anymore • people have to wait for snow to go wooding • don't have to walk through deep snow anymore (easier travel on foot)	• snow up north that stays year round is melting • much less snow now • snow comes later now (not until December) • first snow comes and melts (used to come and stay) • less spring snow • snow melts earlier • snow is much thinner	• hard on blueberries, less blueberries now because no snow to keep bushes warm in winter • don't think you could make snow houses now	• not as much snow • fall snow comes and melts (used to stay) • snow is sugary, thinner, not wet snow anymore • no heavy snow falls in the fall • crust on top of snow is more common now • get more "glitter" now (mild spell and rain on snow, then freezes)	• harder to get around when you're used to • no snow to cool char when you're fishing in the summer • can't make snow house if you need to • hard on caribou for getting food • impacts caribou and partridge and makes walking on snow hard

	Focus Group—Elders (men)		Focus Group—Elders (women) + 1 individual woman		Focus Group—Hunters (men)	
	Reported Changes	Reported Impacts	Reported Changes	Reported Impacts	Reported Changes	Reported Impacts
Ice	• little snow on the ice now • pack ice doesn't seem to come here from north anymore	• hard for seals (nowhere to build breathing holes and dens, hard on pups)	• goes out of the bay much earlier • less snow on the ice, sometimes bare • freeze up later (late Dec.) • break-up takes longer now	• ships come in earlier, don't have to wait as long for supplies • can go out to fishing camps earlier • very hard to travel on ice • hard on seals when there's no snow on the ice • makes thin ice and dangerous in spring for travel • have to be more careful going off on the land	• no good ice till January (about a month later now) • not as solid/thick • takes long time to solidify when it's there • breaks up earlier • cracks form early • ice seems to be saltier • don't get the fog or vapour over the water before freeze-up anymore • takes longer to break-up, because of temperature fluctuations in April • rough ice has changed, not coming in the bay after break-up, no multi-year pieces coming in • rough ice at the ice edge seems to break up fast (quickly) • much less snow on the ice • some ice on rivers is thinner	• dangerous travelling in spring, have to be more careful • can't get out on it during this time, too dangerous • get stuck in community longer time now during break-up, go crazy • pups get attacked by crows etc. • dangerous crossing is you don't know
Glaciers	• glaciers in north are melting, much smaller • cracks in glaciers larger, and more plentiful					
Brooks	• many have dried up	• no drinking water at some camps in summer			• some brooks up north have dried up • some rivers dried up or are very low	• char have to wait for large rain to get up river

	Focus Group—Elders (men)		Focus Group—Elders (women) + 1 individual woman		Focus Group—Hunters (men)	
	Reported Changes	Reported Impacts	Reported Changes	Reported Impacts	Reported Changes	Reported Impacts
Ponds	• some have dried up	• hard to find drinking water sometimes • move camping areas • bake apples grow well in areas of old ponds			• some in north have dried up	
Land	• torn up more from caribou		• caribou paths everywhere now		• land is much drier now • no more "underground ice"	• has impact on berries, much fewer some years and burnt
Caribou	• changed migration route	• harder to get, have to go further	• changed migration (even out on islands now) • meat has changed too • sometimes scarce	• not as much fresh meat in the freezer this year	• some bugs in caribou now • stronger taste to caribou now • marrow used to be thick and greasy, now just stringy • coming out along coast recently • moving back inland now I think • moving farther south • main herd comes out to the coast now (never used to)	• don't eat the liver anymore because of bugs • easier access, but this is changing
Seals	• habitat has changed	• hard on pups, lack of snow on ice, bellies get scratched up			• more square flippers now (bearded seals) • harp seals come through earlier (August now, used to be Sept–Oct) • more sick, skinny seals and more crawlers	
Fox					• gets strong smelling earlier now (February) • fox are fatter now	
Marten					• coming back (disappeared for 15–20 years)	
Rabbits					• disappeared for a few years and are back now	

	Focus Group—Elders (men)		Focus Group—Elders (women) + 1 individual woman		Focus Group – Hunters (men)	
	Reported Changes	Reported Impacts	Reported Changes	Reported Impacts	Reported Changes	Reported Impacts
Moose					• now seen here within last 10 years	
Beaver					• last 20 years started to see beaver again up this far	
Birds					• fewer young geese • fewer ducks • cormorants now seen here within last 10–15 years • partridge come out to coast more	
All animals			• getting smarter, know where nets and tents are of people and stay away • coming closer to people and town	• harder to catch • don't have to go as far to go hunting now	• getting more used to noise	• can get closer to them when hunting
Fish			• no cod anymore (rock cod, tom cod)		• no more fish left • char have changed their diet • char come out of the rivers and go straight out to sea • fish are skinny now too	• hard to get them close to the community now • taste is different • harder to get, must wait for them to come back into the bay
Berries and other plants			• changes with years, some years none b/c it is too hot • more burnt berries and leaves	• really miss berries if there aren't many	• berry leaves are burning more • berries dry up earlier in year • some years, no berries	
Other			• water in harbour is polluted now	• fewer people fishing there • less fish there	• sun burns your skin more • more trees with brown needles • contaminants and worms in wildlife	

Appendix 2: Environmental changes and impacts reported in Kuujjuaq, May 2001

	Focus Group—Hunters and Elders (men)		Focus-Group—Elders (women)	
	Reported Changes	**Reported Impacts**	**Reported Changes**	**Reported Impacts**
Weather	• gets warmer sooner and gets colder sooner • spring comes earlier • warming and cooling are more gradual (take longer) • no real large changes, some years are just different • not as hot for as long as it used to be	• more impact has come from getting older and losing our traditional ways • we don't exercise as much anymore	• gets extremely cold when not expected to • fast changes in temperature • less rain • warmer in summer	• have been stopped when we were going fishing, would have gone fishing more in the past
Ice	• break up is earlier • seems thinner now during the winter	• seals gone earlier in spring	• ice breaks up faster now	• changes travel times and ice safety
Seals	• fewer because of ice leaving earlier • more harp seals here, especially in fall			
Caribou	• more sick caribou seen • more caribou with bad joints and livers seen • thinner now • more sport hunting	• don't eat some parts now and don't eat parts that look bad • limits caribou for Inuit and disturbs them, sometimes diverts them away from Inuit when we are hunting • sometimes hard to get caribou because of this	• liver has white spots now • caribou have returned, more now • marrow is all bloody	• don't need to go as far to go hunting for caribou • not as good tasting
Fish (Lake Trout)	• some lake trout caught are very skinny	• don't eat them all now		
Water	• river by Kuujjuaq is getting shallower every year • some small lakes have dried up	• damage to boats and equipment from low water (related to damming) • limited drinking water on the land sometimes		
Foxes	• fewer now • more rabid foxes now • foxes further from community now	• harder to get in traps		
Muskox	• increasing	• chasing caribou away from some areas		
Beluga whales	• fewer • whales don't go to some locations anymore	• get fewer beluga whales now		
Sun	• the sun is very hot now, dangerous • get sun burns now for first time	• dangerous, it burns your skin now • have to wear cream on your skin now	• sun burns easily now • it is hotter nowadays	• have to be careful with the sun • skin burns now sometimes • stay inside sometimes because of sun • too hot in tents sometimes • plants are dried up • no berries sometimes, sun burns them • cloud berries bloom and get dry quickly

Figure 9-1. Storm clouds developing near Aklavik, Northwest Territories (photo by Jerome Gordon).

9

Putting the Human Face on Climate Change Through Community Workshops:

Inuit Knowledge, Partnerships, and Research

Scot Nickels,[1] Christopher Furgal,[2] Jennifer Castleden,[3] Pitseolalaq Moss-Davies,[1] Mark Buell,[4] Barbara Armstrong,[4] Diane Dillon,[5] and Robin Fonger[5]

It is in the northern regions of the world, such as Arctic Canada, that climate-related changes and subsequent impacts to ecosystem and human health are expected to be greatest. The latest Intergovernmental Panel on Climate Change (IPCC) report suggests that early predictions of the adverse effects of climate change were in fact conservative (Houghton et al., 2001). In Arctic Canada, changes have already been reported in many regions through projects such as the Mackenzie Basin Impact Study, (Cohen, 1997), Voices from the Bay (McDonald et al., 1997), Inuit Observations on Climate Change Project (International Institute for Sustainable Development, 2001), Arctic Climate Change: Observations of Inuit in the Eastern Canadian Arctic (Fox, 2000), Elder's Conference on Climate Change (Nunavut Tunngavik Incorporated, 2001), and Climate Change and Health in Nunavik and Labrador—What We Know From Science and Inuit Knowledge (Furgal et al., 2002). These projects have documented environmental changes as described by Inuit and other indigenous peoples in these regions.

There is a growing concern among Inuit about the impacts on environment, health, and culture from global phenomena such as climate change and contaminants. In a recent tour of Inuit communities organized by the Inuit Tapiriit Kanatami (ITK) to discuss issues related to persistant organic pollutants, like toxaphene and chlordane in the North, communities repeatedly raised issues regarding climate change and the associated human impacts. Community members spoke of rapid

1. Inuit Tapiriit Kanatami, Ottawa, Canada
2. Public Health Research Unit, Centre Hospitalier Universitaire du Québec (CHUQ), Pavillon—Centre Hospitalier Université Laval (CHUL), Quebec, Canada
3. International Institute for Sustainable Development, Winnipeg, Canada
4. Inuvialuit Regional Corporation
5. Inuvialuit Joint Secretariat

changes never experienced before and of new dangers due to variable weather and ice conditions in the area.

As the national organization representing Inuit communities and regions Inuit Tapirrit Kanatami (ITK)[1] is dedicated to supporting the abilities of individuals and communities to address issues affecting them. One issue ITK is currently working on is responding to Inuit concerns about climate change. To date, few data (qualitative or quantitative) have been gathered or organized in a cohesive form to support communities' ability to identify, understand, and communicate their concerns about environmental change, both within and outside of their community. Further, much of the discussion on northern climate change has centered on *documenting* changes and not necessarily *discussing* how communities are adapting or can adapt in the future. One potential tool to support community capacity in dealing with these issues is the development of local monitoring programs that can be used to formulate appropriate response strategies. It is for these reasons, and with this intent, that ITK initiated a project with several communities in the Inuvialuit Settlement Region (ISR)[2] of the Canadian Western Arctic to help them document observations, understandings, and effects of climate-related changes and develop strategies to cope and adapt where possible.

The project used a series of workshops to document observations and concerns of environmental change. Bringing together research and support personnel from across Canada, along with Inuvialuit regional organizational staff, the researchers visited three communities (Tuktoyaktuk, Aklavik, and Inuvik) in January and February 2002 to document Inuit observations, report the effects these observations are having on community residents, and identify and develop potential adaptation strategies or tools. The workshops represent a starting point in this process to collect information to support community, regional, national, and international processes on climate change. In addition, the process is intended to help bring a "human face" to the issue of climate change in the circumpolar Arctic regions.

This chapter is a review of the process involved in the Arctic Climate Change: Observations from the Inuvialuit Settlement Region Project. Describing this recent and in progress work provides a good example of how ITK is addressing Inuit specific issues of climate change, how and why the existing partnerships have developed to organize and conduct such a project, and the value of the participatory methods used. Initial workshops in the first three communities have only recently been completed (from January 25 to February 5, 2002) and thus, only preliminary analysis has occurred to date. Some initial results and authors' impressions from their involvement with the project are outlined in this chapter. Further, this project is a good example of what other papers in this volume are speaking about—"how to learn to act through sharing, listening, and understanding others' ways of observing and 'knowing'" (see introduction to this volume). The examples in this chapter provide good illustrations of how and why we need to take documentation further to include the human face of climate change through community workshops, Inuit knowledge, partnerships, and research.

ITK: Supporting Northern Commmunities' Response to Climate Change

Founded in 1971, the Inuit Tapiriit of Kanatami (ITK) is the national Inuit organization dedicated to supporting the needs and aspirations of all Canadian Inuit (see Figure 9-2). Since its establishment, ITK has broadened its aims and objectives in response to the changing social, economic, environmental, and political challenges facing Inuit. It has done so in a manner that reflects the emerging relationship between Inuit and the rest of Canada and between ITK and the four Inuit regional organizations. With the exception of Labrador and the off shore of Nunavik, the four regions have now signed final land claim agreements.[3]

The ITK Environment Department represents Canada's Inuit primarily on matters of regional and national interest and works directly with technical staffs of the regional organizations. ITK also works closely with the Inuit Circumpolar Conference Canada, the organization representing Inuit in Canada at international fora on environmental issues. In 1997, ITK and the Canadian office of the Inuit Circumpolar Conference (ICC) formalized their relationship. This solidified the links between ITK and Inuit of Alaska, Russia, and Greenland thus assuring a strong international base for cultural, political, economic, and environmental cooperation. The existing structure allows ITK to deal with difficult global scientific and policy issues (e.g., contaminants and climate change) at a local level, while ensuring its influence extends to regional, national and international arenas.

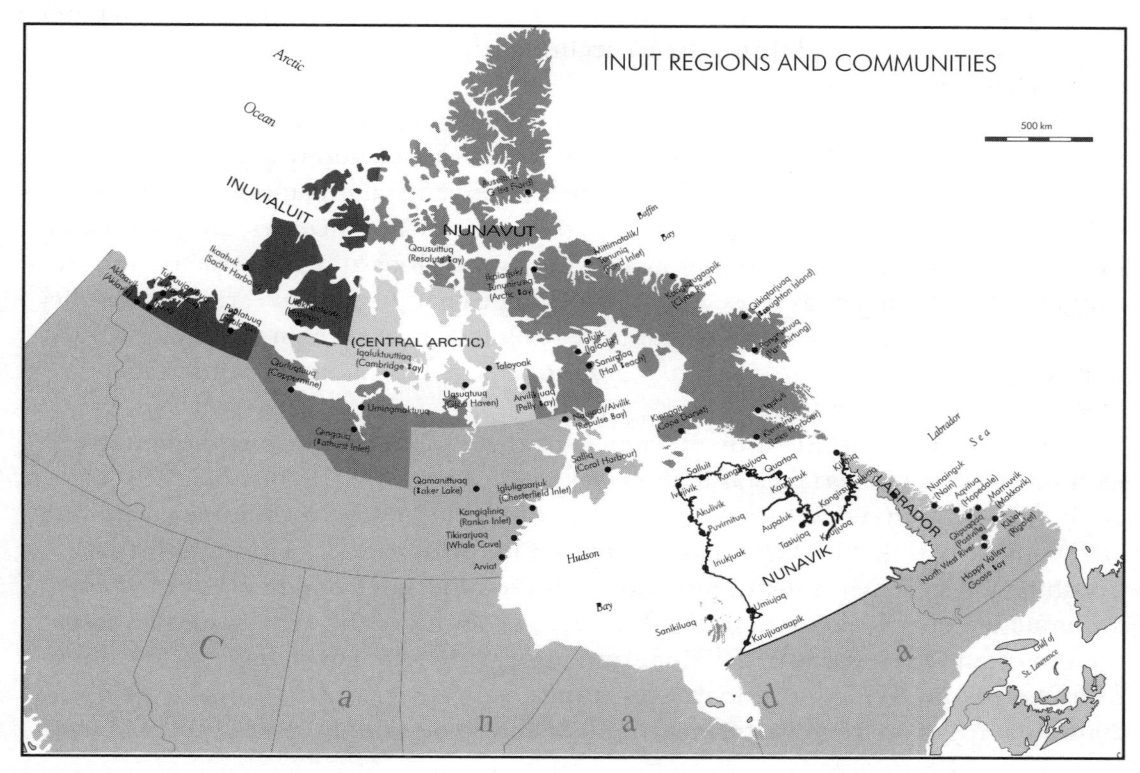

Figure 9-2: Map of Inuit regions in Canada supported by Inuit Tapiriit Kanatami.

In Inuit communities, dependence on the land and sea and the food it provides is central to cultural, spiritual, physical, and mental health, as well as economic prosperity. Change in the quantity or quality of the Arctic ecosystem, and thus traditional food supply, affects all aspects of community life and well-being. It is not unexpected, therefore, that during the past decade, the most important areas of activity for ITK have been issues related to environmental change. Inuit did not create the persistent organic pollutants (POPs), heavy-metals, or radio-nuclides that are entering the Arctic as contaminants. But it is Inuit who bear the immediate and long-term consequences of such substances. Similarly, if significant changes occur in the Arctic lands, air, waters and animals as a result of greenhouse gas induced climate warming, it is Inuit and other northern indigenous peoples who will suffer the most immediate physical, spiritual, economic, and social impacts. Arctic regions serve as clear and early indicators of such global problems. In this respect, Inuit are the "human face" of the effects of environmental change. It should not be surprising, therefore, that Inuit take a very strong position in demanding a tough and effective Canadian policy related to the control of contaminants and greenhouse gas emissions throughout the world.

ITK is supporting Inuit concerns about climate change in many ways. For example, it is arguing for an Inuit-specific discussion on health impacts of climate change so that appropriate health policies are developed and put in place. ITK sees the link between human health and environmental change as an issue of primary importance (see also Furgal et al., this volume). Inuit may know more about extreme weather and how to adapt, survive, and thrive these conditions than any other peoples. However, recent changes may pose new challenges.

Through the link to food and nutrition, climate change may also have impacts on health issues currently facing Inuit such as diabetes, cancer, and respiratory diseases. Characteristics such as a high proportion of children and youth among the total population, current housing shortages in many communities, low incomes, and inequalities in access to health care services put many Inuit populations at higher risk to such environmental health issues as those impacts of climate change. It is also critical for policy makers to be aware of the impacts that climate change may have on the psychological well-being and mental health of Inuit communities due to changes in traditional lifestyles.

Too often, scientific attention on climate change stops with the environmental effects or the plight of certain sentinel species in the Arctic such as the polar bear or caribou. Forgotten seems to be the understanding that in the North, humans are situated at the top of the food chain as they live off the land and sea. Through this relationship with the environment, Inuit are at risk of suffering from the magnified and accumulated effects of environmental change as they move through various levels of the ecosystem. Even the very well done and informative climate change poster series from the Yukon, NWT, and Nunavut territories (Natural Resources Canada, 2000, 2001) that focus on the physical environment and some wildlife species, seem to have forgotten that a majority of their constituents suffer the direct and indirect effects of climate change because of their intimate relationship with the environment.

In addressing these issues, ITK believes that climate change mechanisms cannot be limited to direction from federal/territorial and provincial levels but must include Inuit at the national, regional, and local levels. The scale of predicted Arctic social and environmental changes means that a much higher priority must be given to Inuit voices in the policy debate on climate change (World Health Organization, 2000, see also Fenge, 2001). Inuit must be involved in order to be effective in identifying the reality of these problems in the North, in collecting and organizing the best information to enhance our understanding of these issues, and finally in developing appropriate strategies to adapt, take advantage of, or take action on these issues at local, regional, national, and international scales. It is this belief that provides the impetus for ITK initiatives to support a coordinated approach and enhancement of local capacities to deal with these issues. It is important to ensure funding is available to support participation and capacity building initiatives for Inuit stakeholders. ITK is working towards ensuring that Inuit have the capacity and resources to participate in community, regional, national, and international climate change activities (Inuit Circumpolar Conference, 2001). Further, they are making efforts to support Inuit in making informed decisions regarding the impacts of climate change and to determine the most appropriate mitigation and adaptation actions to minimize any adverse effects on the health and well-being of individuals and communities.

ITK has identified the need for an integrated and intersectoral approach to climate change research, policy development, and communication/education activities as it pertains to human health and the environment. In order to achieve this goal, intersectoral collaboration between environmental professionals (biologists, health researchers, chemists, and environmentalists), health and social service professionals, community organizations such as the Hunters and Trappers Association (HTA) and governments (regional, territorial, and federal) is required. ITK is currently involved in a variety of climate change activities. These initiatives include an ongoing strategic planning process, in partnership with the Inuit Circumpolar Conference (ICC), to develop a long-term Inuit Climate Change Strategy; the development of a joint ITK Health and Environment Department program; and the development of a community monitoring and reporting network on local observations and indicators in the Canadian Arctic.

These activities, in combination with ITK's various other climate change initiatives, will ensure Inuit from all communities:

- are informed of the potential health and environmental effects associated with climate change;
- participate in the federal government decision-making process on climate change;
- assist the federal departments involved to deliver their climate change mandate;[4]
- are able to participate in the scientific and policy initiatives required to define and react to the changes that climate change may bring to the Arctic; and

- are developing the capacity to make informed decisions regarding the impacts of climate change.

In order to fulfill this mandate, ITK has begun to partner with appropriate organizations and individuals to provide the proper technical support in conducting work on climate change and other environment and health issues facing Inuit today. Two of these organizations, the International Institute for Sustainable Development and the Public Health Research Unit at the CHUQ, and three communities (Tuktoyaktuk, Aklavik, and Inuvik) are working with ITK on the Arctic Climate Change: Observations from the Inuvialuit Settlement Region Project. These organizations and communities are briefly profiled below.

International Institute for Sustainable Development

ITK sought to develop professional relationships with key organizations that would complement the goals and aspirations of ITK in responding to the effects of climate change in Inuit communities of Canada. One of these organizations is the International Institute for Sustainable Development (IISD). IISD is a nongovernmental organization headquartered in Winnipeg, Canada, with offices in Ottawa, New York and Geneva. It contributes to sustainable development by advancing policy recommendations on climate change and energy, international trade and investment, economic policy, measurement and indicators, and natural resource management. IISD works with partner organizations, governments, and businesses in Canada and around the world through collaborative projects, knowledge networks, internship programs, and communications to promote a global transition to sustainable development.

IISD's climate change and energy team is particularly interested in monitoring and reporting of current climate change-related impacts. In 1999, IISD partnered with the community of Sachs Harbour and several other local and governmental departments in the Inuvialuit Settlement Region to document local observations of climate change (Inuit Observations on Climate Change, IISD 2001; see Jolly et al., this volume). The project gained international attention from the climate change community and beyond, largely due to the production of a widely disseminated video that demonstrated conclusively that the impacts of climate change are already being felt on the local scale. Perhaps most significantly, the project sought the experience and local knowledge of the people of Sachs Harbour—an approach strongly endorsed by ITK.

The methodology and approach of the IISD project was seen by ITK as able to contribute significantly to future work on understanding the impacts of climate change and promoting adaptation strategies in Canada and internationally. Following their work in Sachs Harbour, IISD expressed strong interest in continuing to work with communities and organizations in the Canadian North on climate change issues. As a result, they are now working with ITK developing the capacity among northern communities to collect and share observations, vulnerabilities, and adaptive strategies on climate change through the community-based local observation network described in this paper.

Public Health Research Unit—CHUQ

The Public Health Research Unit at the CHUQ (Centre Hospitalier Universitaire de Québec) is a diverse organization that includes health and environment professionals and researchers with extensive experience working on issues in the Canadian North and elsewhere around the world. The team is comprised of experts in the disciplines of medicine, biology, anthropology, psychology, toxicology, and sociology among others. Through their experience, senior researchers have a detailed understanding of the relationships between northern indigenous people, the environment, and health. Much of their past work has directly influenced policy related to such issues as contaminants and health in the circumpolar North and elsewhere. Further, some team members are active in an advisory capacity in the North (e.g., Nunavik Nutrition and Health Committee) and are thus in a position to inform policy at regional scales. The team works on issues related to environmental health impacts, assessment, and surveillance in the North and around the world and is an established collaborating centre of the World Health Organization (WHO) and Pan American Health Organization (PAHO) on these issues. The team is currently directing a number of climate and health related initiatives, including work in the regions of Nunavik and Labrador (see Furgal et al., this volume). Their current role in the ISR project presented here is through the involvement of one team member, Chris Furgal, and provision of expert advice and review on health-related issues pertaining to climate change in the Canadian Arctic.

Tuktoyaktuk

Tuktoyaktuk, or Tuk as it is known locally, is geographically located at 69° 26' N latitude and 133° 01' W longitude and is 137 kilometres north of Inuvik in Kugmallit Bay in the Beaufort Sea, east of the Mackenzie Delta (Figure 9-3). The community was incorporated as a Hamlet in 1970. With a population of just under 1,000 people, Tuk is the largest purely Inuvialuit community and is roughly split into 1% Metis, 2% Dene, 9% non-aboriginal, and 88% Inuit. Languages spoken in the Hamlet include Inuvialuktun and English. The traditional name for Tuktoyaktuk, in Inuvialuktun, is *Tuktuujaartuq,* which means "looks like a caribou."

Aklavik

Aklavik is geographically located at 68°12' N latitude and 135° W longitude and is situated in the delta of the Mackenzie River within 90 kilometres of the Beaufort Sea. Due to severe flooding in the area in the 1950s, the government of Canada decided to move the community of Aklavik to present day Inuvik. However, due to lack of good fishing and trapping areas in the new community, many residents refused to move. Since that time, Aklavik's motto has been "Never Say Die." Aklavik gained hamlet status on January 1, 1974. Today, the hamlet has a population of some 750 to 850 depending on the time of year, with less during the hunting season when families leave the community. Languages spoken in Aklavik include Inuvialuktun, Gwich'in, and English. Aklavik's traditional name is Aklarvik, meaning "barren land grizzly place."

Inuvik

Inuvik is geographically located at 68°18'N; 133° 29'W, some 97 kilometers south of the Beaufort Sea. Inuvik gained town status on January 1, 1979. On July 18, 1958, Inuvik, which translates directly as "living place" in Inuvialuktun, was officially inaugurated by proclamation of the fifteenth session of the Council of NWT. Inuvik was the first planned town north of the Arctic Circle, with many people originally living in Aklavik moved to Inuvik because of repeated flooding in that community. Inuvik re-

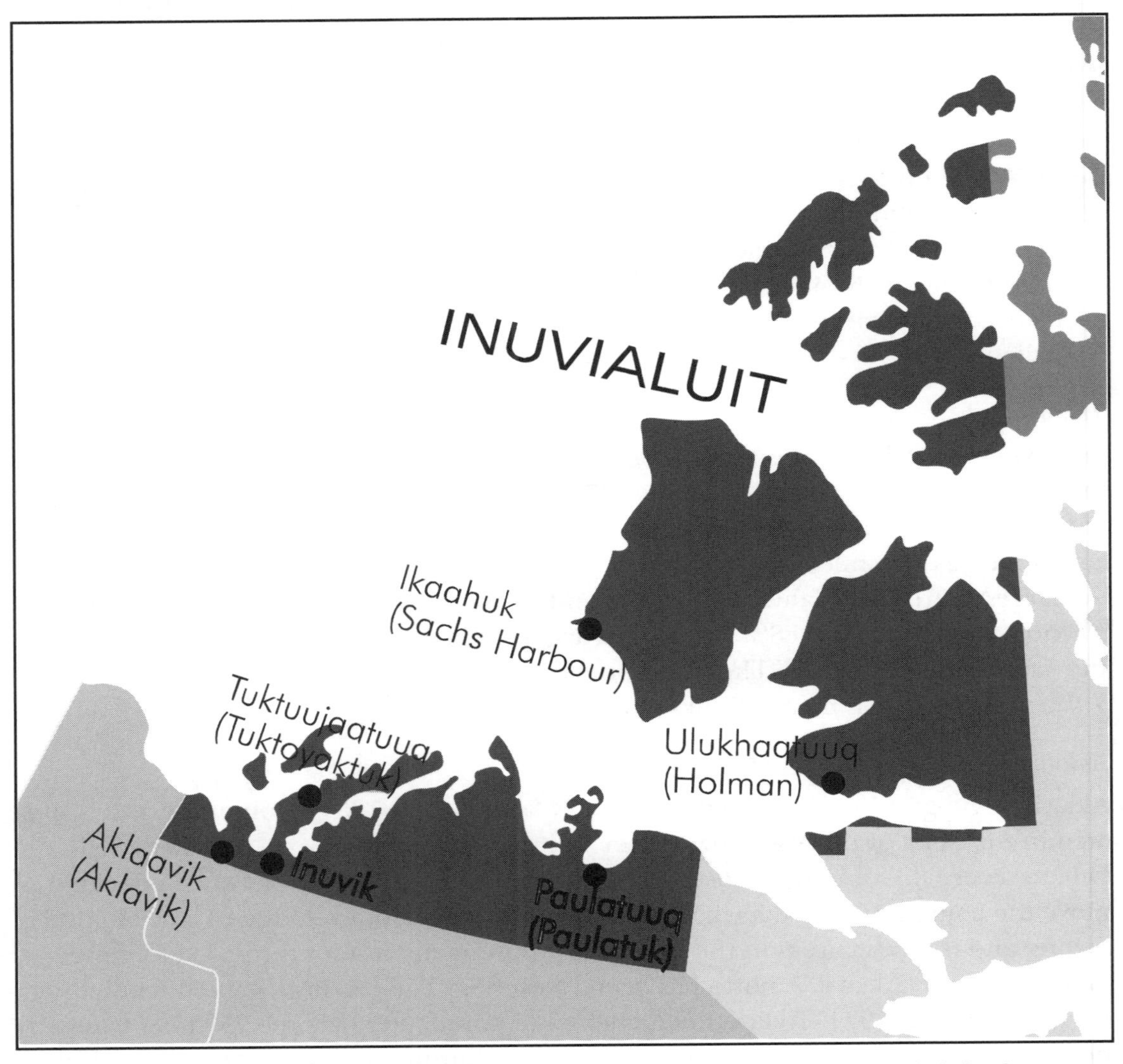

Figure 9-3: Map of communities (Inuvialuit and English names) in the Inuvialuit Settlement Region.

flects this, as it is very much a southern town. Today, Inuvik's population of about 3,296 is equally split into non-Native, Inuvialuit, and Gwich'in peoples. Languages spoken in the community include Inuvialuktun, Gwich'in, North Slavey, English, and French. The town's area is approximately 57 square kilometres, of which about 5% is currently developed.

Methods

The Arctic Climate Change: Observations from the Inuvialuit Settlement Region Project began well in advance of conducting the actual workshops. Consistent, in terms of project fundamentals, was the view that individual communities needed to be visited in order to hear and articulate the wealth of local Inuit knowledge while at the same time gaining an understanding of regional viewpoints. ITK, IISD, and the Inuvialuit Regional Organizations agreed that the Sachs Harbour study (IISD, 2001) should be expanded upon and extended to other communities within the ISR before moving such research to other Inuit regions.

Partnerships developed gradually while ITK sought to involve the necessary expertise. The study format went through several incarnations before the present methodologies were selected. The facilitators agreed on an approach in which all workshop participants—community members, regional representatives, and facilitators—would be co-investigators. The workshops were intended to be a dynamic learning process in which everyone had an opportunity to investigate the issue of climate change and its meaning to the community. This approach was based on the belief that the process needs to allow for multiple exchanges of information and perspectives on climate change.

With limited funds and person hours available, project organizers found it necessary to restrict the number of communities and the research period to a manageable size. With the help of representatives from several Inuvialuit Regional Organizations, three communities were chosen (Tuktoyaktuk, Aklavik, and Inuvik) for this study. These communities were chosen because: (a) they are in close proximity to one another; (b) they are connected by a system of ice roads; and (c) they could accommodate workshops during the appropriate period. Conducting the workshops in late January meant that the team could travel to each community using the ice roads, which at that time were in good condition. Travelling by ice road meant that the team would save time and money, not being dependent on time-consuming and costly commercial flights to communities. The eagerness and ability of the three communities to accommodate the workshops within a two-week period (January 24 to February 6, 2002) meant the project was feasible for the employment schedules of individuals of the team as well.

Of central importance to this project is the methodology of the workshops. This section describes the process used by the research team to develop and adapt workshop methods for use in the project. The stages used to conduct the workshops were: methodology training session with regional representatives; community workshops broken into two days; and the development of community tools in the form of indicator identification (see Table 9-1).

Methodology Training Session: Regional Representatives

Before the training session, the workshop facilitators[5] prepared written material outlining proposed exercises to gather community observations of climate change and develop community indicators. The day before the first community workshop began (January 26, 2002), the project facilitators met with four representatives[6] from Regional Inuit Organizations in the Inuvialuit Settlement Region (ISR) to discuss the workshop methodologies. Regional representatives are the front-line workers for Regional Organizations in their various capacities (i.e., human health, environment) and are an important link between residents in communities and these organizations. There were several reasons for having them participate directly in the workshops, including:

1. Building local capacity to implement and carry out climate change workshops in communities in the region
2. Providing a further opportunity for regional representatives to interact with communities. This will enable communities to get to know their regional representatives and thus enhance regional capacity
3. Streamlining the linkages between local, regional, national, and international projects and programs

This day was used to review the proposed workshop agenda with the regional representatives and discuss the approach, objectives, and desire to have the community set the workshop agenda as much as possible. In keeping with the decision to ensure all participants had an opportunity to shape the objectives, input received from the regional representatives was used to adapt the methods and initial agenda for the workshops. In particular, the regional representatives provided guidance on what to expect from communities and the region in terms of participation and communication.

All workshop materials and methods were prepared as a written draft manual so that regional representatives could use the documentation as a rough training manual while learning the workshop methodology. Also, it was expected that they would use this material to aid in their facilitation of future workshops, since one of our objectives was to have regional representatives assume greater responsibility for the workshops as the workshops progressed. This was also in anticipation of additional workshops to be held at a later date in three other communities in the region (Holman, Paulatuk, and Sachs Harbour), at which the regional representatives would lead the facilitation.

Community Workshops

Workshops were held during a two-day period in each community between January 28, 2002 and February 2, 2002. The first day was used to document observations and produce timelines, while the second day was used to identify effects and coping strategies and conduct future strategic planning. The following sections outline Day 1 and Day 2 activities in detail.

Table 9-1. Primary activities of climate change community workshops (January–February 2002)

Activity	Description
Training session	Four representatives from Regional Inuit Organizations were involved in learning the workshop methodology.
Agenda identification	The participants were broken into small groups after opening introductions. Small breakout groups were asked to respond to five questions to help set the workshop agenda and to gather initial perspectives, desires, etc. for the workshop.
Observation identification	Participants wrote down examples of environmental changes they have experienced that may indicate that the Arctic climate is changing. Each observation or example was written on a separate card. People wrote their own observations, or they were recorded by the project team, or other community members.
Timeline identification	Community members, particularly elders, traced the changes they have experienced (environmental, wildlife-related, social, economic) back through time in living memory, placing them in order of occurrence, providing a historical context for recent observations.
Effects identification	Small groups were asked to review the observations they recorded in the timeline exercise and discuss the associated impacts they are experiencing as a result of these changes. These were discussed from the participants' personal perspective.
Adaptation methods	Participants discussed what could be done by individuals, households, and communities to adapt to observed changes and identified those things already being done by the community to adapt to these changes.
Planning	Small groups identified who should be told about this workshop and the results in order to respond to the issues raised here and be aware of what concerns exist within the community on this issue.
Indicator identification	Following the workshop, observations were translated into measurements that could be used to monitor such changes in and around the community. These indicators were developed based on the various observations documented at the workshop and are intended to provide an initial list of community identified indicators.

Community Workshop Day 1

The first day of each workshop began with brief introductions by the project team and workshop participants. Participants were then divided into two smaller groups to provide a more comfortable discussion environment (Figure 9-4). Several exercises were then used to engage local residents in the discussion. These exercises drew from participatory analysis and planning techniques, already well proven in the Inuit Observations on Climate Change, Sachs Harbour study (see Jolly et al., this volume). These techniques included Participatory Rural Appraisal (PRA)[7] and Objectives Oriented Project Planning (ZOPP),[8] which encourage participation by everyone at a workshop, allowing a community to identify and analyze its own problems.

These techniques used several precise activities to stimulate local resident dialogue. They were (in order of use): agenda identification, identification of observations, and development of a timeline identification (see Table 9-1). Each of these activities is outlined below.

Agenda Identification. In the small groups, participants were asked five questions that would eventually be used to structure the agenda of the workshop. Specifically, these questions were:

1. Why did you come today?
2. What interests you about this topic?
3. What do you hope will be the results of the workshop?
4. Do you have other interests related to this topic that you would like to include in the workshop?
5. Have you attended other workshops that gave you a positive experience? Why? What made it possible?

Figure 9-4. Small group discussions, Tuktoyaktuk, January 28, 2002. Team photo.

Responses to these questions were recorded on flip chart paper and posted on the wall for all workshop participants to see. Small groups were brought together and a short plenary discussion was held to identify common responses and priorities in relation to these questions (Figure 9-5). Using these responses, local residents, facilitated by the team members, developed the agenda for the remainder of the workshop. Without exception, each group began immediately identifying observed changes in their environment, since it was their main interest and reason for coming to the workshop.

Identifying Observations. Given the outcome of the agenda-setting exercise—identification of concern over recent environmental changes—the next task was to identify and record observations of climate and environment-related changes in the local area. Continuing to work in the same small groups, the workshop facilitators recorded each individual observation on a separate index card. Participants were invited to write their own observations on index cards, should they prefer this method to speaking and having someone record their information. This visualization technique allows for less verbal or dominant people to participate in the meeting.

Once each group was satisfied that their observations had been exhausted and adequately recorded, the larger group reconvened and the workshop facilitators provided a summary of the observations that had been recorded by each group. Participants were invited to add and/or clarify their observations at this time.

Timeline Identification. The next activity involved participants arranging their recorded observations along a community timeline. Participants constructed a timeline that drew on their knowledge and oral history of the area to indicate when different phenomena (recorded in the earlier section on individual cards) were first observed. Decadal headings (1930s, 1940s ... 2000) were taped on a large wall in the meeting room, and participants were asked to tape the index cards (observations) to the wall under the appropriate decade (Figure 9-6). To help with recalling periods or incidences of change, participants were also asked to identify significant community events that corresponded to each decade, such as the signing of the Inuvialuit Final Agreement in 1984.

Figure 9-5. Plenary discussion group, Inuvik, Northwest Territories, February 4, 2002. Team photo.

When the workshop group was satisfied with the chronological arrangement of observations, a brief plenary review and discussion was held to ensure that dates and observations corresponded accurately. Participants discussed and made changes to the timeline as appropriate during this plenary and the coffee break that followed. Ultimately, the timeline provided a visual account of historical climate-related changes in the area.

The project team had planned to conduct a strategic planning session to identify and develop actions and reactions to specific changes upon completion of the timeline exercise; however, this exercise was pre-empted as it became clear that the participants were not prepared to discuss planning and adaptation until they had first discussed the impacts or effects the recorded changes had on them individually and as a community. This led to a session focused on discussing the effects of changes as the next step in the workshop.

The first day of the workshop ended and the project team (facilitators and trainers) met in the evening to review the Day 1 events. It was determined that the small group format was much more effective in supporting participation and that the observations themselves were in fact community indicators of change. This implied the need for a slight adaptation to the original workshop plan, as it included a discussion and exercise on "indicators" and developing community indicators. As a result of the review discussion amongst the project team, it was decided that the second day of the workshop would be conducted in the same small groups, with an emphasis on discussing the effects that the changes had on individuals in the community and the community as a whole. "Effects" in this sense would be defined as positive, negative, or no observed effect related to climate and environmental change.

Community Workshop Day 2

Based on the initial agenda set by the community through the opening questions exercise, and review by the workshop facilitators after Day 1, the following agenda for Day 2 was developed:

Figure 9-6. Chronological arrangement of observations providing a visual account of local historical climate-related changes. Tuktoyaktuk, January 28, 2002. Team photo.

- Discussion of effects of observations (positive, negative, or none)
- Strategy/planning session with the following questions:
  1. Looking at the effects, what can you do at the household or individual level to adapt or take advantage of these changes and effects? Or, what are you already doing?
  2. Looking at the community or regional level, what can or should organizations do in response to these changes and their impacts in the communities? Or what are they already doing?
  3. Where or to whom should the results of this workshop be given?

Working in the same small groups, participants reviewed the observations they had recorded the previous day and were asked to identify potential or existing effects of these observed changes. These effects were recorded on flip-chart paper and posted on the wall. They included impacts on both individuals and the community as a whole.

Using the recorded changes and effects identified by the small groups, a final planning exercise was held. Participants were asked the three questions outlined above, related to how they were responding to the changes experienced. An important component of this final planning exercise was asking participants to identify where and to whom the results and workshop report should be sent (i.e., who should be aware of the discussions held and concerns and perspectives raised at this workshop?).

Finally, the workshop concluded with presentations to the entire group. The presentations summarized observations, effects and potential or current coping strategies that are used in the community. Last, participants' ideas about to whom the workshop results should be sent were also discussed (Figure 9-7).

The project team then provided the workshop participants with an overview of current regional, national, and international climate change initiatives taking place. The team suggested potential ways that the information gathered in the workshop could be useful within and outside the community. Specifically discussed was how the information collected in the three community workshops will fit into regional, national, and international programs via the work of IISD, CHUQ, ISR organizations, and ITK as well as through community responses to the workshop report and through international

Figure 9-7. Presentation summarizing community observations, effects, and potential or current coping strategies to the entire workshop group in Inuvik, February 5, 2002. Team photo.

reports and programs such as the Arctic Climate Impact Assessment (ACIA). The purpose of linking the workshop results to other initiatives was to let participants know that they were part of a larger process for addressing climate change in the North and that this workshop was not solely for the sake of collecting more information. It was important for local residents to know that their workshop was in fact adding to an existing "process" for responding to climate change in the circumpolar Arctic.

Finally, the team drew the attention of the workshop participants back to the initial objectives or goals that the group had identified at the beginning of the workshop. A summary of what was addressed, what was not addressed, and what could be responded to outside of this workshop provided an evaluation for the group in meeting its original goals.

Development of Community Tools: Indicator Identification

Upon completion of the three community workshops, individual community observations were examined and translated into measurements that could be used to measure and monitor changes in and around that community. These indicators were developed based on the various observations documented at the workshop and are intended to provide an initial list of community-informed indicators. Translation of observations to indicators was conducted by the research team and was not discussed at community workshops. They will provide the foundation from which future discussions on community and regional monitoring for climate change and associated impacts can begin.

Results and Discussion: Hearing What Communities Have to Say

The first phase of this ongoing research project was recently completed on February 5, 2002, with workshops conducted in the communities of Tuktoyaktuk, Aklavik, and Inuvik. At the time of this writing, less than two months later, only preliminary analysis has been completed. However, we feel that our initial results and impressions, while preliminary, still make a strong contribution to this volume in revealing the extent, scope, and value of understanding Inuit knowledge and experience on the subject of environmental change.

In this section we provide a brief overview of observations and concerns regarding climate-related changes reported from each of the three individual communities. Since a detailed report of workshop results is unavailable at the time of this writing, we try where possible to provide succinct examples from one community, Tuktoyaktuk. Our objective is to describe how observed environmental changes are affecting Inuvialuit individuals, livelihoods, and activities; existing or potential community adaptation or coping strategies; links between these workshops and processes taking place at the regional, national and international levels on climate change; and particular tools that can help to support communities to develop response strategies to these changes. These examples provide good illustrations of what can be learned from putting the human face on climate change through the use of community workshops,

Inuit knowledge, partnerships, and research. Our overview of results follows the sequence of activities conducted in the community workshops.

Observations and Timeline

A large number of observed environmental changes were documented from the community-based knowledge discussed over the course of the workshops. These have been summarized into six categories: changes related to weather; seasons; erosion; permafrost; sea/lake ice; water levels; and wildlife, fish, insects, and plants (Table 9-2). The organization presented here is solely for the purposes of presentation in this chapter and does not reflect any priorities or structure provided by the communities themselves.

In Tuktoyaktuk, examples of changes in the weather patterns along the coast included sudden and intense events with high winds that are unpredictable because they rapidly change direction over time. These storm events were linked to coastal erosion problems that are common in the community today. Heavy wave action from several directions is causing rapid deterioration of the shoreline, resulting in the erection of costly breakwaters and the movement of buildings along this shoreline to safer locations.

In Aklavik, changes in the water levels are linked to the closing up of certain river channels and a decrease in freshwater sources and quality of drinking water. Isolated and extreme weather events, including funnel clouds, have also increased in frequency in the area (Figure 9-8).

Examples of observed changes from the community of Inuvik include adjustments in the migration and travel routes of particular birds and animals as well as the influx

Figure 9-8. Funnel cloud observed outside of Aklavik captured on film (photo by Jerome Gordon).

of certain species never seen before. Inuvik residents are experiencing earlier break-up in spring and later freeze-up in the fall, which affects the length of time people are able to travel along waterways in both summer and winter.

Residents show a remarkable consistency in observations from each community. Three examples of comparable observations voiced in each community in this region are: (a) since the 1980s the weather has become more unpredictable, (b) since the 1980s there are longer, warmer summers and shorter, warmer winters, and (c) since the 1980s there have been unprecedented changes in the travel and migration routes of animals. Recorded observations also reveal differences that exist between communities. For example, an observation from Tuktoyaktuk on the subject of permafrost noted that pingo formation is changing rapidly, with the result that some are actually decreasing in size. This is a community-specific observation articulated by residents of Tuktoyaktuk only. Pingos do not exist in the vicinity of Aklavik and Inuvik and therefore residents did not mention this change occurring in their area.

Not only are there differences in the types of observed environmental changes from one community to the next, but there also exist differences in the extent to which these changes impact individual communities. In Aklavik, for example, sources of fresh drinking water were seen by community members as a key area of concern, whereas in Tuktoyaktuk, erosion was the priority, even though residents observe both of these changes in each of their communities. A further example of this concerns the movement of new species into new areas. All three communities observed new species of animals and birds moving into their areas, but in many cases the specific species discussed by residents varied.

As is clearly outlined in Table 9-2, a great variety of changes are articulated by Inuvialuit who hold an immense understanding of local environmental processes and who indicate concrete evidence for environmental change. For example, residents of each community frequently identified the timing of seasonal changes as a clear indicator of change. To them, earlier freeze-ups and break-ups in the transition seasons (spring and fall) are considered evidence of climate change. The timing of break-up in each community is occurring earlier and at a faster rate than previous decades (since the 1950s) and is also considered an indicator.

The results of this project demonstrate the abundance of local expertise in the communities relating to both historical and current observations. Residents were able to communicate when many of the changes related to weather, seasons, erosion, permafrost, sea and lake ice, water levels, and wildlife, fish, insects, and plants first began to be observed. According to residents in each community, the majority of observations of environmental change began to occur after the 1980s. The number, extent, and types of changes are also increasing with each passing year. Clearly, Inuit knowledge is closely tied to their history of land use and occupancy and current land-based activities. The environmental changes Inuvialuit are speaking about are closely tied to the way these changes impact their daily lives.

Table 9-2. Examples of recent local environmental changes reported in three Inuvialuit communities

Phenomena	Observed Change	Community		
Weather	Weather more unpredictable	T	A	I
	Sudden intense changes (not gradual)	T	A	I
	Isolated "extreme" events becoming more frequent (more variability)	T	A	I
	More thunderstorms with lightning	T		I
	Rapid and unpredictable changes in wind direction and velocity		A	I
	More rain in fall, less snowfall in winter	T	A	I
Changing seasons	Changes most evident in transition seasons (spring	T	A	I
	and fall)	T	A	I
	Longer, warmer summers	T	A	I
	Shorter, warmer winters (less –40 degree days)	T	A	I
	Spring comes earlier	T		
	Kids swim earlier in lakes	T	A	I
	Autumn comes later	T	A	I
	Hotter summer days, for longer duration			
Erosion	Storms eroding banks, exposing permafrost	T		
	Rapid erosion, losing land to the ocean	T		I
	More shoals because of erosion	T	A	I
	Buildings needing to be moved	T		I
Permafrost	Increasing disappearance of permafrost	T	A	I
	Some lakes draining due to melting permafrost		A	
	Increasing landslides and slumping	T	A	
	Increased depth of active layer	T	A	I
	More mud on land	T		
	Pingo formation changing rapidly (decreasing in size)	T		
Sea/Lake ice	Earlier break-up, later freeze-up	T	A	I
	Less multiyear ice and more open water	T		I
	Rate of ice break-up has increased	T	A	I
	Spring break-up along rivers not as loud		A	I
	Less ice along the shores of rivers in spring		A	I
Water and water levels	All lakes are lower	T	A	I
	More algae around lakes	T	A	
	Less freshwater sources—some drinking water sources have disappeared	T	A	
	Freshwater does not taste as good anymore—tastes swampy because it is not moving as it should		A	
Wildlife/fish insects/plants	Less fish and poorer quality—skinnier fish	T		I
	Changing animal travel/migration route	T	A	I
	Changes in the timing of animal/bird movements and activities	T	A	I
	New wildlife/birds/insect/plant species	T	A	I
	Different condition of wildlife/fish	T		I
	Changes in numbers of certain species	T	A	I
	More willows growing, and growing taller	T		I

Letters denote in which community changes were reported in: T = Tuktoyaktuk, A = Aklavik, I = Inuvik.

The results show great scope and detail of local Inuit knowledge on environmental change. This analysis, using preliminary examples of observations, is limited in conveying the depth of local-specific knowledge that residents from particular communities have. Justice is not given here to the rich, local, observational details that can be provided for each community-specific observation pending further analysis. However, these observations do highlight the basic similarities and differences existing within the region, the value of local knowledge as sources of data for local observations of environmental change, and thus the need to involve local people in these processes. Scientific models that presently illustrate or predict environmental change do not account for the fine-tuned intraregional differences outlined here.

This leads us to the next section, which discusses the effects that certain observations of environmental change have on the perspectives of individuals, households, and communities.

Inuvialuit Perspectives on the Effects of Environmental Change

As exemplified in the previous section, Inuvialuit at the three community workshops reported a significant amount of detailed knowledge of observations related to environmental change specific to their locations. Following this discussion and the organization of changes along a community timeline, the workshop process facilitated the discussion of individuals' perspectives of effects these changes are having on them personally, their household or community. Using examples from the Tuktoyaktuk workshop, this section describes the types and variety of effects reported by participants.

Some climate-related changes that are taking place in the Tuktoyaktuk area are reported to have direct impacts on individuals in the community and various aspects of their health and well-being (Table 9-3). For example, the unpredictability of weather is reported to make travel and hunting more challenging and sometimes dangerous as weather systems change more quickly today than before, increasing the risk of being stranded out in a storm or caught in dangerous travel situations. Some direct impacts related to climate changes were reported to be positive in some instances though. Warmer days bring more opportunities for children to enjoy summer activities such as swimming in the Tuktoyaktuk area. Changes were also reported to have effects on various aspects of the physical (e.g., permafrost) and living (e.g., wildlife) environment. Changes in temperature are having negative impacts on the land, such as an increase in the rate of permafrost melting and a decrease in the size of pingos in the area. Warmer temperatures are also bringing more insects to the area which are, in turn, a nuisance to some wildlife species (e.g., caribou), influencing their health and distribution. Warmer surface waters in streams were reported to cause fish kills in some cases. Although this chapter only presents examples of these effects from one of the community workshops, it is important to note that both similar and unique effects were reported at other workshops. Thus, the local perspective of change-induced effect is important to document. All of these changes to aspects of the biotic and

Table 9-3. Examples of observations and reported associated effects at the individual, household, or community level, as reported in Tuktoyaktuk

Observed Change	Associated Effect on Individual, Household, or Community
Weather	
Unpredictability of weather	• Travel difficult to plan and dangerous • Warmer summers means kids can swim more months of the year
Changing seasons	
Hotter summer days	• Worry about tundra fires with extended periods of hot days • Warmer summers means more bugs and more difficult to work on the land • Not able to store country food properly and therefore not able to store for use in winter
Warmer winters	• Less fuel needed for heating homes
Erosion	
Increasing coastal erosion	• Increasing costs of moving buildings • Fear of losing cultural artifacts (e.g., graveyard)
Permafrost	
Increasing disappearance of permafrost	• Was a preferred source of fresh drinking water for Inuit, now disappearing and have to search elsewhere • Pingos used as navigational landmarks disappearing • Caribou get stuck in increasingly muddy areas
Sea/lake ice	
Early break-up	• Longer ice-free period extends shipping season • Duck and goose hunting season is cut shorter. Ducks and geese less available in community • Makes it more difficult to get caribou that are important this time of year • Can get supplies into community earlier
Later freeze-up	• Have to wait for important hunting and fishing season, it shortens the fall/winter hunting season • Have to wait longer to go out to get caribou which is an important source of fresh meat this time of year (diet change) • More travel and activities happen in the dark (late season) and therefore, more difficult and dangerous
Water/water levels	
More algae around lakes	• Humans don't want to drink the water—avoid water • Do not trust the water safety—avoid water
Less freshwater sources	• Need to search for more sources • Must bring water from greater distances (expensive)
Water tastes bad	• Avoid drinking water form certain locations
Water warmer at surface	• Kills fish in nets
Wildlife/Fish/Insects/Plants	
More bugs, particularly mosquitoes and other biting insects	• Increased nuisance • Game more difficult to see • Animals bothered by bugs—unable to eat—in poor condition
More vegetation growing taller	• Hard on Skidoos • Provides shelter, increases bugs, which affect humans and animals
Changing animal travel/migration routes	• Makes hunting more expensive, requires more fuel, gear, and time—high costs means some residents (particularly elders) cannot afford to go hunting

abiotic environment then have effects on the health, lifestyle, and economic activities of individuals, households, and the community.

The indirect effects of climate-related environmental changes reported in the Tuktoyaktuk area impact individuals' cultural, spiritual, physical, and mental health and well-being as well as the economic prosperity of Inuvialuit communities. For example, the warmer temperatures observed in this area are reported to have negative impacts in that they affect the storage of country foods which need cool temperatures during the summer months, yet at the same time, these warmer temperatures mean individuals and families spend less on heating fuel during the winter months. Impacts to community infrastructure, as a result of erosion, are currently costing the community because of the need to build break walls and move buildings. Additionally, environmental changes impacting wildlife are having effects on individuals' diets in the community as some are not able to get certain foods at some times of the year. These complex relationships between the climate, people, and the biotic and abiotic environment are heard in Inuvialuit perspectives on the effects these changes are having on their way of life.

In the Tuktoyaktuk workshop, participants continually linked the changes they are observing to other modifications in the landscape, the surrounding ecosystem and community. For example, residents from each community described the relationship between the presence of ice, wind, wave energy, ice safety, the abundance and health of local marine/terrestrial species and the implications these have on the community's hunting success. It is, therefore, important to note that many of the examples of observations that have been placed into one of the seven categories in Table 9-2 for presentation here, were expressed by community members in a cyclical manner—each observation being connected to another—and actually cross-cut several categories. This understanding of the environment, holistic and interrelated in nature is not surprising, as it is implicit in Inuit knowledge as a worldview. Again, this documentation stresses the necessity to incorporate local observations as sources of data for local predictions and impact assessments, as computer based climate models do not account for such things as the individual level impacts and intra-regional differences evident in this study.

Adaptation and Coping Strategies

Communities in the Inuvialuit Settlement Region are observing changes of various forms in the environment, which are having a wide variety of impacts or effects on wildlife, the community, and individuals. These effects range from those impacting community infrastructure, to individuals' ability to access traditional food and water sources, to direct impacts on individuals' health through such things as exposure to heat and UV radiation never experienced before. However, Inuit in this and other regions of the North have adapted to environmental change for thousands of years, allowing them to survive in the harsh conditions of the Arctic.

In the ISR workshops, residents related physical environmental changes occurring to changes in the local ecology and the adaptive modifications of animals, birds, fish,

Table 9-4. Examples of changes, effects and coping strategies/adaptations community residents are using or recommend should be used in their particular community

Observation	Effect	Coping Strategy/Adaptation
Erosion of the shoreline	Relocation of homes and possibly community considered	Stone breakwalls and gravel have been placed on the shoreline to alleviate erosion from wave action
Warmer temperatures in summer	Not able to store country food properly and therefore not able to store it for use in winter	Community members are travelling back to communities more often in summer in order to store country food. This is expensive since it requires more fuel and time. Not hunting as much for future use since there are few places to store extra meat (government-funded community freezers were recently shut down)
Warmer temperatures in summer	Can no longer prepare dried/smoked fish in the same way, it gets cooked in the heat	People are building thicker roofs on the smoke houses to keep some heat out, tarps and other materials used to shelter country foods from heat
Later freeze-up earlier break-up		Travel to camps and for hunting earlier in the spring as the ice breaks up
Lower water levels and some brooks drying up	Not as much good natural sources drinking water available	Bottled water now taken on trips
Changing water levels and the formation of shifting sand bars	More difficult to plan travel in certain areas	Community members are finding new (usually longer and therefore more costly) routes to their usual camps and hunting grounds. Some residents are now choosing to fly to their destinations, tolerating the added costs
Warmer weather in winters	Animal fur does not prime due to warm weather—fur is shorter and not as thick, changing the quality of the fur/skin used in making clothing, decreasing the money received when sold	Some people do not bother to hunt/trap – others buy skins from the store that are not locally trapped but are usually not as good quality fur/skin and are expensive
Water warmer at surface	Kills fish in nets	Nets are checked and emptied more frequently so fish caught in nets do not perish in the warm surface water and spoil
More mosquitoes and other biting insects	Getting bitten more and more concerned about these insects	Use insect repellent lotion or spray Use netting and screens for windows and entrances to houses
Changing animal travel/migration routes	Makes hunting more expensive, requires more fuel, gear and time—high costs means some residents (particularly elders) cannot afford to go hunting	Initiation of a community program for Elders, younger hunters can provide meat to Elders who are unable to travel/hunt for themselves

insects, and Inuit. Similarly, several adaptive initiatives were reported to have already been implemented in reaction some of the changes documented here (Table 9-4). Berkes and Jolly (2001) identify these types of responses as coping mechanisms or short-term responses to change, such as switching species and adjusting the "where, when, and how" of hunting. They contrast coping mechanisms or short-term responses with adaptive strategies or long-term responses, which might include aspects of traditional knowledge that allow for change in the flexibility of seasonal hunting patterns, hunting strategies, networks for sharing food and other resources, and intercommunity trade. In this project, some coping strategies and adaptations appear to already be in development in this region in response to environmental changes reported.

As identified by the participants in the ISR workshops, different types of environmental change have different and unique impacts, each requiring different types of responses. For example, changing migration patterns of caribou were observed in this region. One effect of this kind of change is that it makes hunting more expensive, requiring more fuel, gear, and time to hunt caribou. This extra expense has meant that some elders in the community can no longer afford to go hunting. To cope with this change, the community has initiated a program for elders, whereby younger hunters provide caribou meat to those elders who are unable to travel these increasing distances to hunt caribou. Variation exists in the application of this coping strategy, since the community distribution program is not a standard region-wide situation. However, the opportunity to share this information between these communities was an occasion for "strategy" sharing or discussion, providing some ideas that have been developed in one community and could be applied to others.

Planning

In the final stages of the community workshops, the project team discussed the link between these workshops and processes taking place at the regional, national, and international level on climate change, with particular emphasis on those processes occuring in the circumpolar North. The purpose of linking the workshop results to other initiatives was to inform participants that they were part of these larger processes addressing climate change and that this workshop was not just for the sake of collecting more information but was in fact adding to the "process" for responding to climate change at various levels. At the local level, participants identified several community and regional organizations and groups that should be aware of the discussions and concerns raised through the workshops since they felt these organizations had a role to play in addressing issues such as climate change. Making these links was the first step in enhancing relationships within and between communities. We believe that the follow-up and response by these organizations will be critical in ensuring community input and mobilization in the region in the future.

The process followed in the workshops (community knowledge directing the agenda and being the focus of the discussions) illustrated to the research team and the participants themselves the extent and value of the local Inuit knowledge and individuals' observations of the environment in understanding and addressing environment and health issues. The process was one of mutual learning among all who attended. As shown here, much of the information on climate changes, effects, and potential responses exists with local people and communities. This is the human face of climate change that ITK seeks to support, through local community networking workshops, as used in this study.

While participants expressed appreciation at the opportunity to get together with other elders and youth to hear the issues and learn from each other, they also stressed the need to further collect detailed specific data. They felt that the workshop format simply "scraped the surface" of the knowledge existing within the community. In regards to this point, participants suggested the need for further studies to discuss, in detail, these issues in order to understand and document the greater breadth of their knowledge. They also suggested several methods that would be most appropriate to conduct this work, including household visits with maps for discussion and one-on-one interviews.

From Observations to Indicators: Developing Tools with Communities

To go beyond simply documenting the various changes that are taking place and what effects these are having on communities and individuals, ITK is interested in supporting communities in developing responses to these changes wherever possible. This requires a variety of approaches to enhance community capacity to respond, cope, and ultimately to adapt in order to take advantage of some scenarios (e.g., positive effects) and minimize impacts of other climate-related scenarios (e.g., negative effects). Since changes and effects related to climate in the North are not taking place in isolation from other forms of change in communities (e.g., political, social, and economic, other forces inducing environmental change), it is essential to accurately identify what changes are taking place, what they are attributed to, and the relationships between these changes and potential effects at the community and individual levels. In order to understand these relationships and the trends in these changes, some form of "monitoring" or observing needs to take place. In many ways, through individuals' time on the land and discussions among each other, this monitoring and oral record is already in place. However, in order to ensure the recording and sharing of this knowledge in a manageable way and to collect information on critical aspects of the environment that are changing, some formalization of this process is helpful to allow collective understanding and action to occur. It is for these reasons that an approach of identifying and selecting community indicators for climate change is gaining attention across the Canadian North. It is argued that some form of organized knowledge or database will enhance community capacity to identify changes and effects and thus understand the situation and relationships better in order to develop response strategies where required.

Since the task of monitoring changes in the environment and their effects on communities and individuals is daunting, measurements that are indicative of key areas of specific interest to the community are chosen as "indicators" of the status of these relationships and their outcomes. Indicators provide clues to matters of larger significance or make perceptible a trend or phenomenon not immediately detectable, and thus their importance extends beyond what is measured. In this context, indicators are:

> an expression of the link between climate, the environment and aspects of community and individual health, targeted at an issue of specific community concern and presented in a form which facilitates interpretation for decision making. (adapted from Briggs et al., 1996)

Following the ISR workshops, observations were translated into measurements that could be used to monitor changes and impacts. These indicators are intended to provide an initial list of community-identified indicators for climate change in the areas of the ISR communities (examples shown here are from Tuktoyaktuk workshop), and provide a starting point for potential future discussions on monitoring. It is expected that the list of indicators will include some traditional or conventional indicators collected through existing monitoring activities by various agencies (e.g., Environment Canada—temperature, precipitation, etc.) as well as some community observation-based indicators (date of freeze-up, date of break-up, qualitative assessments of various changing conditions) specific to the local area or deemed important by community individuals. Table 9-5 presents some examples of observations reported at the Tuktoyaktuk workshop and their associated indicator measurement as developed by the research team.

The type(s) of indicators proposed and retained in a monitoring program must be dictated by the specific goals of the program. Therefore, it will be critical to discuss with communities and stakeholders the intention, interests, and ultimately the purpose they want the indicators and a potential monitoring program based on these indicators to serve (e.g., do they want to simply monitor and report changes? Do they want to be able to predict potentially dangerous situations before they happen? Do they want to identify a link between a change in the environment and a specific impact on the community? etc.).

The International Joint Commission (1991) outlines five such examples of common uses for environmental indicators (for environment and health). They are:

- Compliance indicator: assessment of current condition of environment;
- Change indicator: to document trends or changes;
- Early warning indicator: to anticipate hazardous conditions before impacts occur;
- Diagnostic indicator: to identify causative agents to specify appropriate action;
- Relational indicator: to identify interdependence between indicators.

To ensure that indicators are chosen that help attain specific goals, a variety of scientific based (e.g., reliability, validity) and use-based (e.g., feasibility, cost, catalyst for action, etc.) criteria must be identified and used in the indicator selection process un-

Table 9-5. Examples of potential indicators developed from community observations

Observation	Potential Indicator
Less snowfall = less freshwater	• Less snowfall could be a predictive indicator of less freshwater sources the following spring, therefore measure total precipitation falling as snow/yr., at nearest weather station or measured locally
All lakes are lower	• Max. lake depth/yr. (for a specific lake of importance to the community or representative lakes in the region)
Freshwater not as good anymore—tastes swampy because it is not moving as it should	• Water quality indicators (various) from important natural drinking water sources for community (total coliform counts, etc.)
Less freshwater sources—some drinking water sources not there now	• Monitoring of existence of natural freshwater sources habitually used by community (presence/absence)
Less fish and poorer quality—skinnier, fewer, but larger because they spend more time in the lakes (whitefish)	• Fish stock population survey on important river for community harvesting
Winter temp is warmer now: less –40 degree days	• # of –40°C degree or less days/yr. • Average temp/month (analysis on winter months' data)
Break-up before: late June early July; now early June	• Observation of break-up date recorded each year (data from SAR or community observations **after** "break-up" location and definition is determined
Permafrost is disappearing (going down)	• Annual mean depth of discontinuous permafrost layer (transect measurements in local area)

dertaken in partnership with communities (as in Eyles and Furgal, in press). Further, future discussions with communities will include explicit consideration for issues such as the audience of the information, comparability across scales (community, regional, and national), and the need for core regional indicators and those that are specific to the changes and priorities of individual communities (as in Eyles and Furgal, in press). This, however, is just the beginning. From the set of indicators developed, baseline data on the health of Inuit living in Canada can be developed, and this data will serve as a starting point to measure future health and environment initiatives directed at people living in the North.

Conclusions

In our discussions with Inuvialuit from the communities of Tuktoyaktuak, Aklavik, and Inuvik, it quickly became apparent that residents hold a vast knowledge of the ways climate change is altering the local environment and their communities. The scope of observed changes encompasses many aspects of the environment and ecosystem in the North, from wildlife to permafrost to air and water temperature.

Participants in the community-based workshops conducted as part of this study identified a large number of ways in which the environmental changes affect more than just the physical environment and ecosystem. The effects are felt at the individual, household, and the community level. Mental and physical health, spiritual and cultural well-being, and the economic health of the community were all identified as being impacted by current changes. When migration patterns of geese change near Tuktoyaktuk, residents have new difficulties to contend with at hunting time. They must travel further distances over terrain that may not be as well known, which costs in both time and money. They may have to spend more time away from their community and families, and may not return with enough meat to share with elders. This affects individual health (danger in travelling, less country food in diet), sense of community (residents have to spend more time away from community), and economic well-being (residents spend more on fuel to travel greater distances).

The extent to which the Inuit of the western Arctic identify with the local environment makes it clear that climate change has great impact in this region. Many residents spend a great deal of time on the land. The prevalence of country foods in the Inuit diet further strengthens the human/environment bond in the North (see Furgal et al., this volume). Furthermore, northern communities have a sense of community spirit that is not normally experienced in the south. For this reason, residents are keenly aware of the health of their region and community and are alert to the ways in which climate change affects them. This is seen in the complexity of the links of change residents identified, and the detail provided by participants in their observations and identified effects.

Changes are happening at an alarming rate in the North and are affecting the people who live there in a wide variety of ways. To some extent, it is up to community members to cope with these changes and the ways in which their lives are being affected. In many instances the knowledge of how to adapt to these changes exists in the communities. For example, residents of Aklavik, who are worried about the quality of drinking water when they are out on the land, simply take drinking water with them. In Tuktoyaktuk, the community is already addressing the problems associated with erosion by developing measures to move the town itself if necessary and by using rock to stave off the further erosion of the beach. However, a number of the adaptations or coping mechanisms are not possible without the support of outside institutions. In Aklavik, residents identified better communication with weather stations to facilitate safer travel to the coast; however, without the cooperation of Environment Canada and the U.S. Weather Service, this is not possible.

More efforts are needed on sharing information and facilitating discussion between communities and other organizations about issues such as adapting to climate change. As stated above, much of the knowledge of how to cope with environmental changes already exists in the communities; what is missing is the process by which residents can be involved in deciding their efforts. Some of the adaptations used in individual communities needs to be shared between communities. Residents of Aklavik, for example, are finding new routes to the coast through the maze of rivers that make up the Beaufort Delta region. Residents of Inuvik could use this information as well, since many of these waterways are shared by these two communities.

The series of workshops described here is merely the first step in a larger process that supports communities, like the ones in the Inuvialuit Settlement Region, that are feeling the effects of climate change. By identifying the issues, communities, and stakeholder organizations, they will be better equipped to address them and work together to develop adaptations and coping mechanisms. Furthermore, by developing the capacity in these regions to conduct research, develop monitoring networks, and make links with the regional organizations to which concerns can be articulated and addressed, local residents can begin to take ownership of the issue. Local Inuit will be involved, their concerns will be heard and, most importantly, the residents of communities that have been typically marginalized in Canadian society will have a voice at the regional, national, and international levels.

The development of indicators related to climate change and its effects on the health of Inuit communities (due to the special relationship between the residents of the North and the physical environment) provides a mechanism for quantifiable data collection that will aid in the development of health and environmental policy as it relates to the Inuit world. The set of indicators developed as a result of this project will directly feed, through the ITK and other organizations, into a number of initiatives at the national level in Canada, such as the Inuit Health Information Initiative and the First Nations and Inuit Health Renewal Project.

The Arctic Climate Change: Observations from the Inuvialuit Settlement Region Project has allowed Inuit knowledge to feed into a larger scientific data pool regarding climate change, while at the same time allowing that knowledge to stay in the community. The forging of new partnerships such as those developing between Inuvialuit communities, ITK, IISD, and the CHUQ, and the strengthening of existing ones, is key to the sustainability of this project and support for the communities studied.

Most importantly, through these kinds of workshops, Inuit have a voice on an issue that affects their lives and communities on a daily basis. Not only will their concerns be brought to the attention of national and international governments and NGOs through existing structures and new partnerships, but also Inuit can feel empowered to do something about it. Methods of coping with changes will be better suited, and more likely to be adopted, if it is Inuit themselves who develop them. It is now difficult to ignore the "human face" that has been superimposed upon the climate change issue. This face is now distinguishable in the residents of Inuvialuit communities, and we hope through the use of Inuit knowledge, partnerships, and research that these faces will be recognizable around the world.

Acknowledgements

This paper and the Arctic Climate Change: Observations from the Inuvialuit Settlement Region Project were both made possible by the commitment, support, and generosity of a number of people and organizations. The team would like to thank the following community residents for participating and sharing their experience and knowledge so that all of us could enjoyably learn from each another: Danny A. Gordon, Annie B. Gordon, Carol D. Arey, Jacob Archie, Danny C. Gordon, Pat Kasook, Jerome Gordon, Louisa Kelanik, Jim B. Edwards, Agnes Edwards, David John, Alice Husky, Gary Montgrand, Ruth Furlong, Faye Gordon, Maureen Pokiak, Agnes Felix, Noah Felix, Lena Anikina, Dennis Raddi, Bradly Voudrach, Ernest Pokiak, Emmanuel Adam, Billy Jacobson, Ernest Cockney, William Vaneltsi, Lucy Adams, Ruby McLeod, Sarah Tingmiak, Johnny Banksland, Andy Tardiff, Joe Teddy, Mary Teddy, Walter Elias, Jessie Colton, Elias Aviungana, Maureen Elias, Deva Gordon, Elizabeth Firth, Albert Bernhardt, Agnes Nasogaluak, Catherine Mitchell, Mary Allen, Louie Goose, Roy Goose, Lynn Lau, and Richard Binder.

We would also like to thank the following organizations for encouraging their employees take part in the project training and participate in the workshops: Inuvialuit Regional Corporation, Inuvialuit Joint Secretariat (Inuvik); Hunters and Trappers Committee (Inuvik, Aklavik, Tuktoyaktuk); Inuvialuit Game Council; International Institute for Sustainable Development (IISD); Public Health Research Unit, Centre Hospitalier Université Laval, Quebec; and ITK (particularly the Environment and Health Departments). The following people must be thanked for their planning, ad-

Figure 9-9. Group photo of workshop participants, Inuvik, February 5, 2002. Team photo.

vice, energy, participation, and humour: Debbie Raddi, resource person, Tuktoyaktuk Halmet Office and Eleanor Ross, resource person, Tukotyaktuk Hunters and Trappers Committee.

The project was made possible through financial contributions from the Northern Ecosystem Initiative (Environment Canada), Health Canada, Department of Indian Affairs and Northern Development, Inuvialuit Regional Corporation, Inuvialuit Joint Secretariat (Inuvik), Hunters and Trappers Committees (Inuvik, Tuktoyaktuk), Inuvialuit Game Council, and Inuit Tapiriit Kanatami.

Notes

1. On December 2, 2001, Inuit Tapirisat of Canada announced its name change to Inuit Tapiriit Kanatami (ITK). "Tapirisat" means "we will unite" and, after 30 years of achievements and the signing of four land claims, Inuit felt it was time to acknowledge that Inuit are united, which is "Tapiriit" in Inuktitut. "Kanatami" means "of Canada."
2. The Inuvialuit Settlement Region (ISR) is the westernmost Inuit settled land claim region, which comprises six communities: Tuktoyaktuk, Aklavik, Inuvik, Sachs Harbour, Holman, and Paulatuk.
3. Labrador Inuit Association (LIA) was established in 1975. It represents the 5,000 Inuit of Labrador. LIA currently has an Agreement in Principle with the federal government for a land claims settlement. Makivik Corporation was established in 1978 after the signing of the James Bay and Northern Quebec Agreement. It represents the 9,000 Inuit of Nunavik. Inuvialuit Regional Corporation (IRC) was established in 1984 as part of the Inuvialuit land claim. It represents the 5,000 Inuit of the western Arctic. Nunavut Tunngavik Incorporated (NTI) was established in 1992 as part of the Nunavut Land Claims Agreement. It represents the 21,500 Inuit of Nunavut.
4. Inuit are the largest landowners in Canada next to the crown.
5. Scot Nickels and Pitsey Moss-Davies, ITK; Chris Furgal, CHUQ; Jennifer Castleden, IISD.
6. Mark Buell, health liaison officer, Inuvialuit Regional Corporation; Robin Fonger, HTC technical support officer, Inuvialuit Joint Secretariat; Barbara Armstrong, regional contaminants coordinator, Inuvialuit Regional Corporation; Diane Dillon, administrative support for HTCs, Inuvialuit Joint Secretariat.
7. Robert Chambers of the University of Sussex pioneered the PRA approach over twenty years ago. He has written extensively on its use in promoting local input into project planning and implementation. See for example Chambers, 1997, *Whose Reality Counts? Putting the First Last.*
8. The ZOPP technique was developed by the German development agency GTZ. ZOPP is an acronym for Ziel Orientierte Project Planning (see *ZOPP: An Introduction to the Method.* 1987 Deutsche Gesellschaft Für Technische Zusammenarbeit (GTZ) GmbH, Frankfurt, Germany)

References

Berkes, F., and D. Jolly (Riedlinger). 2001. Adapting to climate change: Social-ecological resilience in a Canadian Western Arctic community. *Conservation Ecology* 5(2): 18. [online] http://www.consecol.org/vol5/iss2/art18

Briggs, D., C. Corvalan, and M. Nurminen. 1996. Linkage methods for environment and health analysis. Geneva, Switzerland: UNEP/US EPA/WHO.

Chambers, R. 1997. *Whose Reality Counts? Putting the First Last.* London: Intermediate Technology Publications.

Cohen, S. J. 1997. What if and so what in northwest Canada: could climate change make a difference to the future of the Mackenzie basin? *Arctic 50* (4): 293–307.

Deutsche Gesellschaft Für Technische Zusammenarbeit (GTZ). 1987. *ZOPP: An Introduction to the Method.* GmbH, Frankfurt, Germany.

Eyles, J., and C. Furgal. In press. Indicators in Environmental Health: Identifying and selecting common sets. In: C. Furgal and P. Gosselin (Eds.), Consensus Conference on Environmental Health Surveillance. Supplement of Canadian Journal of Public Health.

Fenge, T., 2001. The Inuit and climate change. *ISUMA* 2(4): 79–85.

Fox, S. L. 2000. *Arctic climate change: Observations from the Inuvialuit Settlement Region.* In: F. Fetterer and V. Radionov, eds. Arctic climatology project, environmental working group arctic meteorology and climate atlas. Boulder, CO: National Snow and Ice Data Center. CD-ROM.

Furgal, C., Martin, D., Gosselin, P., Viau, A., Labrador Inuit Association (LIA), and Nunavik Regional Board of Health and Social Services (NRBHSS). 2002. *Climate Change in Nunavik and Labrador: What We Know from Science and Inuit Ecological Knowledge.* Final Project Report prepared for Climate Change Action Fund. CHUQ-Pavillon CHUL, Beauport, Québec.

Houghton, J. T., Y. Ding, D. J. Griggs, M. Noguer, P. J. van der Linden, and D. Xiaosu (Eds). 2001. Climate Change 2001: The Scientific Basis. Contribution of Working Group I to the Third Assessment Reort of the Intergovernmental Panel on Climate Change (IPCC). Cambridge, UK: Cambridge University Press.

International Institute for Sustainable Development (IISD). 2001. *Final Report: The Inuit Observations on climate change project.* Winnipeg: IISD.

International Joint Commission. 1991. *A proposed framework for developing indicators of ecosystem health for the Great Lakes region.* Windsor, Canada: International Joint Commission.

Inuit Circumpolar Conference. 2001. Statement by Violet Ford of the ICC. From consultation to partnership: Engaging Inuit on climate change. In *Silarjualiriniq: Inuit in global issues* 7 (January to March): 2–4.

McDonald, M., L Arragutainaq, and Z. Novalinga. 1997. *Voices from the bay.* Ottawa, Ontario: Canadian Arctic Resources Committee and Environmental Committee of the Municipality of Sanikiluaq.

Natural Resources Canada. 2000. *Taking the chill off: Climate change in the Yukon and Northwest Territories.* Geographic Survey of Canada and Aurora Research Institute, miscellaneous report. Map.

Natural Resources Canada. 2001. *Degrees of variation: Climate change in Nunavut.* Geographic Survey Canada, miscellaneous report no. 71. Map.

Nunavut Tunngavik Incorporated. 2001. *Elder's conference on climate change.* Cambridge Bay, Nunavut: Mimeographed.

Riedlinger, D., and Berkes, F. 2001. Contributions of traditional knowledge to understanding climate change in the Canadian Arctic. *Polar Record, 37* (203): 315–328.

World Health Organization (WHO). 2000. *Climate change and human health: Impact and adaptation.* WHO. Protection of the Human Environment.

Figure 10-1. Spring goose hunting camp at Middle Lake, Banks Island, May 2000. After a string of unusually warm springs that saw families travelling to the camp on four wheelers rather than snowmobiles, this spring was unexpectedly cold with more snow than people had seen in years. There was no open water and the geese were three weeks late.

10

Epilogue

Making Sense of Arctic Environmental Change?

Fikret Berkes
University of Manitoba

Are there ways of speaking of global issues such as climate change that accord weight to culturally specific understandings as well as to the universalistic frameworks of science? (Cruikshank, 2001)

Together the essays in this volume document a number of remarkable recent efforts in understanding indigenous knowledge and views on environmental change in the Arctic. These studies are primarily from North America, with another major initiative in progress to collect indigenous observations from Scandinavia and Russia (see Appendix by Mustonen, this volume). This flurry of activity is perhaps surprising, given that the Peterson and Johnson (1995) collection, based on a 1993 conference on climate change in the North, does not mention even one study on indigenous knowledge of climate change. The explanation for such a spectacular burst of activity in less than a decade has to do with the urgency of recent observations and the ability of indigenous peoples to bring their views to the attention of a wider international audience. "The majority of earth's citizens have not seen any significant climate change," whereas "residents of the Circumpolar North, especially the indigenous communities, are already witnessing disturbing and severe climatic and ecological changes" (appendix by Mustonen).

Given that our conventional science faces some fundamental problems in coming to grips with climate change, the emphasis on local observations and indigenous science is entirely appropriate. It is important to understand that the problem of climate change (and the broader issue of environmental change) is intractable by conventional scientific methods. To begin with, conventional disciplines in the sciences and social sciences are inadequate to deal with problems involving the interaction of humans with their environment. These coupled social and ecological systems (social-ecological systems for short) need to be understood and approached as *complex adaptive systems* (Berkes and Folke, 1998).

Environmental change does not lend itself to analysis by conventional approaches. It falls into the class of problems that has been referred to as "wicked problems"—those with "no definitive formulation, no stopping rule, and no test for a solution" (Ludwig, 2001). Ludwig uses climate change as a prime example of wicked problems that cannot be separated from issues of values, equity, and social justice and argues that there will likely never be a final scientific resolution of such problems.

The issue of climate change is not unique in this regard. It is merely one of many issues of social-ecological systems for which the expert-knows-best managerial approach is no longer sufficient. Where there are no clearly defined objectives and a diversity of mutually contradictory approaches, the notion of a disinterested expert no longer makes sense (Ludwig, 2001). Research on social-ecological system problems "must be created through a process by which researchers and local stakeholders interact to define important questions, relevant evidence, and convincing forms of argument" (Friibergh Workshop, 2000). Such research requires *place-based models* because "understanding the dynamic interaction between nature and society requires case studies situated in particular places and cultures" (Friibergh Workshop, 2000).

That brings us to the importance of local observations and to the objective of this volume. This book is aimed at the documentation of indigenous knowledge and perspectives on environmental change. There are three main points I would like to make in this concluding chapter that provides an overview of the status of the field:

1. From a complex systems perspective, local observations and place-based research are crucially important within the larger area of environmental change research.
2. Dealing with indigenous knowledge requires a major shift in thinking about the meaning of knowledge and knowing and in developing new models of community-based research for sharing knowledge.
3. Regarding frontiers of polar social science, arctic environmental change research provides the perfect venue to try to develop approaches needed to deal with the phenomenon of "change" in general.

The first two points provide two separate lines of argument that both feed into a discussion of frontiers of polar social science, with emphasis on the study of change. In exploring these three points, I will refer to the various chapters in this volume, focusing on some of the insights that emerge from the work documented here and elsewhere.

Significance of Local Observations and Place-Based Research

Any of us who have been to a recent conference on climate change is immediately struck by the overwhelming importance of global climate models in the overall discussion. As summarized by McBean et al. (2001), these models have evolved considerably over the years, and are being further improved both in terms of resolution and through the inclusion of new physical parameterizations. They consist of an atmospheric component coupled to ocean and sea ice models and a land surface component. These sophisticated climate change models are used, not to predict weather, but to indicate slow mean change of average weather. They are built on the physical principles that are thought to govern the various components of the climate system.

Before a model is deemed useful, it must be satisfactorily tested against recent climate data.

There is little doubt that these models are important and useful; they have indicated, for example, that climate change effects may be expected to be particularly pronounced in Alaska and the Canadian Western Arctic, projections consistent with many local observations (chapters by Krupnik and Jolly et al., respectively). There is an imbalance, however, in the way the models dominate the discussion in many climate change meetings. Can global climate change models provide the whole answer?

First, I would argue that global models, without local observations of change, are limited in their explanatory power. Second, models, as indicators of average change, do not deal effectively with the problem of uncertainty and do not incorporate the importance of small-scale perturbations that can tip a weather system in one direction or another. Third, models, as indicators of average change, miss out on the major human and ecological impact of environmental change, which is not so much about mean change but about extreme events. I expand on each.

Numerical models are becoming increasingly influential in environmental policy, including climate change debates. However, computer simulations do not, and cannot, provide predictions. As Oreskes et al. (1994) have pointed out, the verification and validation of numerical models of natural systems is impossible. The only propositions that can be verified (i.e., proved true) are those dealing with pure logic or mathematics. These are closed systems, and the components of models in closed systems are based on axioms that are true by definition. By contrast, natural systems are open systems. Our knowledge of them is always incomplete, and we can never be sure that we are controlling all the variables. Models provide a powerful tool for climate change research, but they cannot, with any stretch of the imagination, provide the full story.

Environmental change is a complex systems problem. Complex systems cannot be analyzed at one level alone. One of the major lessons of complex adaptive systems thinking (Levin, 1999; Gunderson and Holling, 2002) is that complex systems phenomena, such as climate change, occur at multiple scales. There are feedbacks across the different levels of the scale, both geographically (local, regional, global) and in terms of social organization (individual, household, community, etc.) (Berkes and Jolly, 2001). No single level is the "correct" one for analysis. Climate change cannot be understood at the global level alone, just as it cannot be understood at the local level alone. Since there is coupling between different levels, the system must be analyzed simultaneously across scales.

How can indigenous observations and traditional knowledge be used to help with the cross-scale problem? Projects involving multiple communities and examining indigenous observations at regional as well as local scales are very significant in this regard because they provide cross-scale insights (chapters by Fox, Thorpe et al., Kofinas et al., and Nickels et al., this volume). Also important are interactive meetings that bring together local experts to try to construct a picture of regional patterns of change. This approach has been used successfully in the Hudson Bay Bioregion project (McDonald et al., 1997), the Beaufort Sea conference (Fisheries Joint Management Committee [FJMC], 2000), the Cambridge Bay workshop of the Nunavut Tunngavik Inc. (Fenge, 2001), the Barrow Symposium on Sea Ice (chapter by Norton), and the Girdwood Workshop on Sea Ice Change (chapter by Krupnik). Such cross-scale observations and the sharing of knowledge between local experts and scientists complement the findings of global change models and help fill in the missing parts of the environmental change story.

Second, the issue about small-scale perturbations also follows from complex systems thinking, and from the idea that complex systems are unpredictable. Uncertainty is inherent in natural systems and in social-ecological systems, in particular. Models of climate change, as Holling (1997: 3) points out, "are full of uncertainties and will always be so. Surprise is inevitable whatever is done or not, and will unfold on a regional stage where adaptive response becomes central." Oppenheimer (2000) examines issues associated with large-scale, linear, and rapid change, with emphasis on climate change. He challenges the idea that climate change can ever be a smooth and linear process. It is much more likely to be a nonlinear process, with thresholds and skips and jumps, involving periods of rapid change and wild swings. He emphasizes the potential for small-scale perturbations to trigger change at regional and global scales, and suggests that anticipating large-scale change might require a specialized observational and modeling program that focuses on small-scale regional events.

If true, this analysis poses a major challenge to conventional global climate change models and underscores the importance of local observations and place-based research. The St. Lawrence Island sea-ice observation project (chapter by Krupnik) and the Barrow historical sea-ice documentation project (chapter by Norton) are very significant in this regard because they provide precisely the kind of specialized observations that are needed to capture small-scale perturbations that can trigger large-scale change. The St. Lawrence project, for example, is set up to capture daily events.

Third, global models indicating average change are limited in their capacity to deal with major social and ecological impacts. It is known to climate change experts that extreme weather events are important from a policy perspective because of the stress they cause (McBean et al., 2001). Hence, adaptation strategies cannot aim exclusively at slow mean change but need to take into account extreme weather events as well. To deal with the issue, statistical estimates on observed and projected changes are produced on parameters such as higher maximum/minimum temperatures and precipitation, to supplement models (McBean et al., 2001). This is a necessary but insufficient set of considerations to deal with the issue. Yes, it is useful to have global and regional estimates of frequency changes in extreme events, but the *actual impacts* of extreme weather occur on the ground, at regional and local scales.

The importance of these local/regional extreme events on arctic wildlife populations is well known to ecologists and anthropologists (Krupnik, 1993). Ice and snow can restrict forage availability for caribou and musk ox. For example, caribou on Coats Island, Nunavut, suffered major declines from winter mortality in years of high snow accumulation and sparse food supply (Gates et al., 1986). Hence, increased snowfall in the Arctic carries high risks to the ecosystem, and extreme events such as winter ice-storms can cause mass starvation. A relatively short-lived weather event can also have cascading population effects (Gunn, 1995). On the human side, the importance of extreme weather events is well known through floods, ice-storms, and hurricanes. Extreme events are probably best appreciated by the insurance industry (Kovacs, 2001)!

Several chapters in this volume, and Norton in particular, provide examples of the importance of extreme events, both in terms of impacts and in terms of causing adaptation problems. Something as simple as warm spells in mid-winter can disrupt entire regional economies by interrupting transportation on ice-roads, as happened in Canada's Northwest Territories in 2000–01 and in northern Manitoba in 2001–02. Such warm spells can also interfere with the local economy by interrupting hunters' travel over ice and by causing safety problems, as documented in the chapter by Jolly et al.

Analyzing the issue of extreme events further, the findings indicate that there may be three related phenomena, as observed at the local level by the indigenous peoples of the Arctic: weather is *more variable,* weather is *less predictable,* and there is an increased frequency of *extreme weather events.* This conclusion is based on combined observations from nine Canadian arctic communities (chapters by Fox, Jolly et al., Thorpe et al.; Kofinas et al.; Jolly et al., 2002), and it is supported by the findings from St. Lawrence Island and Barrow, Alaska (chapters by Krupnik and Norton). The overall effect of these three related changes is potentially very serious for indigenous peoples' lifestyles, nutrition, and safety, as documented in chapters by Furgal et al. and Nickels et al.

My own observations and discussions with local indigenous experts in various parts of the Canadian Arctic and the Subarctic over the years indicate that the issue of predictability is of prime importance in its own right. Northern land-based livelihoods depend on the peoples' ability to predict the weather ("is the storm breaking so I can get out?"), "read" the ice ("should I cross the river?"), judge the snow conditions ("could I get back to the community before nightfall?"), and predict animal movements and distributions. A hunter who cannot make the right judgment about what to hunt and where, cannot stay a hunter for long!

Impacts of environmental change are stripping arctic residents of their considerable knowledge, predictive ability, and self-confidence in making a living from their resources. This may ultimately leave them as strangers on their own land, as indicated by the chapters by Fox and Norton, in particular. The Fox chapter has a section on the emotional and cultural impacts of climate change that provides insights regarding the mechanisms of such impacts. Similarly, the Bradley chapter describes the impacts of change on one key area—traditional navigation skills. Northern peoples are experts on adapting to conditions that outsiders consider difficult, but it is a question of the speed and magnitude of change, as opposed to how fast people can learn and adapt.

The evidence from many statements by elders and experienced hunters cited in this volume confirm that current environmental change is beginning to stress the ability of Northern indigenous peoples to adapt. Rapid change requires rapid learning, and unpredictability superimposed on change interferes with the ability to learn. Extreme events and higher variability are definitely related to lower predictability. I am not sure if variability *per se* is so important, but I am quite sure that predictability is an area of environmental change research that should be pursued in its own right and analyzed in greater detail. Such work can only be undertaken through place-based research. The analysis of extreme events and predictability requires local-level research; global change models can inform such research, but these models have little to say directly on the question.

Indigenous Knowledge and New Models of Community-Based Research

Place-based research, especially in the Arctic, requires working with indigenous peoples and dealing with local and traditional environmental knowledge. This is not so easy to achieve. Some international programs, such as Arctic Climate Impact Assessment (ACIA) launched by the Arctic Council, aim to incorporate traditional knowledge into their work, a laudable idea. But ACIA leaders and others advocating the use of local observations may be underestimating the difficulty of such an approach. In this section, I deal with two challenges posed by indigenous knowledge and community-based research. The first challenge concerns the nature of knowledge and the relationship between Western science and indigenous knowledge. The second and a very urgent challenge is the need to develop new models of community-based research that do justice to local observations and facilitate sharing of knowledge.

Carrying out place-based research requires a major shift in the scientific philosophy and planning, as well as in our view of knowledge—away from expert-knows-best science, and towards accepting indigenous knowledge (and civil science in general) as a source of knowledge that complements science. Indigenous knowledge has been used in a number of areas of environmental management, from wildlife co-management to environmental assessment. However, the relationship between Western science and traditional knowledge has remained controversial. There are both similarities and differences between traditional science and Western science. Both kinds of knowledge are ultimately based on observations of the environment and both result from the same intellectual process of creating order and making sense of disorder. But they are different in a number of substantive ways. Traditional ecological knowledge is often an integral part of a culture and tends to have a large social context. Elsewhere, I have characterized traditional ecological knowledge as a knowledge-practice-belief complex (Berkes, 1999).

The conflict between science and traditional knowledge is in part related to claims of authority over knowledge. In the Western positivist tradition, there is only one kind of science—Western science. Knowledge and insights that originate outside institutionalized Western scholarship are not easily accepted, and scientists tend to dismiss understandings that do not fit their own. Scientists tend to be skeptical; therefore, they demand evidence when confronted with traditional knowledge. However, some aspects of this knowledge-practice-belief complex do not easily lend themselves to scientific verification. For example, Cruikshank (2001) points out that Athapascan and

Tlingit elders' belief in a sentient "land that listens" and glaciers *who* may be offended by noise, does not fit well with the narratives of geophysical science!

Insomuch as indigenous knowledge remains poorly appreciated by scientists and by Western society in general, skepticism works both ways. Traditional knowledge holders are skeptical of book learning and tend to dismiss scientists who do not have extensive first-hand knowledge of a specific area. I remember the story once told by a caribou biologist. He was in Baker Lake, in Canada's Nunavut Territory, for the first time, and he made the mistake of announcing himself as an expert on caribou. The local Inuit were incredulous. "You mean," they said, "you know about our caribou here too?" From their response, it was clear to the biologist that what he knew about the caribou, which is universal knowledge according to the positivist tradition, carried not one bit of weight in Baker Lake. His knowledge was not considered legitimate unless it was specific to the local area, obtained largely first-hand, and in apprenticeship with a local knowledge-holder. The Baker Lake Inuit had no trouble with Western science, as long as it fit *their* understanding of the caribou and it was obtained at least in part through the kind of learning process that *they* considered legitimate, that is, through first-hand observations and guided by the teachings of elders. In this regard, their attitude hardly differed from that of scientists confronted by another system of knowledge.

Thus, the question of how to deal with indigenous knowledge is anything but settled. At the risk of oversimplification, one could say that there are two main positions. The first position claims that the use of indigenous knowledge offers just another information set from which data can be extracted to plug into scientific frameworks of understanding. Many scientists have no trouble with the notion of local knowledge, as long as it is verifiable and as long as the local knowledge stays in the realm of observable information—and does not wander into the realm of belief. In particular, biologists who work in the North are particularly appreciative of indigenous northerners as experts in natural history. This is in contrast to geophysical scientists (including climatologists) who have little contact with indigenous knowledge and tend to be rather dismissive (Bielawski, 1992).

The second position, often associated with post-modern philosophers of science, is that Western science is but one knowledge system among many, even though it also happens to be the dominant knowledge system by far. There is a related point here, which is quite obvious to indigenous peoples themselves but not always clear outsiders. When information is extracted from traditional knowledge and plugged into scientific frameworks, it is taken out of its cultural context and its integrity is threatened (Nadasdy, 1999). Anthropologists have questioned the validity of subsuming indigenous knowledge within theoretical frameworks that rely on Western notions of subjectivity and causality (Cruikshank, 2001).

It is this second position that guides research in virtually every case described in the chapters in this volume.

There is a good reason that chapters in this volume deal with indigenous knowledge respectfully, as knowledge in its own right, rather than as "data" and observations to be tested against the standard of science. For many indigenous groups, the social and cultural aspects of traditional knowledge are very significant, and this is one of the reasons why dealing with traditional ecological knowledge has become politically volatile, just as its importance for environmental change research is beginning to be appreciated. In many indigenous areas, researchers no longer have a free hand to conduct their work independently from the people themselves. Indigenous groups are beginning to assert control over their knowledge systems, and they have started to ask the question of who benefits from traditional knowledge research. As indigenous knowledge has become a symbol for many groups to regain control over their culture, reclaiming their tradition has become a major strategy in many parts of the world, including the North, for movements of cultural revitalization (Berkes, 1999).

Indigenous knowledge has become very political, but this need not be a barrier to the use of traditional environmental knowledge to inform climate change studies. As this volume demonstrates, traditional knowledge and Western science need not be thought of as opposites. Rather, it is useful to emphasize the potential complementarities of the two and to look for points of agreement rather than disagreement. The use of traditional knowledge contributes to conceptual pluralism, and expands the range of approaches and information sources needed for problem solving (Berkes and Folke, 1998).

In the area of climate change research, it is useful to recap the findings of Riedlinger and Berkes (2001) and the chapter by Jolly et al. This chapter suggests that traditional environmental knowledge and science can be brought together through five areas of convergence, that is, potential areas of collaboration and communication. These relate to the use of traditional knowledge (a) as local scale expertise, (b) as a source of climate history and baseline data, (c) in formulating research questions and hypotheses, (d) as insight into impacts and adaptation in arctic communities, and (e) for long term, community-based monitoring. This is a promising framework that can help guide practical collaborative initiatives. But the question remains: how can we achieve real collaboration in these potential areas of convergence?

This brings us to the second challenge in dealing with indigenous peoples and traditional knowledge—to develop new models of community-based research. Why do we need new models of place-based research at this time? Very simply, because we can no longer use the old models of extracting information from communities, for ethical reasons and for practical political reasons alike. In fact, there is little choice but to develop models of participatory research that can help capture local observations accurately, in a way that incorporates their worldview and values as well.

The contribution of this volume is perhaps most significant in demonstrating the feasibility of new models of place-based, participatory research. The chapters not only illustrate that participatory research is quite possible in the area of environmental change, but also that local participation can greatly *improve* the quality of scientific research. Some of the chapters provide explicit models, but in others, there is insufficient detail to characterize how the reported collaborative research was structured—

perhaps a task for a follow-up workshop. Suffice to say, there is no magic way of constructing new models of community-based research, and there certainly is more than one way to do it.

Figure 10-2 is based on the project described in the chapter by Jolly et al. and illustrates one way of carrying out research that can create a partnership of indigenous knowledge and Western science. There are a number of general features of the partnership arrangement sketched in the figure. The model creates a forum in which the agendas of the partners are made transparent and common objectives discussed. Objectives, research approaches, and rules of conduct are all determined jointly. The research process has both a science and a local knowledge component, with provision for the two to learn from one another. There is continual feedback to the community in the form of preliminary results, and there is feedback to the research team in the form of revised approaches and verification. The results of the project are shared as previously agreed upon, and the overall results are deposited with the community in culturally appropriate ways. The community vets publications, and the local experts receive credit for their contributions as a way of acknowledging their authority over their knowledge.

Frontiers of Polar Social Science: Studying Change

I have argued that place-based research and local observations have a crucial role to play in research on environmental change. Such approaches to climate change are not model-driven but are culture-specific, historically informed, and geographically

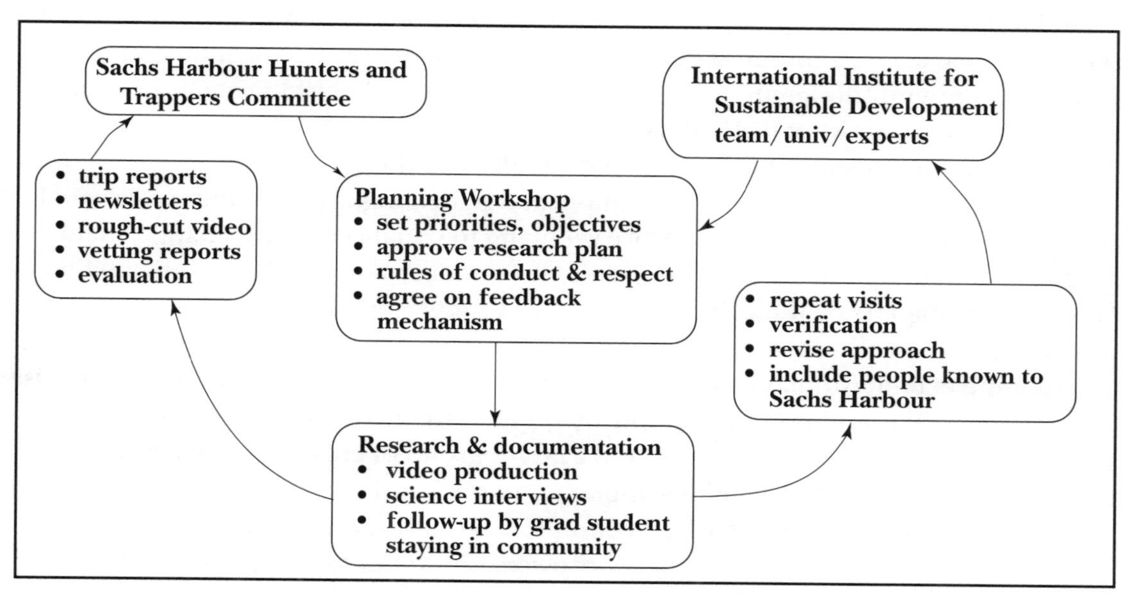

Figure 10-2. A partnership model for research that can combine indigenous knowledge and Western science. The model involves the creation of a "forum" in which the two partners (Sachs Harbour Hunters and Trappers Committee and IISD) interact as equals, for sharing across knowledge systems.

rooted. A major lesson of complexity theory is that scale is important; local and regional scales have to be addressed simultaneously with the global scale. I have also argued that the appreciation of traditional environmental knowledge is key to unlocking the secrets of environmental change at the local level. However, this requires shifting our conventional approaches to knowledge, and developing new models of community-based research for sharing knowledge between indigenous systems of knowing and Western science.

Creative ways of combining indigenous knowledge and science, in ways that do justice to both, have potential to improve our understanding of climate change. These may include the use of daily diaries (chapter by Krupnik), participant observation (chapter by Fox), elder-youth camps (chapter by Thorpe et al.) and expert-to-expert interviews (chapter by Jolly et al.) Achieving improved understanding of indigenous knowledge and perspectives depends on building new participatory models of research in which indigenous peoples are not the "objects" of research but equal partners. On that point, the lesson from this volume is loud and clear*: indigenous knowledge of environmental change can be treated as knowledge in its own right and not merely as a source of "data" for science.*

Partnership approaches are significant also because they pose a fundamental challenge to expert-knows-best science. They show the way to other disciplines on how research can be created through the interaction of scholars and stakeholders. In the case of this volume, the stakeholders are indigenous groups observing, and being affected by, arctic environmental change, and they hold a considerable amount of relevant knowledge. The larger issue is that civil science produced by nonspecialists and stakeholders is a valid input into decision-making for complex environmental problems. There are no "designated" experts on these social-ecological problems; researchers need to interact with stakeholders to define the key questions, to participate in the research, and to interpret the findings (Friibergh Workshop, 2000).

This kind of research is particularly important in dealing with the phenomenon of change in complex systems. Here we are referring to social-ecological change characterized by nonlinear relationships, multiple scales, uncertainty, and unpredictability. Many of the chapters deal specifically with *climate change;* others deal with environmental change but not climate change specifically (chapters by Thorpe, Norton, Bradley, and to some extent, Kofinas et al., and Krupnik). However, the weight of the volume is on climate change, and not for example on arctic contaminants. Environmental change being observed is, in turn, part of a larger group of challenges that arctic peoples face, such as oil and gas development, mining and hydroelectric developments, and the broader set of social, economic, and cultural change.

The subject of environmental change provides a number of challenges to social science research. A key one is related to the study of change—describing change, understanding change, and dealing with impacts of change. Scholars have focused on many dimensions of *change* (e.g., Mitchell, 2002). The urgent nature of the *climate change* issue is such that research is forced to focus, not on the static and the descriptive, but on the dynamic and the analytic. This makes the climate change issue the perfect venue for the study of change. Thus, this current "frontier in polar social science" has the potential also to become a leader in "frontiers of social and natural science" in general.

As the various chapters show, *describing* change based on indigenous knowledge and perspectives is not such a straightforward issue. Recording hunters' and elders' views and presenting them may be relatively easy using standard social science research methodologies, but the difficulty comes in the interpretation. As Huntington explains in his Preface, interpreting indigenous knowledge and perspectives requires something more—an understanding of "what people look for, how and why they look for it, and to what use they put the resulting information." A logical starting point in developing an approach to the study of change from an indigenous perspective is story-telling, a methodology that has the additional advantage of being consistent with the learning style and culture of many indigenous groups.

Moving to the *understanding* of change, Krupnik and Jolly in their Introduction point out that we, as scholars and Western scientists, have to learn through listening to and sharing other ways of observing and knowing. Indigenous understandings "are often articulated as community-based assessments of change. Such assessments translate global-scale processes of change into local-scale evidence."

In learning traditional knowledge through sharing, a logical step to link indigenous views to Western science may be scenario-building. This methodology has the advantage of being consistent with climate change research in the geophysical sciences—those, however, tend to be quantitative. In the present volume, there are few examples of the scenario approach, except perhaps the chapter by Kofinas et al. However, one can imagine incorporating scenarios of qualitative change into the broader studies of climate change, or incorporating quantitative scenarios based on global climate change models in arctic social science research.

Moving from the study of change to the study of *impacts* of change, it is worth repeating Krupnik and Jolly's point that arctic indigenous people have a special stake in modern studies of global change! As Kusugak's Foreword puts it, "the Inuit see themselves as part of the ecosystem and want to be included, not as victims, but as people who can help." Impacts on arctic indigenous peoples include some changes that are relatively easy for southerners to comprehend, such as the appearance of species never seen in living memory. But they also include changes that only a few scientists can appreciate, such as the loss of multiyear sea ice and permafrost melt causing lakes to drain. But further, impacts include changes that an outsider can hardly imagine: fermenting caribou meat spoiling because of temperature change. Again borrowing from Kusugak, boating seasons are longer and berry picking is great. But all the knowledge that people have built about exactly which phase of the moon to go seal hunting at the edge of ice becomes useless when the ice cover changes or the ice gets thinner.

To link indigenous observations to Western science, I will briefly comment on three approaches that have been used in the area of climate change: risk analysis, vulnerability analysis, and resilience analysis. These three approaches are not used to any extent in the chapters in this volume, except for example in assessing risk in travel over ice (chapter by Norton), and their full potential remains to be developed.

Risk analysis is perhaps the best known approach, and it looks at the impacts of statistical increases in various risk-producing events, such as hurricanes and ice-storms, that have costs, for example, for the insurance industry (Kovacs, 2001). The vulnerability analysis, on the other hand, focuses not on physical events but on the combination of physical events *and human conditions* that make a group of people vulnerable to impacts, such as the location of settlements in flood-prone areas. At the most basic level, vulnerability is a measure of exposure to stress. Hence, Adger (1999) and Kelly and Adger (2000) see vulnerability mainly as an issue related to human welfare and suggest focusing on solutions that aim to reduce such vulnerability.

The resilience approach takes the flip side of vulnerability and looks at factors that provide resilience or absorptive capacity in the face of perturbations or stress (Gunderson and Holling, 2002; Berkes and Jolly, 2001). Resilience has the advantage of looking at feedbacks at multiple scales. I consider the resilience approach to be particularly suitable for the study of change because it deals with the flexibility of a system's responses to stress, and it focuses on a system's capacity for learning, self-organization, and adaptation. This is significant for arctic peoples; the wisdom of knowledge-holders is an adaptive quality (chapter by Norton).

Consider, for example, that the well-known adaptations of indigenous peoples to living in the highly variable arctic environment (Krupnik, 1993) confers resilience to environmental change impacts, but up to a limit. Short-term responses such as switching species and adjusting the "where, when, and how" of hunting (chapters by Jolly et al., Thorpe et al., Fox, and Furgal et al.), provide additional means by which arctic people can deal with change.

Adaptations are probably best studied at the local level (e.g., chapters by Fox, Jolly et al., Nickels et al.) and at the regional level (e.g., chapter by Furgal et al.). But the chapters also indicate that societal responses and adaptations may be occurring at multiple levels. By focusing on arctic communities that are already witnessing change and trying to adapt to change, the chapters in this volume provide insights that will be of interest, not only in the Arctic, but elsewhere.

The study of environmental change can focus on responses and adaptations that occur across scale: at the level of the individual, the household, the community, and the region. Just as the vulnerability approach focuses on solutions that aim to *reduce vulnerability,* the resilience approach focuses on solutions that aim to *increase resilience.* Resilience across scale may be increased through self-organization, learning, and according to complexity theory (Levin, 1999), by any measure that can provide for a tighter coupling between levels and for speeding up feedbacks among them. How might these cross-scale linkages contribute to learning and self-organization and hence to the resilience of arctic peoples in the face of climate change?

Co-management arrangements that have been developing in recent years in a number of areas of the arctic are important in this regard. Because co-management connects local-level institutions and government agencies, we hypothesize that it has the potential to contribute to learning and self-organization across scale and hence to increased resilience (Berkes and Jolly, 2001). This is a highly fertile area for polar social science. In the Arctic, a number of vertical linkages have been created across levels of organization in recent years, all the way to the Arctic Council established in 1996, as well as horizontal linkages across areas.

Co-management is only one linkage mechanism among many in a rapidly globalizing world. There are several other mechanisms for cross-scale interaction and institutional interplay now in place. These include, for example, the Arctic Environmental Protection Strategy of 1991 and the Arctic Climate Impact Assessment (ACIA) initiated by the Arctic Council in 2000 (Fenge, 2001). The International Human Dimensions Programme on Global Environmental Change (IHDP) has been exploring issues of institutional interplay, from the local to the international level, in two pilot areas of the world, the Arctic and Southeast Asia. Other linkages are created in ways that were not possible only a few years ago. For example, indigenous hunters and scientists have been interacting in meetings such as the Beaufort Sea 2000 Conference (FJMC, 2000) and the Snowchange Project (appendix by Mustonen), learning from one another and finding ways in which traditional knowledge and science can share observations.

Conclusions

Residents of the Circumpolar North are witnessing disturbing environmental changes. Words fail when trying to make sense of rapid changes that go beyond the realm of experience. Weather has become difficult to predict, as expressed by one Alaska elder with the evocative phrase, "the Earth is faster now" (which is taken as this volume's title). In a similar vein, an elder from the Beaufort Sea community of Aklavik told his dream to university researcher Melissa Marschke (personal communication). He dreamt of fish that came to feed on water bugs on the underside of sea-ice. But now, he dreamt, the ice floes were melting and disappearing and there was no place for the little fish to eat and sleep. Making sense of arctic environmental change is not easy, except perhaps through the discourse of arctic peoples who see these changes in their day-to-day lives.

However, there are impediments to learning from traditional knowledge. The conventional practice of modernist science puts up huge barriers to the study of local knowledge and civil science. Is "such peoples' science" valid? To paraphrase Cruikshank (2001), are there ways of generating and sharing knowledge that accord weight to local understandings as well as to universal science?

Part of the answer is that conventional science is often humbled by its inability to deal with complex social-ecological issues and Ludwig's (2001) "wicked problems." Complex systems problems facing humanity are not adequately addressed by the familiar scientific approach of developing and testing hypotheses serially. Because of nonlinearity, complexity, and long time lags, new methodologies are needed. These, according to the findings of the Friibergh Workshop (2000), may build upon lessons of case studies and work with the local people to produce knowledge "that is both scientifically sound and rooted in social understanding."

To develop new methodologies based on case studies, we need models of place-based or community-based research that can help capture what people are observing, in a way that reflects their world view and values. Chapters of this volume indicate that there may be a number of common features of the models that work, but also that there are a number of different ways in which place-based research can be structured.

Perhaps the key to understanding indigenous knowledge and perspectives is to get the relationship right between traditional knowledge and Western science. I approach the question through an indigenous metaphor (Erasmus, 2002). Kaswentah is the wampum belt recording of the eighteenth-century treaties between the Iroquois and the Europeans. Kaswentah shows the path of two boats, an indigenous canoe and a European sailing ship, travelling together on the river of life. The two retain their own identity and integrity, but are linked to one another by strands of truth, respect, and friendship.

Acknowledgements

The editors deserve much appreciation for their boundless optimism and energy to pull this volume together. For helpful comments and criticisms on this paper, I thank Igor Krupnik, Dyanna Jolly, Alan Diduck, Melissa Marschke, and the Natural Resources Institute seminar group.

References

Adger, W. N. 1999. Social vulnerability to climate change and extremes in coastal Vietnam. *World Development 27:* 249–269

Berkes, F. 1999. *Sacred Ecology: Traditional Ecological Knowledge and Resource Management.* Philadelphia: Taylor & Francis.

Berkes, F., and Folke, C. eds. 1998. *Linking social and ecological systems. Management practices and social mechanisms for building resilience.* Cambridge: Cambridge University Press.

Berkes, F., and Jolly, D. 2001. Adapting to climate change: Social-ecological resilience in a Canadian western arctic community. *Conservation Ecology 5* (2): 18. [online] URL: http://www.consecol.org/vol5/iss2/art18

Bielawski, E. 1992. Inuit indigenous knowledge and science in the North. *Northern Perspectives 20* (1): 5–8.

Cruikshank, J. 2001. Glaciers and climate change: Perspectives from oral tradition. *Arctic 54:* 377–393.

Erasmus, G. 2002. Why can't we talk? The third annual LaFontaine-Baldwin lecture, Vancouver.

FJMC. 2000. *Beaufort Sea 2000. Renewable resources for our children.* Conference Summary Report. Inuvik, NT: Fisheries Joint Management Committee. Available online http://www.fjmc.ca

Fenge, T. 2001. The Inuit and climate change. *Isuma, Canadian Journal of Policy Research 2* (4): 79–85.

Friibergh Workshop. 2000. *Sustainability science.* Statement of the Friibergh Workshop on Sustainability Science. Friibergh, Sweden. Available online http://sustsci.harvard.edu/keydocs/friibergh.htm

Gates, C.C., Adamczewski, J., and Mulders, R. 1986. Population dynamics, winter ecology and social organization of Coats Island caribou. *Arctic 39:* 216–222.

Gunderson, L. H. and Holling, C. S., eds. 2002. *Panarchy: Understanding Transformations in human and natural systems.* Washington, DC: Island Press.

Gunn, A. 1995. Responses of arctic ungulates to climate change. In: Peterson, D. L., and Johnson, D. R., eds. *Human ecology and climate change: People and resources in the Far North.* Washington, DC: Taylor & Francis, pp. 89–104.

Holling, C. S. 1997. Regional responses to global change. *Conservation Ecology 1* (2): 3. [online] URL: http://www.consecol.org/vol1/iss2/art3

Jolly, D., Fox, S., and Thorpe, N. 2002. Inuit and Inuvialuit knowledge of climate change. In: J. Oakes, R. Riewe, K. Wilde, A. Dubois, and A. Edmunds, eds. *Native voices in research.* Winnipeg: Department of Native Studies Press, University of Manitoba, pp. 170–180.

Kelly, P. M., and Adger, W.N. 2000. Theory and practice in assessing vulnerability to climate change and facilitating adaptation. *Climatic Change 47:* 325–352.

Kovacs, P. J. E. 2001. Increase community protection from extreme weather events. *Isuma, Canadian Journal of Policy Research 2* (4): 57–61.

Krupnik, I. 1993. *Arctic adaptations: Native whalers and reindeer herders of northern Eurasia.* Hanover and London: University Press of New England.

Levin, S. A. 1999. *Fragile dominion: Complexity and the commons.* Reading MA: Perseus.

Ludwig, D. 2001. The era of management is over. *Ecosystems 4:* 758–764.

McBean, G., Weaver, A., and Roulet, N. 2001. The science of climate change: What do we know? *Isuma, Canadian Journal of Policy Research 2* (4): 16–25.

McDonald, M., Arragutainaq, L., and Novalinga, Z. 1997. *Voices from the bay: Traditional ecological knowledge of Inuit and Cree in the Hudson Bay bioregion.* Ottawa: Canadian Arctic Resources Committee and Sanikiluaq, NWT.

Mitchell, B. 2002. *Resource and environmental management.* Second Edition. Harlow, UK: Prentice-Hall.

Nadasdy, P. 1999. The politics of TEK: Power and the "integration" of knowledge. *Arctic Anthropology 36:* 1–18.

Oppenheimer, M. 2000. White papers. Anticipating rapid change: Insights from non-linear geophysical systems. http://lsweb.la.asu.edu/akinzig/nsfmeet.htm

Oreskes, N., Schrader-Frechette, K., and Belitz, K. 1994. Verification, validation, and confirmation of numerical models in the earth sciences. *Science 263:* 641–646.

Peterson, D. L., and Johnson, D. R., eds. 1995. *Human ecology and climate change: People and resources in the Far North.* Washington, DC: Taylor & Francis.

Riedlinger, D., and Berkes, F. 2001. Contributions of traditional knowledge to understanding climate change in the Canadian Arctic. *Polar Record 37:* 315–328.

Figure 11-1. A view of Finnmark region in Norway, coast of the Barents Sea.

Appendix: Snowchange 2002

Indigenous Views on Climate Change: A Circumpolar Perspective

Tero Mustonen
Tampere Polytechnic, Tampere, Finland

Snowchange is a multiyear education-oriented project to document indigenous observations of climate change in the northern regions, coordinated by the Tampere Polytechnic, Department of Environmental Management and Engineering in Tampere, Finland.

The objective of the initiative is three fold:

- to collect and document the experiences of indigenous peoples related to climate change,
- to present them in a manner that will be effectively heard by the peoples and decision makers of the south, and
- to enhance and support the indigenous participation and work in climate change issues.

The project began in early 2001. Stage one is expected to conclude in 2004. However, plans exist to continue the work on long-term basis, leading for example to the establishment of monitoring networks around the North.

The Snowchange project operates at three levels: at the regional level, the majority of documentation work takes place in the European North, meaning the Sami communities of Scandinavia and the Kola Sami in Murmansk region of the Russian Federation.

At the national level, the project promotes indigenous viewpoints in the media, educational spheres and research community in Finland, Russia, and Scandinavia.

At the circumpolar level, the project engages and works with various partners in the North American Arctic and Russian Arctic, including Siberia and the Far East. The Snowchange website functions as an information portal and capacity building resource for communities, research and educational organizations, and other stakeholders in the north.

Principal Partners

- Indigenous communities around the Circumpolar North, especially the European North and Russia
- Tampere Polytechnic, Finland
- the Arctic Territories GLOBE Program
- Aurora Research Institute
- Murmansk Humanities Institute, Russia
- Arctic Climate Impact Assessment, Alaska, USA
- International Arctic Science Committee
- Russian Association of the Indigenous Peoples of the North (RAIPON)

Many of these partnerships were formed during a series of visits to Canada and the United States, to places such as Whitehorse, Yukon, and Inuvik and Tuktoyaktuk in the Northwest territories, as well as communities in Alaska.

Documenting Indigenous Perspectives

The first documentation work took place in September 2001 among the Sami people of Finland, in the reindeer-herding village of Purnumukka in northern Finland. Pentti Nikodemus, a Sami reindeer herder, expressed his concerns for the sudden variation of climate conditions and the effect of unpredictability. He described sudden, extremely cold weather, close to –50°C in the region in the previous winter, which prevented the use of motor transportation in reindeer herding. Pentti commented that this was a good occasion to use the reindeer as a means of transportation. As well, the late and heavy snowfall has been troubling the Purnumukka region.

After the Purnumukka visit, community interviews were conducted with the Kola Sami in the town of Lovozero (Luujavre) in the Murmansk Region, Russia. Additional interviews were carried out in the regional capital, city of Murmansk.

During the interviews, the Kola Sami expressed concern over weather-related changes and changes in the traditional cycle of seasons that are affecting their life. For example, Larisa Avdeeva stated that sudden periods of above-zero temperatures followed by a quick freeze overnight make it difficult for the reindeer to access the lichen. As well, local reindeer herders had witnessed the arrival of new species of insects, plants, and birds, which in the past were common only in the more southern parts of Russia. In addition, because of the late freeze-up, movement is more difficult in the tundra. The people interviewed stated that there are many other concerns in addition to climate change, such as the state of the Russian society, lack of resources, and other information. But a definite impact to the traditional lifestyle had been seen because of the climate change. All in all, the September 2001 visit yielded approximately thirty-five hours of good-quality material.

In 2002, documentation of indigenous observations continued in various areas across the northern zone. Kaisu Pulli, a member of the Snowchange staff, travelled to Inuvik in Arctic Canada to help the Inuvialuit Settlement Region in its documentation work. She will be stationed in that region until August 2002.

A visit to the Sami community of Utsjoki (Ochejohka), in northern Finland was made in March 2002, to continue the documentation process. It was organized together with Elina Helander, a Sami researcher from the University of Lapland, and members of the Kaldoaivi Reindeer Herders Association. Initial results indicate concerns by the Sami about unpredictability in weather; ice rain that freezes overnight in fall; erosion caused by new, strong winds; and lateness of snowfall.

Documentation will also continue in the spring and summer 2002 in the Murmansk region of arctic Russia, coordinated by Sergey Zavalko from the Murmansk Institute of Humanities. Materials will be made available in English in fall 2002.

Together with the Russian Association of the Indigenous Peoples of the North (RAIPON) in Moscow and Russian researcher Dr. Tatjana Vlassova (Institute of Geography, Russian Academy of Sciences), documentation work will be expanded to the Nenets Autonomous Okrug. With additional resources, we hope to incorporate two or three additional Siberian native communities, over the 2002–2003 period. Finally, Snowchange work will also take place with the Tahltan Nation in northern British Columbia, Canada over the course of 2002.

Communicating Indigenous Observations: The Snowchange Website

The project relies on a ground-up approach to material collection and sharing. In practice, this means that local participants and communities are trusted to prioritise which local observations of climate change are most relevant. Open-ended semi-directive interviews will be used to collect and document such observations. Oral stories and observations in plain language are encouraged: submissions can be made via email, tape, CD, minidisk, written paper, or electronic format. While only selected materials will be published in the project publications, all materials collected will be included in an on-line database and advertised in various international fora.

The project will be producing a series of publications. These publications will be posted on a special Snowchange website, providing a cost-effective, readily accessible, and highly visible record of climate change across the Circumpolar North. The internet address for this website is www.snowchange.org.

The use of a project website means that there are no practical limits to the amount of information capable of being stored, and it will be possible to update and include information beyond the initial project time. In the course of 2002 the website will be developed to encourage nonlinear, experimental "observational education of climate change," as opposed to linear, hierarchical flow.

Snowchange 2002: An International Indigenous Workshop on Climate Change

In addition to documentation, various conferences will be organized on the local level to distribute and engage the public, media, research community and local people in the project. The first conference, Snowchange 2002: An International Indigenous Workshop on Climate Change, took place February 22–24, 2002, in Tampere, Finland.

The workshop brought together seventy representatives from the indigenous northern communities and organizations. Participants came from the Sami Council; the Kola Sami community of Lovozero, Russian Association of Indigenous Peoples of the North (RAIPON); the Tahltan Nation of the Yukon Territory; the government of Nunavut (Department of Sustainable Development); from the Inuvialuit Settlement Region and the Gwich'in Nation in Northwest Territory, Canada; from Inuit Tapirisat of Canada office in Ottawa; and from the Haida Nation in British Columbia. There were also many representatives of various northern research and environmental organizations, such as Tampere Polytechnic, World Wide Fund for Nature, Friends of the Earth, the International Institute of Sustainable Development, the Russian Academy of Sciences and the Arctic Centre (University of Lapland).

Along with partnerships, discussions and media coverage, the first Snowchange conference participants drafted a Declaration on Traditional Ecological Knowledge (see below). This declaration will be taken to the United Nations' Framework Convention on Climate Change and to the minister of environment of Finland, Ms. Satu Hassi.

Second Snowchange Conference, February 2003 (Murmansk, Russia)

The next Snowchange Conference is scheduled to take place in Murmansk, Russia, in February 2003. It will be hosted by the Murmansk Humanities Institute together with local indigenous organizations and other partners.

Draft Declaration on Traditional Ecological Knowledge

Adopted by the Representatives and Participants of Snowchange Workshop, February 2002—Tampere, Finland

Indigenous Traditional Ecological Knowledge is a Viable Source of Information in Scientific Assessments on Climate Change and Should be Recognised as an Equal Tool of Research

While the threat of global climate change has received a tremendous amount of attention worldwide, very little has actually been done to address the root causes of this change. This is in part because the majority of the Earth's citizens have not seen any significant climate changes thus far, and may not see any until major ecological damage has already been done. The residents of the Circumpolar North, especially the indigenous communities are already witnessing disturbing and severe climatic and ecological changes.

The United Nations Framework Convention on Climate Change and the Kyoto Protocol should be ratified as soon as possible. There is as well a need for an "Arctic Message on Climate Change," for example through the work of the Arctic Council and other relevant international fora.

But more is needed. Our modern needs, the ways of spending and consuming will have to change. We need to choose again and choose well.

The changes threaten to increasingly undermine the North's ecology, economy, society, and culture.

It is felt that traditional ecological knowledge (TEK) and the first teachings of the indigenous peoples of the world should be recognised as equal tools of research, methodology, assessment, and observation in the global climate change work as the methods of the scientific community.

Traditional ecological knowledge has been discussed in various indigenous fora. It is local, relevant, and valid.

Traditional ecological knowledge is a system of knowledge that builds on generations of people living in close relationship with nature. Thus indigenous climate change assessments and observations build on countless generations of knowledge, since time immemorial.

Traditional ecological knowledge carries within itself systems of classifications and empirical observations about the local environment and a system of self-management that governs sustainable resource use. Knowledge brings responsibility.

Indigenous peoples around the world use the traditional ecological knowledge, and especially the elders of the different peoples, the carriers of this vast system of knowledge, accumulate the traditional ecological knowledge through direct relations with the environment on the local level. Traditional ecological knowledge accumulates and adapts knowledge in a holistic manner. It is a system of holistic ecology.

In the scientific assessment of the global climate change, mostly only western science data has been seen as valid in the past. It is not enough. Regional examples, such as the work of the Arctic Climate Impact Assessment (ACIA) and the Mackenzie Basin

Impact Study (MBIS) have shown good regional recognition of the validity of traditional ecological knowledge.

There is an urgent need to integrate traditional ecological knowledge respectfully, equally, and responsibly with western science to create a fusion of hybrid way of thinking and perceiving the relationship between human and the nature.

Currently, such recognition of traditional ecological knowledge has not been made.

Various international fora, such as the Intergovernmental Panel on Climate Change and others are urged to validate traditional ecological knowledge as a recognised vehicle of knowledge in the assessment, research, and other scientific work on climate change. More broadly, the same recognition is needed worldwide in all environmental and resource management work.

There is a need of peace, power, and righteousness in finding answers to the challenges of the climate change. The initial impacts of the climate change will be on the indigenous peoples, and therefore there is a clear need for a direct participation to bring awareness. There is a need to develop an active, ongoing circumpolar indigenous community-based monitoring network to share information and to develop local adaptations to climate change.

The representatives of the Snowchange workshop feel that the urgency of climate change requires immediate action from the scientific community in multiple forums to recognise traditional ecological knowledge as an equal system of knowledge.